Focus

On

Developmental

Mathematics

H&H Publishing Company, Inc.
Clearwater, Florida
(813) 442-7760

Robert Hackworth
Joseph Howland
Robert Alwin

Focus On Developmental Mathematics

by
Robert Hackworth, Joseph Howland, Robert Alwin
of St. Petersburg Junior College, Clearwater Campus

Copyright 1989
H&H Publishing Company, Inc.
1231 Kapp Drive
Clearwater, Florida
telephone (813) 442-7760

Editorial/Production Supervision
Robert Hackworth, Joseph Howland

Editor
Karen H. Davis

Design & Marketing
Tom Howland

Production Assistants
Sally Marston, Mike Ealy

Business Management Assistant
Janie Howland

ISBN 0-943202-32-9

Library of Congress Card Catalog Number 89-080839

Printing is the lowest number: 10 9 8 7 6 5 4 3 2

Preface

Focus on Developmental Mathematics was written for those who enter higher education with serious deficiencies in their mathematics backgrounds. These are special students with some unique needs. Many of these students have proved their competence in other academic subjects or other tasks. Most of them have some degree of anxiety about learning mathematics. Nevertheless, mathematics is a requirement for their academic ambitions and they are willing to confront their fears again.

The text seeks to give its student readers a maximum of assistance. First, the text presents its mathematics in a simple, logical way. That is not to say that it plays down the nature of mathematics itself. On the contrary, the subject is embraced by many mathematicians because of its logic and its adherence to logical principles.

Second, the text goes beyond the mathematics and presents ways to study the subject effectively. Many of the students in this course have learned to do mathematics in ways that are so ineffective that failure is a natural consequence. This text, beginning with its introduction on study skills, provides its student readers with methods for studying that will maximize the efforts of the student.

Professors who use this text will also find that it offers them numerous benefits in their teaching of the course. The professor can skim through the FOCUS statements of a unit and quickly determine the content of each section along with its expected level of rigor. The numerous exercises which will be completely explained later in this Preface provide ample opportunity for assignments particularly suited to either individuals or classes. Answers for most of the student exercises relieve the professor from the burden of being the source for that type of feedback and, where answers are not included, the individual problems are keyed to the particular chapter and unit where the student can look for help. An extensive testing program with multiple forms of chapter tests, cumulative tests, and course finals gives the professor an opportunity for testing in a variety of ways, easily giving make-up exams, and implementing a mastery testing program.

Among the unique features of this text is the fact that its introduction is on study skills for mathematics. This introduction can be covered in class or assigned as homework. In either case, the student will have a source to refer to when the actual problems in studying mathematics occur. There are many myths about mathematics and how it is best studied. Developmental students can greatly improve their progress in a mathematics course by addressing the ways to study the subject as well as the mathematics itself.

Another unique feature of this text is the fact that it is written to a specific learning theory. Its title, in fact, comes from that learning theory. FOCUS is the idea that a student always needs to know exactly what needs to be attended to at any moment. Concurrently, the student needs to know that portion of the problem that can be ignored (at least for the moment). Both the writing and the formatting are designed on the basis of giving the student clearly what is to be focused on at any given moment. The design of the text does this by stating in a short sentence the particular focus of any presentation and placing this sentence in an oval on the left side of the page. The text beside the oval FOCUS statement gives a clear example or description of the particular idea that needs to be learned at that time.

Mathematicians will be pleased to find that this text carefully defines the words for the course. Each definition is displayed prominently with a grey background. Each definition is written as an "if and only if" statement. Students are strongly encouraged to clearly establish the meaning of all words and symbols the text defines. Students are also discouraged from memorizing other material in the course. The constant message of the text is encouragement for "figuring out" the processes and the rules that might govern those processes. In other words, the text makes a great effort to show the student those aspects of mathematics that need to be placed in long-term memory while making the plea that the remainder of the course is best learned in terms of its inherent logic.

Each unit ends with an exercise that is separated into three sections: Part A, Part B, and Part C. Part A problems are intended to continue the instructional process by reviewing the foci of the unit and enhancing the material already presented. All answers for Part A problems are given at the back of the text. Instruction is incomplete without feedback.

Part B sections include the traditional drill and practice problems that other mathematics texts use. Answers for all Part B problems are also in the back of the text.

Part C sections review all the material previously learned in the course. For example, a Part C exercise in Chapter 4 will contain problems from Chapters 1-3. These exercises are aimed at counteracting the tendency of many students to learn mathematics just long enough to get through a test. This text demands at the end of each unit that the student show a knowledge of all previous material. This constant review improves the likelihood that the student will begin to integrate new concepts and skills with those previously learned. As this integration occurs, the student's memory greatly improves. Part C problems do not have answers in the back of the text, but each problem is keyed to show the Chapter and Unit in which it was presented. In a way, each Part C section is a final examination for the course to that point in the book.

Word problems are considered of great importance in this text. They are found in every chapter and are included whenever the skill being learned can be enhanced by that type of problem. From the first time that word problems are introduced, every Part C exercise will include some.

Many people have aided us in the development of this text. Our students and colleagues from St. Petersburg Junior College, Clearwater Campus, have played the most crucial role and we wish to publicly thank them for that support. At H&H Publishing, we have had the wonderful assistance of Karen Davis as our editor and Tom Howland as our chief graphics developer.

To the students and faculty who use this text we express the hope that you will let us know your praises and your criticisms. We have tried to do this text perfectly, but know that with your help we can make it even better.

Robert Hackworth
Joseph Howland
Robert Alwin

Clearwater, Florida June 1989

Contents

Focus on Study Skills for Mathematics

Starting college and taking an Arithmetic/Algebra Course is a downer. The student is studying, one more time, material that was not learned before. That is discouraging, and the negative feelings of some students interfere greatly with opportunities for success.

Many students in these situations think something is inherently wrong with them. They tend to give up before they start and make statements like:

> "I've never been good at math."
>
> "I don't have a math mind."
>
> "I don't like math."
>
> "My mother can't do math either."

All of these statements, and others like them, really mean that the student doesn't believe he/she can be successful in a math course. Not surprisingly, once a student accepts that belief the likelihood of doing well in mathematics is slight indeed.

Attitudes of Students Who Do Well in Mathematics

Students that do well in mathematics believe they will succeed and then do the common sense things needed to support that belief.

1. They attend class without fail.

2. They listen carefully and take good notes.

3. They always do all of their assignments.

4. They show all their work and they arrange it very neatly so they can refer to it later and read it easily.

5. They ask questions — lots of questions.

6. They work hard because nothing worthwhile is ever gained without a struggle.

Attitudes of Students Who Don't Do Well in Mathematics

Students who expect to fail in mathematics don't believe they can succeed and frequently do such silly things that their belief comes true.

1. They frequently miss class and make little effort to find out what was missed.

2. They don't listen carefully and rarely take notes that would help them remember.

3. They give up on any assignment that is difficult and skip some problems because they seem easy.

4. They write far less than their teacher would and they write so small that hardly anyone could read it anyway.

5. They hardly ever ask questions and are usually afraid that they will ask a "stupid" question. There really is no such thing as a "stupid" question.

6. They believe they are destined to do poorly anyway so they don't work hard. What's the use??

Today Is a New Day for You and Mathematics

The first objective you need to accomplish in this course is to establish the belief that you can, and will, succeed. This is not an easy task. You need to carefully consider your past experiences in mathematics and make plans to change those behaviors which did not work for you in other courses. Today is a new day in your mathematics life, a chance to start over and do it right, and an opportunity to find the easy ways to study and learn mathematics.

The authors of this book do not believe you are destined to fail. In fact, our many years experience with students like you have convinced us that there is nothing biologically or intellectually deficient about you. More commonly, our students have come to us with beliefs about mathematics and study habits in the subject which have made the subject very hard to learn. The aim of this section of the book is to present efficient and effective ways of dealing with your learning of mathematics.

Many study skills have been preached constantly at you over the years. You have probably been told:

"Read mathematics carefully."

"Keep paper and pencil handy so you can constantly try the problems."

"Arrange your work neatly so you or someone else can find any errors."

"Do all your assignments and never get behind."

All of these statements are excellent advice. If you have ignored them in the past, make certain that you follow them in this course.

In this section, however, we want to go beyond some of the past advice you have been given and expand on four major themes for organizing your study of mathematics. Our aim here is to encourage you to change the ways you study mathematics. It is the way you have studied in the past that is the problem. The mathematics will be far easier if you find new ways to study it.

The four themes for organizing your study of mathematics are from the work of Claire E. Weinstein, Professor of Educational Psychology, University of Texas at Austin. These four themes are:

1. Work and study in quality environments.

2. Learn what is to be memorized and what must not be memorized.

3. Keep your mind active.

4. Learn to evaluate your own work and study habits.

1. Always work and study in a quality environment

A student does better work, stays at it longer, and feels better about the learning when that work is done in a setting in which the student is comfortable and physically relaxed. Your body and your mind should not be the source of any distractions. You want to be able to devote full energies, physical and mental, to your study.

Focus on building a quality study environment

A good restaurant earns its reputation on much more than the quality of its food. A restaurant is judged as much on its atmosphere (ambiance) as on its menu. Customers expect fine food to be served in a comfortable environment.

Make a list of the physical attributes of a fine restaurant. Some of these attributes may be different than you want in a good study. The important new idea to incorporate in your behavior is the selection, by you, of an environment that will allow you to study more effectively. Don't settle for less in your study environment than you demand in your other activities. Otherwise, you will not be your best when studying.

Physical Attributes of Quality Environments

Physical attributes such as lighting, temperature, and noise level are obvious factors to be carefully regulated by the student. Yet these obvious factors are sometimes ignored by students who are having trouble in a subject. Serious study cannot occur while working the drive-in window at MacDonald's or watching the Super Bowl on television. Although these examples may seem extreme, you need to look at your own study sessions and make certain they eliminate all possible physical distractions.

Focus on sounds in your study environment

The sounds in your study environment would be a good place for you to start altering the way you work. Complete silence is not necessarily the best condition, but loud rock and roll is definitely not helpful. You should try to find sound that is soft enough not to be intrusive and with a slow beat. Try playing classical musical at such a low volume that you cannot pay attention to it. That music will over-ride other sounds in your environment that might be distracting.

Psychological Attributes of a Quality Environment

Besides the obvious physical attributes of your learning environment, but equally important, are qualities which are psychological in nature. Fear of mathematics or anxiety over mathematics interferes greatly with the study habits of some students. Those fears or anxieties must be dealt with prior to a study session if they are to be overcome. Your mind, as well as your body, must be free from distractions. A mind that is full of fear or worry has no capacity to also learn mathematics. If your mind is distracted by other thoughts find ways to clear it before studying. Other-wise, it will make your study time inefficient and distasteful. The result will be a lessening of your efforts and eventual failure.

Focus on relaxation techniques

Relaxation techniques can be learned and they will lower the level of fear or anxiety. You can learn methods like Benson's Relaxation Technique from books in the school library. You can also ask for some personal assistance from a school counselor.

All students should learn, as a minimum, to pay attention to their breathing. Do not attempt to alter your breathing. Just pay attention to it and its rhythm. That simple focusing will, by itself, have beneficial results in any anxiety producing situation. Try it. The more you try it, the better will be your results.

Play the Role of a Scholar

A final aspect of your environment needs close attention. Education is more than simply learning a list of facts. It is a different lifestyle with a value system that respects books, music, art, and beauty from an altered perspective. Put good literature, classical music, fine art, and flowers into your environment. Live with these elements in your study environment.

Imagine the private study of a learned professor at Yale and try to create the same study conditions for yourself. Playing the role of a scholar is helpful in becoming a good student.

2. Learn what part of mathematics needs to be memorized and those portions of the subject that must be figured out.

A major problem for many students is the belief that all mathematics should be memorized. This is false and trying to memorize everything always leads to failure. If you have been memorizing much in mathematics, that is not only inefficient – it is destructive. Read this section carefully and develop new ways of dealing with mathematics information.

Memorize Definitions and Symbols

In general, the **only** material that should be memorized in a mathematics course is the meaning of words and symbols. A good mathematics student must know the meaning of a word like "percent" and its symbol "%." **Memorize words and symbols** because their meanings are crucial to working out problems that contain them.

Focus on the meaning of words and symbols

Words and symbols play a crucial role in mathematics which is often overlooked by students. At the same time, words and symbols do not have meanings that can be figured out or discovered. Someone, your teacher or your textbook, needs to give you clear definitions for all the words that are relevant and clear descriptions of any new symbols that are introduced.

Words like "percent" and symbols like the radical sign, $\sqrt{}$, must be clearly understood before problems that contain them can be learned. Such words and symbols must be memorized. Don't skip past them and attempt to do problems without first understanding them.

Do Not Memorize Rules, Laws, or Processes

In general, do not memorize the steps involved in working a problem. **Do not memorize** rules or laws. Such memorization clutters your mind unnecessarily and tends to misdirect all your study. Blindly memorizing a "bunch of rules" is not the same as learning mathematics. Find reasons why rules, laws, processes, or steps work. If you understand why, there will be no need to memorize more than words or symbols.

Focus on not memorizing rules or processes

Treat rules or laws in mathematics just as you treat the legal laws in your community. Nobody memorizes all the state laws. Instead, the good citizen tries to make sense of them and learns how they apply to his/her needs.

Rules and laws of mathematics also need to make sense. Your study of them needs to uncover the logic that supports them. If you learn the reasons for different steps in a problem then you will find it is both unnecessary and unwise to learn the steps themselves!

Focus

Finally, and very importantly, you need to direct your attention to the particular idea or skill that is to be learned. **FOCUS** on each new idea or skill. That is the theme of this text and every effort is made, by the book's format and presentation, to constantly alert you to the new material that must be the object of your attention.

You need to **FOCUS** specifically on the purpose of each new idea or skill as it is presented. Do not be distracted by other parts of a problem or its accompanying material.

Focus on the objective of each lesson

Each new idea or skill in mathematics will be presented to you in a context which contains many other ideas with other skills. It is essential that you recognize both the new ideas and the older ones. You must learn to concentrate on the new ideas. This is what is meant by **focusing**.

Poor mathematics students generally try to learn with only a vague idea of the particular new skill being taught at any moment. Good mathematics students practice the habit of giving their undivided attention to the one new aspect of a problem that needs to be learned while treating the other parts of the problem as background information.

3. Keep your mind active

Learning is an activity of the mind and no one is learning when they aren't thinking.

> **There is no way to divorce learning from thinking.**

You can improve your ability to learn by constantly making certain that you are thinking. Good learning of mathematics occurs when the student is thinking about the mathematics. Poor learning of mathematics occurs when the student's mind is elsewhere.

Focus on thinking while learning

It may come as quite a surprise, but even the best learners encounter difficulties at trying to maintain their thinking. All students, good and bad, have their minds wander. Good students have learned methods for constantly bringing their minds back to the activity. Poor learners have frequently seemed pleased when they could do a lot of problems without thinking.

If you find yourself doing mathematics without thinking, stop and bring yourself back to learning.

Inner Dialogues Keep Your Mind Active

An excellent way to keep your mind active is to carry on a conversation with yourself as you study. This "inner dialogue" might be structured by always asking the following two questions:

> **"What is the meaning of this problem?"**
>
> **"Why is the problem worked this way?"**

Neither of these questions is simple. Both of them force the learner to think and that is their purpose. The first question will keep the student on track in constantly knowing the meaning of words and symbols. The second question will emphasize the need to understand the reasoning behind a problem rather than memorizing the steps used to solve it.

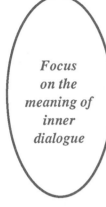

Focus on the meaning of inner dialogue

An "*inner dialogue*" is a form of talking to oneself and such conversations are valuable. It is also valuable to talk with someone else when studying. Find another student who will participate in a common learning experience. Again, the two major questions you can build into your study sessions are:

"What is the meaning of this problem?"

"Why is the problem worked this way?"

You want to constantly look at both meaning and reasoning. The questions, whether in an "inner dialogue" or with a learning "buddy," will structure your thinking and keep you on track for learning mathematics effectively.

4. Learn to evaluate your own learning

Many students believe that it is the task of the teacher to give them a grade and, consequently, accept little responsibility for making accurate judgements about their own progress in a mathematics course. When such students are asked, "How did you do on the test?" they are likely to give very vague or inaccurate responses. This is a characteristic that is drastically different from good students. When good students are asked, "How did you do on the test?" they are likely to be very specific in their comments and will accurately predict the grade they will receive.

Focus on knowing how well you are learning

Most math programs and texts provide students with opportunities for evaluating the extent to which learning has occurred. For example, most texts contain answers for many of the problem sets. Students who check the answers to problems carefully and restudy those problems that are missed gained valuable insight into both the mathematics and their learning of the mathematics.

To make any evaluation of your work pay its best dividends be certain that enough is written to provide a clear trail of your thinking on each problem. Finding wrong answers is not nearly as helpful as finding the point in a problem where you went wrong.

Two Kinds of Knowledge

Knowing how well you are learning in a mathematics course is very different from knowing the mathematics. In other words, there are always two types of knowledge for the mathematics student:

1. Knowledge of the mathematics, and

2. Knowledge about how well the mathematics has been learned.

Knowledge about the mathematics is not enough. The most effective student needs to have a knowledge about his/her learning of the mathematics. In studying mathematics, always pay attention to both the subject and the learning of that subject.

A simple example involves the checking of a problem. The student who does a problem, but skips the check of the problem is accepting responsibility for the mathematics, but ignoring the equally important responsibility for evaluating their learning.

Focus on knowing about yourself

The best of students make mistakes, misunderstand directions, and encounter real difficulties with some topics. But these students have developed mechanisms for telling them when those errors in learning occur.

Knowing when the required material has not been learned gives direction and meaning to further study. Excellence is achieved for the mathematics student when he/she knows the mathematics and also knows that he/she knows the mathematics.

Failures in Learning Are Correctable

A major difference between good and poor mathematics students is the fact that poor students do not know whether they know the mathematics. After studying for a mathematics test, poor students are unable to determine correctly whether they know or don't know the material. This is truly tragic because it is almost always correctable. The tragedy is that poor students generally do not take advantage of the opportunities they have for accurately assessing how much they know.

Use your textbook, your instructor, and other sources for help on and off campus. If you look for them there are ample opportunities for you to determine how much and how well you know the mathematics. Problem sets, review exercises, practice tests, supplementary teaching aids (television, computers, etc.) and learning centers are usually available. Good students take advantage of them. Poor students may not realize their value in the learning process and either ignore them or give them only slight attention.

If you want to improve your knowledge of mathematics, begin by raising your awareness of the ways you use to learn mathematics. All students have troubles, at times, learning the mathematics. Good students realize that and take the necessary steps to overcome the difficulties.

Learn the mathematics, but use every tool available to accurately tell you how well you have learned the mathematics.

A good relationship with yourself, your text book and your instructor will make learning mathematics a very important part of your total educational and real life experiences.

1
Addition and Subtraction of Counting Numbers

Unit 1:   BASE 10 NUMERATION SYSTEM

The first two chapters of this text review the whole numbers and their arithmetic operations. Accordingly, we begin by defining whole numbers. Remember, in mathematics the words that are defined for you are of special importance. Do not overlook them.

In fact, you need to understand the meaning of every defined word so well that you can accurately describe it whenever needed.

Definition	The Set of Whole Numbers

A number is a **whole number** **if and only if** it is in the set $\{0, 1, 2, 3, \ldots\}$.

The three dots in the set braces (. . .) indicate that the numbers continue forever in the same pattern. This means that 4, 5, 6, etc. are also whole numbers.

Focus on the meaning of whole number

The smallest number in the set of whole numbers, $\{0, 1, 2, 3, \ldots\}$, is zero (0).

There is no largest whole number because the three dots (. . .) indicate that the count goes on forever.

Forty-nine (49) is a whole number.

Eight hundred seventy-three (873) is a whole number.

Five million, six hundred forty-two thousand, eight hundred fifteen (5,642,815) is also a whole number.

Counting – The Prerequisite Skill

The one major skill needed to do whole number arithmetic is the ability to count (assign a number name) to each element in a set. Conversely, those who have trouble doing arithmetic usually have trouble counting and, in fact, can greatly improve their ability to do arithmetic simply by practicing their counting skills in a great variety of circumstances.

The box at the right has a number of stars. ☆ ☆ ☆ ☆ ☆ ☆ ☆

Three ways of showing the number of stars are:

The word: seven

The Arabic numeral: 7

The tallies: l̶H̶l̶ ll

Focus on the counting process

The word "seven" is needed to express, orally or in writing, the idea of seven. The symbol "7" allows us to express the idea of seven with a simple numeral. But the tally " l̶H̶l̶ ll" really describes the idea of seven. It even "looks like" the meaning of seven because a person can count the tallies.

We need the word "seven" and the numeral "7" to easily express the idea, but we must always remember the concept of seven which is best indicated by " l̶H̶l̶ ll ."

Addition and Counting

The relationship between counting and addition is crucial to an understanding of arithmetic. To "add 2 and 7" means to "combine the tallies of l l and l̶H̶l̶ ll." The combined count of the tallies gives l̶H̶l̶ llll which is expressed as the symbol 9.

We write 2 + 7 = 9 and that means:

l l added to l̶H̶l̶ ll gives l̶H̶l̶ llll.

Focus on addition being a form of counting

Addition is a form of **counting**. To add 5 and 3 is the same as counting 5 and then counting 3 more. To add 5 + 3 think of 5 objects (❀ ❀ ❀ ❀ ❀) and 3 objects (❀ ❀ ❀).

Count the total number of objects. The total number is 8.

Numerals for Numbers Greater than Ten

The addition of 7 + 5 is shown at the right.

This is twelve tallies

Or one group of ten plus 2

Or 1 ten, 2 ones.

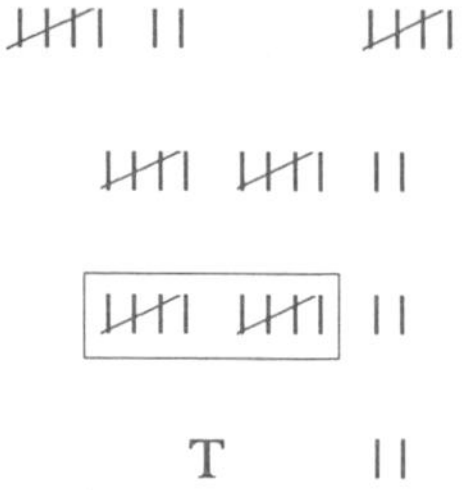

The addition of 7 + 5 gives 12 ones or 1 ten, 2 ones. The numeral "12" has both of these interpretations.

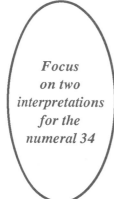

Focus on two interpretations for the numeral 34

The figure at the right shows thirty-four tallies.

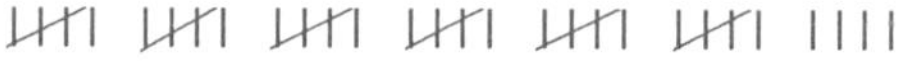

If the tallies are placed in groups of ten, the figure shows three groups of ten with 4 tallies left over.

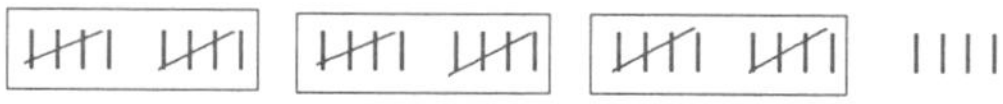

There are two ways to interpret the numeral 34. One interpretation would call it thirty-four ones. A second interpretation would make it three tens, four ones. Either interpretation is correct and both are necessary for understanding arithmetic operations.

Place Values

The **digits** of our base numeration system are 0, 1, 2, 3, 4, 5, 6, 7, 8, 9.

Every digit in a whole number has a place value. The **place value** of a digit is determined by its column.

The figure at the right shows the **place values** of the digits in 8,943,621. 6 is in the **hundred's** place and is called the **hundred's** digit. 4 is the **ten-thousand's** digit.

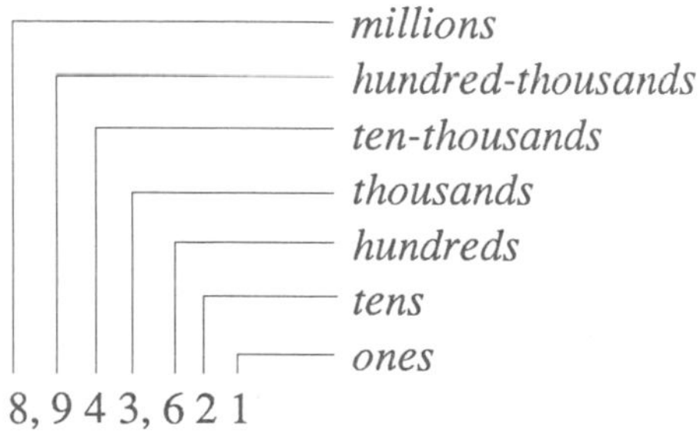

Focus on the place value of digits

In the numeral 5,632 the digit "2" means 2 **ones**, the digit "3" means 3 **tens**, the digit "6" means 6 **hundreds**, and the digit "5" means 5 **thousands**.

In the numeral 4,157,986 the digit "5" means 5 **ten-thousands** and the digit "1" means 1 **hundred-thousands**.

Adjacent Columns in a Place Value Numeral

The numeral 758 means

    7 hundreds, 5 tens, 8 ones.

Adjacent columns of a place value numeral are related by multiples of ten. Each ten is equal to ten ones, each hundred is equal to ten tens, each thousand is equal to ten hundreds, etc.

Focus on the relationship between adjacent column values

If the digit in the hundreds column of 458 is reduced by one, the number of tens is increased by ten.

458 means 4 hundreds   5 tens   8 ones
                    or
458 means 3 hundreds   15 tens   8 ones
                    or
458 means 2 hundreds   25 tens   8 ones
                    or
458 means 1 hundred   35 tens   8 ones

Focus on different interpretations of 758

The numeral 758 can be read and written as:

758 = 7 hundreds, 4 tens, 18 ones   700 + 40 + 18

                    or

758 = 6 hundreds, 15 tens, 8 ones   600 + 150 + 8

Unit 1 Exercise

This exercise reviews the preceding unit. The exercise is divided into three parts. Part A reviews the foci of Unit 1. Part B offers opportunities to practice the skills and concepts of Unit 1. Part C contains problems that review your previous work in this text. All answers for Parts A and B are at the back of the book.

Part A: Reviewing the foci of Unit 1.

1. Add 9 + 4 by counting 4 more than 9.

2. Add 15 + 8 by counting 8 more than 15.

3. Add 16 + 5 by counting 5 more than 16.

4. To add three numbers (addends) add the first two, get an answer (sum), and then add the third addend to the sum.
   6 + 3 + 4 =

5. In the numeral 6,589,143 the digit "6" means 6 millions and the digit "9" means _____.

6. The ten-thousand's digit in 4,367,598 is 6 and the hundred-thousand's digit is ____.

7. A hundred is 10 tens. A thousand is 10 _____.

8. Ten-thousand is 10 thousands. A hundred thousand is 10 _____.

9. 800 has
   8 hundreds, 0 tens, 0 ones or
   7 hundreds, 10 tens, 0 ones or
   7 hundreds, 9 tens, _____ ones.

10. 6,000 has
    6 thousands, 0 hundreds or
    5 thousands, 10 hundreds or
    5 thousands, 9 hundreds, _____ tens.

Part B: Drill and Practice

1. 6 + 3 =

2. 8 + 7 =

3. 5 + 9 =

4. 4 + 4 =

5. 9 + 9 =

6. 8 + 4 =

7. What is the million's digit in 6,498,075?

8. What is the hundred's digit in 5,938,407?

9. What is the one's digit in 8,697,753?

10. What is the ten's digit in 45,687?

11. What is the hundred's digit in 8,697,453?

12. Write a three-digit numeral with one's digit 5, ten's digit 9, and hundred's digit 7.

13. Write a three-digit numeral with hundred's digit 8, ten's digit 6, and one's digit 0.

14. Write a three-digit numeral with ten's digit 7, one's digit 8, and hundred's digit 9.

15. 96 means 9 tens, 6 ones or
    8 tens, _____ ones.

16. 65 means 6 tens, 5 ones or
    5 tens, _____ ones.

17. 917 means 9 hundreds, 1 ten, 7 ones or
    8 hundreds, _____ tens, 7 ones.

18. 6,400 has 6 thousands, 4 hundreds or
    5 thousands, _____ hundreds.

19. 900 has 9 hundreds, 0 tens, 0 ones or
    8 hundreds, _____ tens, 10 ones.

20. 700 has 7 hundreds, 0 tens, 0 ones or
    6 hundreds, 9 tens, _____ ones.

Part C: Enhancing your learning.

An ability to do arithmetic well is closely correlated to an ability to count. Practice counting continually by looking for a variety of situations. Some possibilities are listed below.

1. Use a calendar, count the number of months that have 31 days.

2. Use an alphabetical list of the United States, count the number of states after Florida.

3. On your way to math class, count the number of steps you take from the door of the building to the door of your classroom.

4. A Leap Year has 366 days and occurs every four years (1980, 1984, 1988, etc.). How many Leap Years will there be between 1950 and 1999?

Unit 2:   ADDING IN COLUMNS

Adding Like Terms

If asked to add

    5 apples + 4 cows

the student should reply that it is impossible to add apples and cows. It doesn't make any sense.

Numbers must have the same label to be added and this fact explains addition in columns.

To add 146 + 32

146 means	1 hundred	4 tens	6 ones
+ 32 means	+	3 tens	2 ones
	1 hundred	7 tens	8 ones

The addition of 146 and 32 gives 178 which is 1 hundred, 7 tens, 8 ones.

Focus on writing addition in a vertical form

The problem 56 + 413 means

	5 tens	6 ones
+ 4 hundreds	1 ten	3 ones

The same problem can be written in vertical columns as shown at the right. Notice that digits with the same place value (label) are in the same vertical column.

```
    5 6
+ 4 1 3
```

Addition in Columns

To add   56   the columns are added from right to left.
      + 23

First 6 ones + 3 ones are added.

$$\begin{array}{r} 5\,6 \\ +\,2\,\mathbf{3} \\ \hline 9 \end{array}$$

Then 5 tens + 2 tens are added.

$$\begin{array}{r} 5\,6 \\ +\,\mathbf{2}\,3 \\ \hline 7\,9 \end{array}$$

The answer is 79 which is 7 tens, 9 ones.

*Focus
on adding
in columns*

The addition of 6,542 + 325 means

	6 thousands	5 hundreds	4 tens	2 ones
+		3 hundreds	2 tens	5 ones

In columns, the addition is shown at the right. The addition begins with the ones column and proceeds from right to left.

$$\begin{array}{r} 6\ 5\ 4\ 2 \\ +\ \ \ 3\ 2\ 5 \\ \hline 6\ 8\ 6\ 7 \end{array} = 6{,}867$$

A First Look at the "Carrying" Process

The addition of 7 + 9 is completed vertically as

$$\begin{array}{r} 7 \\ +\,9 \\ \hline 16 \end{array}$$

The answer, 16, has two interpretations:

      16 means   16 ones
                  or
      16 means 1 ten, 6 ones

The "carrying" process for column addition depends upon the second interpretation.

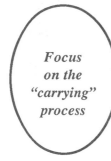

Focus on the "carrying" process

The addition of 800 + 500 may be completed in columns as shown at the right.

$$\begin{array}{r} 8\,0\,0 \\ +\ 5\,0\,0 \\ \hline 1\,3\,0\,0 \end{array} = 1{,}300$$

There are two interpretations for the answer, 1300:

1,300   means   13 hundreds,   0 tens,   0 ones
1,300   means   1 thousand,   3 hundreds,   0 tens,   0 ones

The second interpretation shows how "1 is carried" to the thousands column.

Vertical Addition Requiring "Carrying"

To add 4,093 and 2,395 the numerals are arranged vertically so that **digits with the same place values are in the same column.**

$$\begin{array}{r} 4\,0\,9\,3 \\ +2\,3\,9\,5 \\ \hline \end{array}$$

Each column from right to left is separately added. The answer to the ones column is 8.

$$\begin{array}{r} 4\,0\,9\,3 \\ +2\,3\,9\,5 \\ \hline 8 \end{array}$$

The answer to the tens column is 18. Since 18 requires two digits, we interpret 18 as:
    18 tens   means   1 hundred, 8 tens

$$\begin{array}{r} 1 \\ 4\,0\,9\,3 \\ +2\,3\,9\,5 \\ \hline 8\,8 \end{array}$$

Now the addition in the hundreds column involves the digits 1, 0, and 3. The answer in the hundreds column is 4.

$$\begin{array}{r} 1 \\ 4\,0\,9\,3 \\ +2\,3\,9\,5 \\ \hline 4\,8\,8 \end{array}$$

The addition is completed by adding the digits in the thousands column. The answer in the thousands column is 6.

$$\begin{array}{r} 1 \\ 4\,0\,9\,3 \\ +2\,3\,9\,5 \\ \hline 6\,4\,8\,8 \end{array} = 6{,}488$$

"Carrying" is necessary whenever the sum of a column exceeds nine. When this occurs use "carrying" so that a single digit appears in each column of the answer.

The example of this focus shows the addition of 546 + 308 + 439

<table>
<tr><td></td><td></td></tr>
</table>

1. The addition of the ones column digits
   gives 23 ones which is interpreted as
          23 ones equals 2 tens, 3 ones
   This interpretation means that "2" is
   carried to the tens column and 3 is recorded
   in the ones column.

```
    2
  5 4 6
  3 0 8
+ 4 3 9
      3
```

2. The addition of the tens column
   now involves four digits:
        2 + 4 + 0 + 3
   The answer to the addition in the
   tens column is 9.

```
    2
  5 4 6
  3 0 8
+ 4 3 9
    9 3
```

3. The addition of the hundreds column
   gives 12 hundreds which is interpreted
   as:
          1 thousand, 2 hundreds

```
1   2
  5 4 6
  3 0 8
+ 4 3 9
1 2 9 3
```

4. The sum of the thousands column is 1.

The final answer is 1,293. 1,293 is read
"one thousand two hundred ninety three."

Focus on addition using "carrying"

Unit 2 Exercise

This exercise reviews the preceding unit. The exercise is divided into three parts. Part A reviews the foci of Unit 2. Part B offers opportunities to practice the skills and concepts of Unit 2. Part C contains problems that review your previous work in this text. All answers for Parts A and B are at the back of the book. Each problem of Part C is accompanied by a notation «CU» that refers to the Chapter (C) and Unit (U) in which that type of problem is studied.

Part A: Reviewing the foci of Unit 2.

1. Add 7 + 4 + 6 by first adding 7 and 4.

2. To add 56 + 23 the addends are   56
    first written in vertical form as   + 23
    shown. Write 63 + 25 in
    vertical form.

3. Write 725 + 42 in vertical form.

4. Add the problem at the right   63
    by adding the columns from   + 25
    right to left.

5. Add 438 + 927 using vertical addition and
    show the "carrying" marks.

6. Add 9,036 + 5,281 using vertical addition
    and show the "carrying" marks.

7. Add 543 + 628 by writing the addends
    in vertical columns and show the
    "carrying" marks.

8. Add 905 + 1,483 + 384 using column
    addition and showing all carrying.

9. Add 674 + 8,905 + 655 using column
    addition and showing all carrying.

Part B: Drill and Practice

Add.

1.   5
      + 3

2.   2
      + 9

3.   3
      + 6

4.   4
      + 5

5.   1
      + 8

6.   3
      + 7

7.   12 + 3 =

8.   13 + 9 =

9.   17 + 4 =

10. 16 + 7 =

11. 14 + 2 =

12. 11 + 7 =

13. 6 + 5 + 8 =

14. 4 + 9 + 6 =

15. 2 + 8 + 7 =

16. 6 + 1 + 8 =

17.   4
       3
      + 8

18.   2
       6
      + 9

19.   6
       3
      + 5

20.   7
       4
      + 9

21.   437
     + 551

22.   618
     + 279

23.   84
     + 14

24.   814
     + 152

25. 487 + 252 =

26. 654 + 359 =

27. 495 + 358 =

28. 294 + 6,308 =

29. 8,638 + 958 =

30. 426 + 5,270 + 8,654 =

31. 7,179 + 869 + 7,658 =

32. 6,845 + 97,403 + 6,510 =

33. 44,509 + 81,654 + 3,124 =

34. 4,326,913 + 58,496 =

35. Add 67,315 + 412,684 + 5,094 + 678

36. 575,903 + 96,318 + 4,867 + 718 + 987,865

Part C: Review. Answers for the problems of Part C are not given. However, the notation «C,U» refers to the Chapter (C) and Unit (U) in which problems of the same type were presented.

1. What is the one's digit in 8,697,753? «1,1»

2. What is the ten's digit in 45,687? «1,1»

3. Write the numeral with 7 as its hundreds digit, 6 as its ones digit, and 5 as its tens digit. «1,1»

4. Write the numeral with 5 as its one digit, 4 as its hundreds digit, and 3 as its tens digit. «1,1»

5. 813 means
   8 hundreds, 1 ten, 3 ones or
   7 hundreds, _____ tens, 3 ones.
   «1,1»

6. 5,800 has
   5 thousands, 8 hundreds or
   4 thousands, _____ hundreds.
   «1,1»

Unit 3:   THE DISTANCE BETWEEN WHOLE NUMBERS

THE NUMBER RAY

The figure below is a number ray. The dot at the left end of the number ray shows the position of zero (0). The positions of the numbers 0, 1, 2, 3, 4, 5, 6, 7, 8, and 9 are indicated by marks on the number ray.

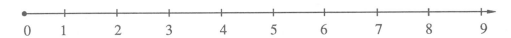

The arrow at the right end of the number ray indicates that the ray continues to the right and has positions for the other whole numbers, 10, 11, 12, etc.

Focus on the position of a number on the number ray

The size of a number is determined by its position on the number ray. The further to the right a number is, the greater its size. Of the numbers 19 and 23, 23 is the greater because it is further to the right on the number ray than 19.

Definition		Greater Than
A number x is **greater than** a number y	**if and only if**	on the number ray, x is further from zero than y.

Greater Than and Less Than

5 is to the right of 2 on the number ray. 5 is also further from zero than 2.

This means that 5 is larger **(greater)** than 2. The symbol for **greater than** is >. The statement that 5 is greater than 2 is written in symbols as 5 > 2.

When 5 is greater than 2, then 2 is **less than** 5. The symbol for **less than** is <. The statement that 2 is **less than** 5 is written in symbols as 2 < 5.

Focus on the meaning of greater than

6 is to the right of 4 on the number ray and is therefore **greater than** 4.

15 is to the right of 8 on the number ray. 15 is **greater than** 8.

Since 6 > 4, then 4 is less than 6.   4 < 6. Similarly, 8 < 15.

Distance on the Number Ray

The figure below shows that 7 is 3 units from 4.

Notice that the distance between two whole numbers is found by counting the number of spaces from the smaller to the larger.

Focus on the distance between whole numbers

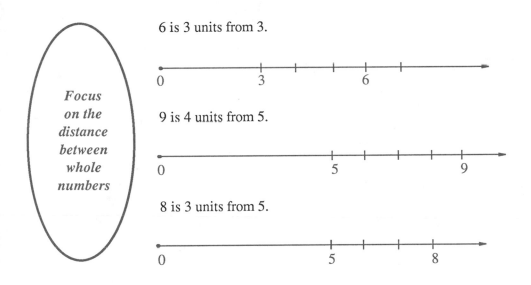

6 is 3 units from 3.

9 is 4 units from 5.

8 is 3 units from 5.

Subtraction and Distance

The question, "How far is it from 5 to 7?" is equivalent to asking, "What is the distance from 5 to 7?"

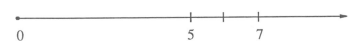

The subtraction problem 7 − 5 = _?_ has the answer 2. The distance from 5, the smaller number, to 7, the larger number, is 2.

Definition   **Subtraction**

The subtraction of two whole numbers x and y $(x - y)$ is the number z	**if and only if**	z is the number that can be added to y to give x.
$(x - y) = z$	**if and only if**	$(y + z) = x$

Subtraction of Whole Numbers

By the definition of subtraction, $8 - 3 = \underline{?}$ asks the question:

What number can be added to 3 to give 8?

The answer to the question is 5. Therefore, $8 - 3 = 5$.

Also by the definition, $4 - 7 = \underline{?}$ asks the question:

What number can be added to 7 to give 4?

There is no whole number answer for that question. For whole number subtraction problems that have no answer, we will use a capital letter X to indicate there is no whole number answer.

Focus on the subtraction of two whole numbers

The subtraction question

$$13 - 8 = \underline{?}$$

asks the question: What number can be added to 8 to give 13? The answer to the question is 5. Therefore, $13 - 8 = 5$.

In the subtraction question $1 - 6 = \underline{?}$, the 6 is greater than the 1 and the question $6 + \underline{?} = 1$ has no whole number answer. Therefore $1 - 6 = X$.

Arranging a Subtraction Problem in Vertical Columns

The subtraction problem 76 − 23 has been written vertically as shown
at the right. Notice that digits with the same place value have been
aligned vertically. The answer to 76 − 23 is 53 because 23 + 53 = 76.

$$\begin{array}{r} 7\ 6 \\ -2\ 3 \\ \hline 5\ 3 \end{array}$$

The subtraction problem 56 − 93 has been written vertically as shown
at the right. Notice again that digits with the same place value have been
aligned vertically. The answer to 56 − 93 is X because there
is no whole number that can be added to 93 to give 56.

$$\begin{array}{r} 5\ 6 \\ -9\ 3 \\ \hline X \end{array}$$

Column Subtraction

The subtraction problem 8,437 − 205 = ? is
written vertically at the right.

$$\begin{array}{r} 8\ 4\ 3\ 7 \\ -\ \ 2\ 0\ 5 \\ \hline 8\ 2\ 3\ 2 \end{array} = 8,232$$

The subtraction is completed by separately
subtracting each column from right to left.
Always begin any vertically arranged subtraction
problem by subtracting the column at the far right.

To subtract 5,384 − 262 the problem is first
written vertically as shown at the right.

$$\begin{array}{r} 5\ 3\ 8\ 4 \\ -\ \ 2\ 6\ 2 \end{array}$$

Subtraction begins in the ones column.
4 − 2 = 2 because 2 + 2 = 4.

$$\begin{array}{r} 5\ 3\ 8\ 4 \\ -\ \ 2\ 6\ 2 \\ \hline 2 \end{array}$$

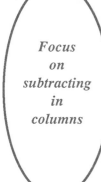

*Focus
on
subtracting
in
columns*

Now the subtraction occurs in the tens column.
8 − 6 = 2 because 6 + 2 = 8.

$$\begin{array}{r} 5\ 3\ 8\ 4 \\ -\ \ 2\ 6\ 2 \\ \hline 2\ 2 \end{array}$$

The subtraction moves to the hundreds column.
3 − 2 = 1 because 2 + 1 = 3.

$$\begin{array}{r} 5\ 3\ 8\ 4 \\ -\ \ 2\ 6\ 2 \\ \hline 1\ 2\ 2 \end{array}$$

Finally, the subtraction finishes with the hundreds
column. 5 − 0 = 5 because 0 + 5 = 5.

$$\begin{array}{r} 5\ 3\ 8\ 4 \\ -\ \ 2\ 6\ 2 \\ \hline 5\ 1\ 2\ 2 \end{array} = 5,122$$

5,384 − 262 = 5,122.

Borrowing Process

The problem 483 − 219 requires "borrowing." Although 219 is less than 483, the subtraction in the ones column is not possible without borrowing because 9 is greater than 3.

$$\begin{array}{r} 4\ 8\ 3 \\ -2\ 1\ 9 \\ \hline X \end{array}$$

To make the subtraction in the ones column possible, 483 is re-written

483 means   4 hundreds, 8 tens, 3 ones

or

4 hundreds, 7 tens, 13 ones

$$\begin{array}{r} 4\ {}^{7}\!8\ {}^{1}3 \\ -2\ 1\ 9 \\ \hline \end{array}$$

The subtraction in columns now can be completed.
The ones column gives 4.   13 − 9 = 4
The tens column gives 6.   7 − 1 = 6
The hundreds column gives 2.   4 − 2 = 2

$$\begin{array}{r} 4\ {}^{7}\!8\ {}^{1}3 \\ -2\ 1\ 9 \\ \hline 2\ 6\ 4 \end{array}$$

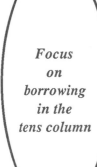

Focus
on
borrowing
in the
tens column

The problem 419 − 245 requires "borrowing" in the tens column.   The subtraction begins with the ones column where the answer is 4.   9 − 5 = 4

$$\begin{array}{r} 4\ 1\ 9 \\ -2\ 4\ 5 \\ \hline 4 \end{array}$$

To make the subtraction in the tens column possible,
4 hundreds, 1 ten   is rewritten as
3 hundreds, 11 tens
The answer for the tens column is 7.   11 − 4 = 7

$$\begin{array}{r} {}^{3}\!4\ {}^{1}1\ 9 \\ -2\ 4\ 5 \\ \hline 7\ 4 \end{array}$$

The subtraction in columns can now be completed by subtracting the digits in the hundreds column.
3 − 2 = 1

$$\begin{array}{r} {}^{3}\!4\ {}^{1}1\ 9 \\ -2\ 4\ 5 \\ \hline 1\ 7\ 4 \end{array}$$

419 − 245 = 174

Borrowing In Two Columns

The problem 6,518 − 2,544
requires borrowing in two
columns.

$$\begin{array}{r} 6\ 5\ 1\ 8 \\ -\ 2\ 5\ 4\ 4 \\ \hline \end{array}$$

The subtraction begins with the
ones column where no borrowing
is necessary.   8 − 4 = 4

$$\begin{array}{r} 6\ 5\ 1\ \mathbf{8} \\ -\ 2\ 5\ 4\ \mathbf{4} \\ \hline \mathbf{4} \end{array}$$

The subtraction in the tens column
requires borrowing because 1 − 4 = X.
  5 hundreds, 1 ten = 4 hundreds, 11 tens
          11 − 4 = 7

$$\begin{array}{r} 6\ \ ^4\!5\ ^1\!1\ 8 \\ -\ 2\ \ 5\ \ \mathbf{4}\ 4 \\ \hline \mathbf{7}\ 4 \end{array}$$

The subtraction in the hundreds column
now requires borrowing because 4 − 5 = X.
  6 thousands, 4 hundreds   becomes
        5 thousands, 14 hundreds
            14 − 5 = 9

$$\begin{array}{r} ^5\!6\ ^{14}\!5\ ^1\!1\ 8 \\ -\ 2\ \ \mathbf{5}\ \ 4\ 4 \\ \hline \mathbf{9}\ \ 7\ 4 \end{array}$$

The subtraction in the thousands column
completes the problem. 5 − 2 = 3
  6,518 − 2,544 = 3,974

$$\begin{array}{r} ^5\!6\ ^{14}\!5\ ^1\!1\ 8 \\ -\ \mathbf{2}\ \ 5\ \ 4\ 4 \\ \hline \mathbf{3}\ \ 9\ \ 7\ 4\ =\ 3{,}974 \end{array}$$

In general, work all subtraction from right to left and only borrow when it becomes necessary.

$$\begin{array}{r} 9\ 3\ 1\ 6 \\ -\ 5\ 6\ 0\ 9 \\ \hline \end{array}$$

The subtraction begins with the ones column and borrowing is necessary. $6 - 9 = X$

$$\begin{array}{r} 9\ 3\ {}^{0}\cancel{1}\ {}^{1}6 \\ -\ 5\ 6\ 0\ 9 \\ \hline 7 \end{array}$$

The subtraction in the tens column requires no borrowing.
$$0 - 0 = 0$$

$$\begin{array}{r} 9\ 3\ {}^{0}\cancel{1}\ {}^{1}6 \\ -\ 5\ 6\ 0\ 9 \\ \hline 0\ 7 \end{array}$$

The subtraction in the hundreds column now requires borrowing because $3 - 6 = X$.
  9 thousands, 3 hundreds becomes
    8 thousands, 13 hundreds

$$\begin{array}{r} {}^{8}\cancel{9}\ {}^{1}3\ {}^{0}\cancel{1}\ {}^{1}6 \\ -\ 5\ 6\ 0\ 9 \\ \hline 7\ 0\ 7 \end{array}$$

The subtraction in the thousands column completes the problem.

$$\begin{array}{r} {}^{8}\cancel{9}\ {}^{1}3\ {}^{0}\cancel{1}\ {}^{1}6 \\ -\ 5\ 6\ 0\ 9 \\ \hline 3\ 7\ 0\ 7 \end{array} = 3{,}707$$

Focus on borrowing from two columns

$$9{,}316 - 5{,}609 = 3{,}707$$

Borrowing In Several Columns

The problem 8,000 − 6,257 represents a special case in borrowing.

```
  8  0  0  0
− 6  2  5  7
```

As usual, the subtraction begins with the ones column. Borrowing is necessary but with zeros in the ten and hundreds column the first borrowing possible is from the thousands column.

```
  7  1
  8  0  0  0
− 6  2  5  7
```

Borrowing is still needed for the ones column but the first place where there is something to borrow is now the hundreds column.

```
     9
  7  1  1
  8  0  0  0
− 6  2  5  7
```

Borrowing is still needed for the ones column but now there is something in the tens column.

```
     9  9
  7  1  1  1
  8  0  0  0
− 6  2  5  7
```

Subtraction in all the columns is now possible without further borrowing.

```
     9  9
  7  1  1  1
  8  0  0  0
− 6  2  5  7
  1  7  4  3
```

8,000 − 6,257 = 1,743

Focus on borrowing from several columns

To subtract 6,000 − 3,179 it is helpful to see that 6,000 is:

6 thousands, 0 hundreds, 0 tens, 0 ones,	6 0 0 0
or	
5 thousands, 10 hundreds, 0 tens, 0 ones.	5 ^{1}0 0 0
or	
5 thousands, 9 hundreds, 10 tens, 0 ones.	5 9 ^{1}0 0
or	
5 thousands, 9 hundreds, 9 tens, 10 ones.	5 9 9 ^{1}0

The four problems below show the successive borrowing described above that is needed to subtract 6,000 − 3,179.

```
  6  0  0  0      5 ¹0  0  0      5  9 ¹0  0      5  9  9 ¹0
− 3  1  7  9    − 3  1  7  9    − 3  1  7  9    − 3  1  7  9
                                                 2  8  2  1
```

The final answer is 2,821.

Unit 3 Exercise

This exercise reviews the preceding unit. The exercise is divided into three parts. Part A reviews the foci of Unit 3. Part B offers opportunities to practice the skills and concepts of Unit 3. Part C contains problems that review your previous work in this text. All answers for Parts A and B are at the back of the book. Each problem of Part C is accompanied by a notation «CU» that refers to the Chapter (C) and Unit (U) in which that type of problem is studied.

Part A: Reviewing the foci of Unit 3

1. The figure below shows a number ray. Which capital letter (A) or (B) indicates the position of 3?

2. On the number ray below, which capital letter shows the position of 10?

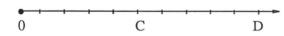

3. On the number ray below which capital letter shows the position of 7?

4. On the number ray below which capital letter shows the position of 5?

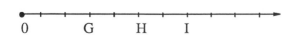

5. How far is 7 from 1?

6. How far is 8 from 3?

7. How far is 8 from 8?

8. How far is 9 from 8?

9. Find the distance from 5 to 8 and answer the subtraction problem 8 − 5 = _____.

10. Find the answer to 4 − 1 = _____ by finding the distance form 1 to 4.

11. Find the answer to 12 − 5 = _____ by finding the distance from 5 to 12.

12. Find the distance from 4 to 6 and find the answer to 6 − 4 = _____.

13. Find the distance from 5 to 14 and find the answer to $14 - 5 =$ _____.

14. Which problem below has X as its answer?
$4 - 1 =$ _____ or $1 - 4 =$ _____

Write the following problems vertically and find the answer.

15. $7 - 2 =$

16. $12 - 6 =$

17. $14 - 8 =$

18. $437 - 125 =$

19. $286 - 143 =$

20. $637 - 407 =$

21. $155 - 31 =$

22. In subtracting $546 - 127 =$ _____ which column requires "borrowing"?

23. In subtracting $546 - 127 = ?$, 546 must be rewritten as 5 hundreds, 3 tens, and _____ ones.

24.   546 $5\,{}^{3}\!4\,{}^{1}6$
    $-\,127$ becomes $-1\;\;2\;\;7$

Find the answer (difference).

25. Will the problem at the right require borrowing?

$$\begin{array}{r} 6157 \\ -\,4032 \\ \hline \end{array}$$

26. Will the problem at the right require borrowing?

$$\begin{array}{r} 4512 \\ -\,2042 \\ \hline \end{array}$$

27. What column requires borrowing in the problem at the right?

$$\begin{array}{r} 4512 \\ -\,2042 \\ \hline \end{array}$$

28. Show the "borrowing" needed to do the problem at the right.

$$\begin{array}{r} 4512 \\ -\,2042 \\ \hline \end{array}$$

29. Is borrowing necessary to subtract in the problem at the right?

$$\begin{array}{r} 4663 \\ -\,4381 \\ \hline \end{array}$$

30. What column of $4,663 - 4,381$ requires borrowing?

31. Show the borrowing needed
    for the problem at the right

    $$\begin{array}{r} 4663 \\ -\,4381 \\ \hline \end{array}$$

32. Subtract 4,663 − 4,381.

33. Subtract 4,765 − 1,538 and clearly
    show any borrowing.

34. Subtract 6,184 − 2,048 and clearly
    show any borrowing.

35. Show all your work in subtracting:
    9,516 − 4,267

36. Show all your work: 4,682 − 3,947

37. To subtract the problem shown
    at the right, 5,000 must be
    rewritten for the borrowing
    process. Show all the work and
    find the answer to 5,000 − 2,043.

    $$\begin{array}{r} 5000 \\ -\,2043 \\ \hline \end{array}$$

Part B: Drill and Practice

1. Is 28 greater than 12?

2. Is 45 greater than 63?

3. Is 102,456 greater than 97,314?

4. Is 57,645 greater than 9,999?

Subtract.

5. 7 − 4

6. 3 − 1

7. 8 − 0

8. 6 − 4

9. 13 − 8

10. 14 − 4

11. 16 − 9

12. 9 − 7

13. 18 − 9

14. 12 − 8

15. 6 − 10

16. 9 − 4

17. 4 − 11

18. 8 − 5

19. 4 − 2

20. 0 − 8

21. 7 − 5

22. 14 − 9

23. 14 − 7

24. 8 − 15

25. 6 − 8

26. 16 − 8

27. 12 − 8

28. 989 − 101

29. 183 − 291

30. 397 − 104

31. 855 − 914

32. 4,512 − 2,042

33. 5,436 − 4,254

34. 4,738 − 1,835

35. 1,747 − 1,651

36. 5,618 − 4,906

37. 9,453 − 2,281

38. 4,927 − 3,182

39. 6,738 − 4,973

40. 4,664 − 1,186

41. 6,805 − 2,241

42. 2,843 − 147

43. 4,000 − 1,564

44. 200 − 1,547

45. 50,406 − 15,698

46. 63,819 − 25,947

47. 47,832 − 26,278

48. 40,340 − 6,186

49. 50,035 − 19,478

Part C: Review. Answers for the problems of Part C are not given. However, the notation «C,U» refers to the Chapter (C) and Unit (U) in which problems of the same type were presented.

1. What is the one's digit in 4,870,576? «1,1»

2. What is the ten's digit in 63,509? «1,1»

3. Write the numeral with 6 as its hundreds digit, 4 as its ones digit, and 0 as its tens digit. «1,1»

4. Write the numeral with 9 as its one digit, 7 as its hundreds digit, and 4 as its tens digit. «1,1»

5. 542 means 5 hundreds, 4 tens, 2 ones or 4 hundreds, _____ tens, 2 ones. «1,1»

6. 8,300 has 8 thousands, 3 hundreds or 7 thousands, _____ hundreds. «1,1»

7. 6,795 + 246 + 7,906 «1,2»

8. 4,103 + 36,090 + 8,531 «1,2»

9. 67,440 + 53,816 + 8,975 «1,2»

10. 3,932,080 + 47,913 «1,2»

11. 52,731 + 325,814 + 3,094 + 125 «1,2»

12. 512,324 + 47,923 + 9,712 + 917 + 798,586 «1,2»

Unit 4:   WORD PROBLEMS INVOLVING ADDITION AND SUBTRACTION

Addition/Subtraction Requires the Same Label

In both addition and subtraction problems, it is extremely important to understand that the numbers being added/subtracted must have the same labels.

Adding or subtracting in columns is possible only because the digits in those columns have the same labels (place values).

Adding 5 houses and 3 dollars is impossible because the labels are different.

Subtracting 8 cows from 11 apples is impossible because the labels are different.

Definition   Addition Terms

The numbers in the addition problem at the right are **addends**. 5,469, 395, and 4,067 are **addends**.

The answer to an addition problem is its **sum** or **total**. 9,931 is the **sum** of the addition at the right.

$$
\begin{array}{r}
5469 \\
395 \\
+\,4067 \\
\hline
9931
\end{array}
$$

Focus on the meaning of addend and sum

The answer to an addition problem is its **sum**. The word **total** is also used for the answer to an addition problem.

In the problem at the right, the numbers 543, 289, 1,038, and 63 are addends. The sum or total is 1,933.

$$
\begin{array}{r}
543 \\
289 \\
1038 \\
+\quad 63 \\
\hline
1933
\end{array}
$$

Clue Words for Addition Problems

When arithmetic problems are stated in words, the student must develop an ability to determine when the problem requires an addition of the numbers.

Although there are no rules with word problems that must always be followed, some words frequently are good clues that addition is necessary.

The words "sum" and "total" are answers for addition problems.

The terms "more than," "increased by," and "plus" also might indicate that addition is the operation to be used.

Focus on solving a word problem requiring addition

Word Problem: A rancher has three pastures measuring 4,352 acres, 518 acres, and 47 acres. Find her total pasture land.

The clue word in the problem is **total**. Since "total" is the answer to an addition problem, the word indicates the numbers should be added.

$$4,352 + 518 + 47 = 4,917$$

Her total pasture land is 4,917 acres.

Definitions   Subtraction Terms

The numbers in a subtraction problem are the **minuend** and **subtrahend**. The answer is the **difference**.

In the problem at the right the **minuend** is 5,867; the **subtrahend** is 2,514; and the **difference** is 3,353.

5867	minuend
− 2514	subtrahend
3353	difference

Focus on the names of numbers in a subtraction problem

The numbers in a subtraction problem are the **minuend, subtrahend,** and **difference.**

$$
\begin{array}{r}
4653 \\
-3197 \\
\hline
1456
\end{array}
$$

minuend
subtrahend
difference

In the problem above, the **minuend** is 4,653, the **subtrahend** is 3,197, and the **difference** is 1,456.

Focus on checking a subtraction computation

To check a subtraction problem, find the sum of the subtrahend and the difference. The sum must be equal to the minuend.

$$
\begin{array}{r}
9 \\
-2 \\
\hline
7
\end{array}
\qquad
\begin{array}{r}
2 \\
+7 \\
\hline
9
\end{array}
$$

To check $9 - 2 = 7$, add 2 and 7. The sum of 2 and 7 is 9 and this checks the computation.

$$
\begin{array}{r}
413 \\
-296 \\
\hline
117
\end{array}
\qquad
\begin{array}{r}
296 \\
+117 \\
\hline
413
\end{array}
$$

To check any subtraction, add the subtrahend and the difference to see if it equals the minuend.

Solving Word Problems Requiring Subtraction

There are clue words which often indicate subtraction is the way to solve a word problem.

The word "difference" is the answer to a subtraction problem.

Subtraction is also normally used to solve problems which ask "how much more" or "how many more."

The terms "decreased by," "less than," "diminished by," or "minus" often indicate a problem that will be solved using subtraction.

Focus on a word problem solved using subtraction

Word Problem: Paul is 18 years old and his father is 41. What is the difference in their ages?

The word **difference** is the clue to solving the problem because "difference" is the answer to a subtraction.

$$41 - 18 = 23$$

23 years is the difference in their ages.

Unit 4 Exercise

This exercise reviews the preceding unit. The exercise is divided into three parts. Part A reviews the foci of Unit 4. Part B offers opportunities to practice the skills and concepts of Unit 4. Part C contains problems that review your previous work in this text. All answers for Parts A and B are at the back of the book. Each problem of Part C is accompanied by a notation «CU» that refers to the Chapter (C) and Unit (U) in which that type of problem is studied.

Part A: Reviewing the foci of Unit 4.

1. Sam Jones owes two bills of $512 and $87. Find the sum of money owed by Sam.

2. It makes no sense to add 3 cats and 4 dogs because the numbers have different labels - cats and dogs. Is it sensible to add 57 apples and 12 peaches?

3. Is it sensible to subtract 47 Fords from 53 horses?

4. To add two numbers they must have the same label. Is it possible to find the total of 19 cows and 483 cows?

5. If Sara has $15 and Mike has $38, how many more dollars does Mike have?

6. If Dan weighs 158 pounds and Joe weighs 141 pounds, how much more does Dan weigh?

7. The **subtrahend** of the problem at the right is 3,197 and the **minuend** is _____.

    $$\begin{array}{r} 4819 \\ -\,3197 \\ \hline 1622 \end{array}$$

8. The **minuend** of $5 - 3$ is 5 and the **subtrahend** is 3. What is the **minuend** of $56 - 47$?

9. In an addition problem the numbers being added are called **addends**. In a subtraction problem the numbers are the _____ and _____.

10. The answer to an addition problem is called its **sum**. What is the sum of 476 + 51?

11. Would subtraction be used to solve? Farmer Smith sold 930 acres from his 2,536 acre ranch. How many acres were left?

12. In an addition problem should the numbers have the same label?

13. In both addition and subtraction problems the numbers must have the same labels. Does it make sense to add 5 dogs and 3 miles?

14. Does it make sense to subtract 47 cows from 58 pigs?

15. Does it make sense to subtract 97 boxes from 482 boxes?

Part B: Drill and Practice

1. What is the **subtrahend** of 963 − 673?

2. The **minuend** of 6,715 − 2,438 is _____.

3. The **subtrahend** of 4,693 − 1,946 is _____.

4. In the problem 17 − 9 = 8 the minuend is _____.

5. In the problem 47 − 35 = 12 the difference is _____.

6. In the problem 56 − 11 = 45 the subtrahend is _____.

7. An addition problem has two or more addends. Does a subtraction problem have two or more minuends?

8. In the problem 2 + 4 = 6, 4 is a/an _____.

Subtract and check.

9. 573 − 268 =

10. 917 − 432 =

11. 465 − 199 =

12. 600 − 362 =

13. If Sam worked 25 days in January and 18 days in February, how many more days did he work in January?

14. Carol had 33 vacation days coming. After a 15 day vacation, how many more days were left?

15. A sculptor started a statue from a block of granite weighing 942 pounds. She estimated chipping off 150 pounds. How much granite was left?

16. If a nurse earns $147 a week and a cab driver earns $205 a week, how much more does a cab driver earn each week?

17. If a man has read 387 pages of a 520 page book, how many pages are left?

18. It is 486 miles from Town A to Town B and only 195 miles from Town A to Town C. How much further is it from Town A to Town B than from A to C?

19. Farmer A has 853 acres and Farmer B has 290 acres. What is the difference in the size of their farms?

20. The words _____ and _____ mean the answer to an addition problem.

21. Bill loaded the following amounts of concrete onto a truck: 100 lbs, 50 lbs, 25 lbs, 93 lbs. Find the total.

22. Find the sum of Julie Black's investments if she has separately invested $46,724 and $9,764.

23. Numbers should not be added if they have _____ (the same, different) labels.

24. Joan has 18 more college credits than Larry. If Larry has 12 credit hours, then how many does Joan have?

25. Mark increased his savings by $34 dollars. If he had $413 in savings, what is his new amount?

26. A druggist prepared 45 prescriptions on Monday, 32 on Tuesday, and 54 Wednesday. What was the total number of prescriptions those three days?

27. A nurse worked 20 days in June, 24 days in July, 19 days in August, and 12 days in September. What was the sum of days worked those four months?

Part C: Review. Answers for the problems of Part C are not given. However, the notation «C,U» refers to the Chapter (C) and Unit (U) in which problems of the same type were presented.

1. What is the hundreds digit in 45,876?
    «1,1»

2. What is the thousands digit in 763,509?
    «1,1»

3. Write the numeral with 5 as its hundreds digit, 8 as its ones digit, and 3 as its tens digit. «1,1»

4. 689 means   6 hundreds, 8 tens, 9 ones or
                5 hundreds, ___ tens, 9 ones.
        «1,1»

5. 53,816 + 246 + 2,506   «1,2»

6. 47,913 + 36,090 + 2,345   «1,2»

7. 800 − 247   «1,3»

8. 50,086 − 15,358   «1,3»

9. 64,289 − 25,947   «1,3»

10. 46,332 − 24,158   «1,3»

11. 340 − 6,186   «1,3»

12. 70,045 − 19,478   «1,3»

CHAPTER 1 TEST

This test reviews the objectives of the chapter. The student is expected to know how to do **all** of these problems before finishing the chapter. Answers for this test are at the back of the book.

«1,U» shows the unit in which the problem is found in the chapter.

1. 4 + 5 =   «1,1»

2. 6 + 8 + 3 =   «1,1»

3. What is the ten's digit of 4,369?   «1,1»

4. What is the ten thousand's digit of 714,638?   «1,1»

5. 743 + 269 =   «1,2»

6. 538 + 894 =   «1,2»

7. 267 + 481 =   «1,2»

8. 913 + 936 =   «1,2»

9. The distance from Town A to Town B is 435 miles and from Town B to Town C is 658 miles. Find the sum of the two distances. «1,4»

10. A carpenter earned $319 one week and $67 the next week. What were the total earnings during the two week period?   «1,4»

11. 124 + 183 + 567 =   «1,2»

12. 924 + 484 + 326 =   «1,2»

13. Arrange the following numbers in size from smallest to largest.

78,   215,   6,032,   139,   394,   76   «1,3»

14. 9 − 5 =   «1,3»

15. 8 − 3 =   «1,3»

16. 5 − 7 =   «1,3»

17. 14 − 9 =   «1,3»

18. 1,749 − 532 =   «1,3»

19. 3,667 − 3,415 =   «1,3»

20. 8,009 − 413 =   «1,3»

21. 7,000 − 3,692 =   «1,3»

22. It is 371 miles from Town A to Town B and only 165 miles from Town A to Town C. How much further is it from Town A to Town B than from A to C?   «1,4»

23. Farmer A has 758 acres and Farmer B has 299 acres. What is the difference in the size of their farms?   «1,4»

2
Multiplication and Division of Counting Numbers

Unit 1:   MULTIPLES OF WHOLE NUMBERS

The Multiplication of Whole Numbers

Multiplying whole numbers can be described in terms of repeated addition.

Multiplying 3 times 5 is the same as adding three 5's.

     $5 + 5 + 5$ is the same as 3×5

The numbers in a multiplication problem are called **factors**.
The answer to a multiplication problem is called its **product**.

> **Definition**   **Multiples**
>
> The number y is   **if and**   there is a whole number
> a **multiple** of n   **only if**   k such that n x k = y.

Multiples of 6

The list 0, 6, 12, 18, 24, 30
shows six **multiples** of 6.

Each number in the list is the product of a whole number and 6.

$$0 \times 6 = 0$$

0 + 6 = 6	1 x 6 = 6
6 + 6 = 12	2 x 6 = 12
12 + 6 = 18	3 x 6 = 18
18 + 6 = 24	4 x 6 = 24
24 + 6 = 30	5 x 6 = 30

After 30 the next **multiple** of 6 is 36 because 30 + 6 = 36.

*Focus on
the multiples
of a
number*

The list 0, 5, 10, 15, 20, 25, 30, 35, 40, 45, 50
shows **multiples** of 5.

The first number in the list is 0 and each other
number is 5 more than the preceding one.

$$0 \times 5 = 0$$

0 + 5 = 5	1 x 5 = 5
5 + 5 = 10	2 x 5 = 10
10 + 5 = 15	3 x 5 = 15
15 + 5 = 20	4 x 5 = 20
20 + 5 = 25	5 x 5 = 25
25 + 5 = 30	6 x 5 = 30
30 + 5 = 35	7 x 5 = 35
35 + 5 = 40	8 x 5 = 40
40 + 5 = 45	9 x 5 = 45
45 + 5 = 50	10 x 5 = 50

The list clearly shows that the multiples of 5 can be found by multiplication **or** by addition. If a multiplication fact is forgotten, do not guess. **Figure it out.** Start with some known fact and add until the desired result is achieved.

Definition   **Consecutive Multiples**

Numbers x and y
(x > y) are   **if and**   x − y = n
consecutive   **only if**
multiples of n

6 and 9 are **consecutive multiples** of 3 because 9 − 6 = 3, but 12 and 21 are not consecutive multiples of 3 because their difference is not equal to 3.

---

*Focus on
the use
of consecutive
multiples*

If a multiplication fact is forgotten or unknown the use of a consecutive multiple often helps in figuring out the desired product.

For example, to find 12 x 7 it may be helpful to know that (12 x 7) and (11 x 7) are consecutive multiples of 7. Consequently,

1. It is relatively easy to see that 11 x 7 = 77

2. Therefore, 12 x 7 will be 77 + 7 or 84.

As another example, the product of 9 x 6 will be 9 more than the product of 9 x 5 because the products are consecutive multiples of 9. Therefore,

$$9 \times 6 = (9 \times 5) + 9$$
$$= 45 + 9$$
$$= 54$$

The Multiplication Facts (Zero Through Nine)

In this table the basic multiplication facts for the whole numbers 0 through 9 are shown.

The table of "facts" is shown at the right. Use the table sparingly. Try to figure out its entries using the ideas of multiples and consecutive multiples.

X	0	1	2	3	4	5	6	7	8	9
0	0	0	0	0	0	0	0	0	0	0
1	0	1	2	3	4	5	6	7	8	9
2	0	2	4	6	8	10	12	14	16	18
3	0	3	6	9	12	15	18	21	24	27
4	0	4	8	12	16	20	24	28	32	36
5	0	5	10	15	20	25	30	35	40	45
6	0	6	12	18	24	30	36	42	48	54
7	0	7	14	21	28	35	42	49	56	63
8	0	8	16	24	32	40	48	56	64	72
9	0	9	18	27	36	45	54	63	72	81

Focus on multiplying by 0

Whenever 0 is multiplied by a whole number the answer is 0.

$$0 \times 7 = 0$$
$$0 \times 4 = 0$$
$$0 \times 0 = 0$$
$$5 \times 0 = 0$$
$$2 \times 0 = 0$$

Any whole number multiplied by zero gives 0.

Also, whenever zero is a number (factor) in a multiplication problem the answer is 0.

Focus on multiplying by 1

$$1 \times 8 = 8$$
$$1 \times 6 = 6$$
$$1 \times 3 = 3$$

When one (1) is multiplied with any whole number the answer will be that same whole number.
$$1 \times 4 = 4$$

Do Not Guess in Multiplication

In multiplying do not guess the answer. Make a list of multiples when
you are not certain. If you don't know the answer to a multiplication you
shouldn't guess. Figure out the answer. Take your time and be correct.

Multiplying Factors with Two or More Digits

The multiplication of factors with two or more digits requires the use of basic
multiplication facts with speed and accuracy. If you are unsure of any
multiplication facts keep reviewing them until they are mastered.

$$
\begin{array}{r}
6\ 3\ 0\ 2\ 1 \\
\times \quad\quad 2 \\
\hline
1\ 2\ 6\ 0\ 4\ 2 = 126{,}042
\end{array}
$$

To multiply 63,021 by 2, each
digit of 63,021 is multiplied by 2.
This process is shown at the left.

2 x 1 one   = 2 ones

2 x 2 tens   = 4 tens

2 x 0 hundreds = 0 hundreds

2 x 3 thousands = 6 thousands

2 x 6 ten thousands = 12 ten thousands or
1 hundred thousand and 2 ten thousands

"Carrying" in Multiplication

$$
\begin{array}{r}
2 \\
3\ 7 \\
\times\ \ 4 \\
\hline
1\ 4\ 8 = 148
\end{array}
$$

The multiplication example shown
at the left requires "**carrying**."
The "carrying" in multiplication is
explained below.

4 x 7 ones = 28 ones or 2 tens, 8 ones
The carrying occurs here. The 2 tens
are carried to the tens column.

4 x 3 tens plus 2 (carried) tens = 14 tens or 1 hundred, 4 tens

Focus on the use of "carrying" in multiplication

```
    3
  5 4 1
x     8
43 2 8  = 4,328
```

In multiplying 541 by 8, each digit of 541 is multiplied by 8. Because 8 x 4 tens = 32 tens, carrying is necessary in the hundreds column and the thousands column.

8 x 1 one = 8 ones

8 x 4 tens = 32 tens = 3 hundreds, 2 tens (3 hundreds are carried)

8 x 5 hundreds plus 3 (carried) hundreds = 43 hundreds or
4 thousands, 3 hundreds

Focus on carrying in the thousand's column

```
    1
  6 5 3 2
x       3
19 5 9 6  = 19,596
```

In the problem at the left, "carrying" is necessary in the thousand's column and the ten thousands column.

3 x 2 ones = 6 ones

3 x 3 tens = 9 tens

3 x 5 hundreds = 15 hundreds = 1 thousand, 5 hundreds
(1 thousand is carried)

3 x 6 thousands plus 1 (carried) thousand = 19 thousands or
1 ten thousand, 9 thousands.

"Carrying" in Three Columns

```
  3 1
  4 6 3
x     5
23 1 5  = 2,315
```

The multiplication shown at the left requires "carrying" in three columns.

5 x 3 ones = 15 ones = 1 ten, 5 ones (1 ten is carried)

5 x 6 tens plus one (carried) ten = 31 tens
          (31 tens = 3 hundreds, 1 ten) 3 hundreds are carried

5 x 4 hundreds plus 3 (carried) hundreds = 23 hundreds
       (23 hundreds = 2 thousands, 3 hundreds)

Focus on carrying "in your head"

The "carrying" in multiplication is best done in your head. For the problem at the right no carrying marks are shown. The multiplication required "carrying" but was completed without writing those "carrying" marks.

A skill at multiplying without writing the carrying marks is needed later. It needs to be practiced.

```
5 7 8 3
x     6
3 4 6 9 8  = 34,698
```

Focus on writing answers, using commas to help in reading answers

To multiply 1,347 by 9 the format shown at the right should be used, including the proper placing of the comma in the product.

```
1 3 4 7
x     9
1 2 1 2 3  = 12,123
```

Unit 1 Exercise

This exercise reviews the preceding unit. The exercise is divided into three parts. Part A reviews the foci of Unit 1. Part B offers opportunities to practice the skills and concepts of Unit 1. Part C contains problems that review your previous work in this text. All answers for Parts A and B are at the back of the book. Each problem of Part C is accompanied by a notation «CU» that refers to the Chapter (C) and Unit (U) in which that type of problem is studied.

Part A: Reviewing the foci of Unit 1.

1. The list 0, 5, 10, 15, 20, 25, 30 shows **multiples** of 5. The next number after 30 in the list is _____.

2. The list 0, 5, 10, 15, 20, 25, 30, 35 shows **multiples** of 5. The next number after 35 in the list is _____.

3. The list 0, 5, 10, 15, 20, 25, 30, 35, 40 shows nine **multiples** of _____.

4. The list 0, 5, 10, 15, 20, 25, 30, 35, 40 shows nine multiples of 5. After 40 what would be the next two multiples of 5?

5. The list 0, 2, 4, 6, 8, 10 shows six multiples of 2. After 10, what would be the next multiple of 2?

6. The list 0, 2, 4, 6, 8, 10, 12 shows seven multiples of _____.

7. Make a list, starting with 2, of eight multiples of 2. _____

8. 9, 12, 15 is a list of multiples of 3. What are the next five multiples of 3?

9. The list 8, 12, 16, 20 shows four multiples of _____.

10. 8, 12, 16, 20. After 20 what is the next multiple of 4?

11. 8, 12, 16, 20 is a list of four multiples of 4. After 20, what are the next five multiples of 4?

12. Start with 2 and make a list of ten multiples of 2.

13. Start with 3 and make a list of ten multiples of 3.

14. Start with 4 and make a list of ten multiples of 4.

15. Start with 5 and make a list of ten multiples of 5.

16. Multiples of 6 are 0, 6, 12, 18, 24, 30. After 30, what are the next five multiples of 6?

17. The numbers in the list 21, 28, 35, 42, 49 are multiples of _____.

18. The list 0, 3, 6, 9, 12, 15 shows six multiples of 3. After 15, what would be the next multiple of 3?

19. The numbers 30, 35, 40 are three consecutive multiples of 5. The numbers 16, 20, 24 are three consecutive multiples of _____.

20. 5 x 1 = 5
    2 x 1 = 2
    7 x 1 = _____.

21. Multiply 1,302 by 3 by multiplying each digit of 1,302 by 3.

$$\begin{array}{r} 1302 \\ \times\quad 3 \\ \hline \end{array}$$

22. Multiply 1,304 by 2 by multiplying each digit of 1,304 by 2.

$$\begin{array}{r} 1304 \\ \times\quad 2 \\ \hline \end{array}$$

23. Multiply without writing any carrying symbols.

$$\begin{array}{r} 4156 \\ \times\quad 4 \\ \hline \end{array}$$

24. Multiply without writing any carrying symbols.

$$\begin{array}{r} 2023 \\ \times\quad 6 \\ \hline \end{array}$$

Part B: Drill and Practice

Multiply.

1. 0 x 8

2. 0 x 3

3. 1 x 9

4. 4 x 7

5. 6 x 9

6. 9 x 7

7. 8 x 6

8. 3 x 8

9. 5 x 9

10. 9 x 4

11. 6 x 6

12. 32, 40, 48 are three consecutive multiples of 8. 24, 30, 36 are three consecutive multiples of _____.

13. 21, 28, 35 are three consecutive multiples of _____.

14. 27, 36, 45, 54 are four consecutive multiples of _____.

15. 18, 21, 24, 27 are four consecutive multiples of _____.

16. Make a list of ten multiples of:
    a) 2
    b) 5
    c) 7
    d) 8

17. 12, 18, 24, 30, 36 are consecutive multiples of _____.

18. 36, 45, 54, 63 are consecutive multiples of _____.

19. 12, 15, 18, 21 are consecutive multiples of _____.

20. After 28, what is the next multiple of 7?

21. After 32, what is the next multiple of 4?

Multiply.

22. 5 x 8

23. 7 x 3

24. 9 x 4

25. 8 x 0

26. 8 x 7

27. 7 x 4

28. 4 x 9

29. 3 x 7

30. 4 x 8

31. 8 x 6

32. 6 x 7

33. $\begin{array}{r} 5173 \\ \times\ \ 2 \\ \hline \end{array}$

34. $\begin{array}{r} 4536 \\ \times\ \ 7 \\ \hline \end{array}$

35. $\begin{array}{r} 6154 \\ \times\ \ 8 \\ \hline \end{array}$

36. $\begin{array}{r} 3157 \\ \times\ \ 6 \\ \hline \end{array}$

37. $\begin{array}{r} 4936 \\ \times\ \ 5 \\ \hline \end{array}$

38. $\begin{array}{r} 8159 \\ \times\ \ 9 \\ \hline \end{array}$

39. $\begin{array}{r} 9317 \\ \times\ \ 3 \\ \hline \end{array}$

40. $\begin{array}{r} 4939 \\ \times\ \ 4 \\ \hline \end{array}$

41. 2607
    x   7

42. 4883
    x   9

Part C: Review. Answers for the problems of Part C are not given. However, the notation «C,U» refers to the Chapter (C) and Unit (U) in which problems of the same type were presented.

1. What is the hundreds digit in 5,316? «1,1»

2. What is the thousands digit in 478,509? «1,1»

3. Write the numeral with 6 as its hundreds digit, 2 as its ones digit, and 5 as its tens digit. «1,1»

4. 429 means   4 hundreds, 2 tens, 9 ones or
                3 hundreds, ____ tens, 9 ones.
        «1,1»

5. 51,316 + 746 + 2,406   «1,2»

6. 53,813 + 57,090 + 5,045   «1,2»

7. 1,300 – 287   «1,3»

8. 47,086 – 33,358   «1,3»

9. 64,469 – 25,917   «1,3»

10. 46,332 – 54,158   «1,3»

11. Bob loaded the following amounts of hay onto a truck: 180 lbs, 75 lbs, 28 lbs, 45 lbs. Find the total.   «1,4»

12. Find the sum of Mr. Smart's income if he has separately earned $35,314 and $8,464. «1,4»

13. Charles is reading a 520 page book. He is currently on page 347. How many pages are left to be read?
        «1,4»

Unit 2:   MULTIPLYING COUNTING NUMBERS

Two-Digit Multipliers

To multiply 67 x 485, the
465 is multiplied by each
digit of 67. This
multiplication is shown
at the right.

```
      4 8 5
    x  6 7
    3 3 9 5   =  7 ones x 485
    2 9 1 0   =  6 tens x 485
  3 2 4 9 5   =  32,495
```

The numbers 3,395 and 2,910 are **partial products**. Because the 6
in 67 is the ten's digit, the partial product 2,910 is placed directly
under the 6.

The partial products are added in columns to give the final
answer, 32,495. This final answer is the **product** of 67 x 485.

*Focus on
multiplying
larger
numbers*

To multiply 54 x 914, the 914
is multiplied by each digit of
54. The example at the right
shows the completed
multiplication.

```
      9 1 4
    x  5 4
    3 6 5 6
  4 5 7 0
  4 9 3 5 6  =  49,356
```

Multiplying by a Numeral with Three Digits

In the example at the left, 619 is multiplied
by 408. Notice that three partial products
are required; one for each digit of 408.
Notice that the partial product when multiplying by zero is zero.

```
      6 1 9
    x 4 0 8
      4 9 5 2
      0 0 0 0
    2 4 7 6
  2 5 2 5 5 2
```

4952 = 8 ones x 619
0000 = 0 tens x 619
2476 = 4 hundreds x 619

252,552 is the final product of 619 and 408.

Focus on multiplication

The example at the right shows the multiplication of 9,306 and 839.

```
    9 3 0 6
   x 8 3 9
   8 3 7 5 4
   2 7 9 1 8
  7 4 4 4 8
  7 8 0 7 7 3 4  = 7,807,734
```

Multiplication Terminology and Applications

In a multiplication problem the numbers being multiplied are **factors**. For example, in the problem 6 x 9 = 54, the numbers 6 and 9 are **factors**. 54 is the **product**.

When two numbers are multiplied using a vertical arrangement of the **factors**, the **factor** on top is the **multiplicand** and the factor on the bottom is the **multiplier**.

```
    4 3 8      multiplicand
   x 6 9       multiplier
  3 9 4 2      partial product
  2 6 2 8      partial product
  3 0 2 2 2    product
```

438 x 69 = 30,222

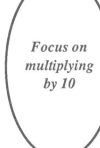

Focus on multiplying by 10

When one factor of a multiplication problem is 10, the product will always be the other factor with a zero affixed.

10 x 98 = 980

10 x 45 = 450

10 x 398 = 3,980

10 x 46,512 = 465,120

Focus on multiplying by 100

To multiply a number by 100, just affix two zeros to the number.

$$546 \times 100 = 54,600$$

$$15 \times 100 = 1,500$$

$$3,858 \times 100 = 385,800$$

Solving Word Problems Involving Multiplication

Frequently, a word problem will be solved by multiplication and some words are good clues that multiplication is needed.

The word "product" is the answer to a multiplication problem and the words "factor," "multiplier," and "multiplicand" are numbers in a multiplication problem.

Another good clue word for word problems requiring multiplication is "each."   If the word "each" appears in a word problem then it is wise to consider the likelihood that the numbers should be multiplied.

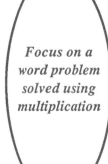

Focus on a word problem solved using multiplication

Word Problem: Find the product of 14 and 25.

The word "product" is the answer for a multiplication problem so the product of 14 and 25 is found by multiplying the two numbers.

```
    1 4              2 5
    2 5              1 4
    7 0            1 0 0
  2 8                2 5
  3 5 0            3 5 0
```

The answer is 350. Note: The multiplicand and multiplier may be interchanged without changing the product.

Unit 2 Exercise

This exercise reviews the preceding unit. The exercise is divided into three parts. Part A reviews the foci of Unit 2. Part B offers opportunities to practice the skills and concepts of Unit 2. Part C contains problems that review your previous work in this text. All answers for Parts A and B are at the back of the book. Each problem of Part C is accompanied by a notation «CU» that refers to the Chapter (C) and Unit (U) in which that type of problem is studied.

Part A: Reviewing the foci of Unit 2.

1. Numbers in a multiplication problem do not have to have the same label. Do numbers in an addition problem have to have the same label?

2. Numbers in a multiplication problem may have different labels. Can numbers in a subtraction problem have different labels?

3. The numbers of a multiplication problem are the multiplicand and the multiplier or both may be called _____.

4. The answer to a multiplication problem is its _____.

Complete the following multiplication.

5.
$$\begin{array}{r} 874 \\ \times\,53 \\ \hline 2622 \\ 4370 \end{array}$$

6.
$$\begin{array}{r} 537 \\ \times\,39 \\ \hline 4833 \end{array}$$

7.
$$\begin{array}{r} 2031 \\ \times\,457 \\ \hline 14217 \\ 10155 \end{array}$$

8.
$$\begin{array}{r} 4915 \\ \times\,250 \\ \hline 0000 \\ 24575 \end{array}$$

9. $10 \times 57 = 570$
    $10 \times 93 = $ _____

10. $10 \times 170 = 1,700$
    $10 \times 580 = $ _____

11. $100 \times 710 = 71,000$
    $100 \times 320 = $ _____

12. $10,000 \times 38 = 380,000$
    $10,000 \times 14 = $ _____

Part B: Drill and Practice

Multiply.

1.
$$\begin{array}{r} 65 \\ \times\ 8 \\ \hline \end{array}$$

2.
$$\begin{array}{r} 35 \\ \times\ 6 \\ \hline \end{array}$$

3.
$$\begin{array}{r} 3175 \\ \times\ \ \ 7 \\ \hline \end{array}$$

4.
$$\begin{array}{r} 4076 \\ \times\ \ \ 9 \\ \hline \end{array}$$

5.
$$\begin{array}{r} 234 \\ \times\ 18 \\ \hline \end{array}$$

6.
$$\begin{array}{r} 205 \\ \times\ 73 \\ \hline \end{array}$$

7.
$$\begin{array}{r} 3852 \\ \times\ 602 \\ \hline \end{array}$$

8.
$$\begin{array}{r} 3417 \\ \times\ 450 \\ \hline \end{array}$$

9.
$$\begin{array}{r} 408 \\ \times\ 73 \\ \hline \end{array}$$

10.
$$\begin{array}{r} 897 \\ \times\ 38 \\ \hline \end{array}$$

11.
$$\begin{array}{r} 471 \\ \times\ 11 \\ \hline \end{array}$$

12.
$$\begin{array}{r} 1635 \\ \times\ \ \ 9 \\ \hline \end{array}$$

13.
$$\begin{array}{r} 1483 \\ \times\ 58 \\ \hline \end{array}$$

14.
$$\begin{array}{r} 359 \\ \times\ 104 \\ \hline \end{array}$$

15.
$$\begin{array}{r} 4382 \\ \times\ 53 \\ \hline \end{array}$$

16.
$$\begin{array}{r} 3007 \\ \times\ 19 \\ \hline \end{array}$$

17. 327 x 10

18. 785 x 100

19. 3,215 x 10

20. 453 x 1,000

21. A druggist received 28 cartons with 24 jars of vitamins in each carton. How many vitamin jars were there?

22. Bob Jones is paid $204 each week. How much does he earn in 15 weeks?

Part C: Review. Answers for the problems of Part C are not given. However, the notation «C,U» refers to the Chapter (C) and Unit (U) in which problems of the same type were presented.

1. What is the hundreds digit in 51,876? «1,1»

2. What is the thousands digit in 609,509? «1,1»

3. Write the numeral with 2 as its hundreds digit, 7 as its ones digit, and 6 as its tens digit. «1,1»

4. 352 means   3 hundreds, 5 tens, 2 ones or
                2 hundreds, _____ tens, 2 ones.
   «1,1»

5. 93,816 + 746 + 2,406   «1,2»

6. 40,513 + 56,090 + 2,145   «1,2»

7. 7,300 – 447   «1,3»

8. 40,086 – 18,358   «1,3»

9. 14,289 – 25,947   «1,3»

10. 47,882 – 34,158   «1,3»

11. 5836   «2,1»
    x   4

12. 6159   «2,1»
    x   8

13. Numbers should not be added if they have _____ (the same, different) labels.   «1,4»

14. Frank has 52 more college credits than Jean. If Jean has 15 credit hours, then how many does Frank have?   «1,4»

Unit 3:   MULTIPLES AND DIVISION

Associating Multiples With Division

Multiples of 6 are: 0, 6, 12, 18, 24, 30, 36, 42, 48, 54, etc.

The **greatest multiple of 6** that can be subtracted from 34 is 30.

34 – 30 = 4   and   30 = 6 x 5

30 is **less than** 34 and it is the **greatest multiple of 6** that
is **less than 34**.

Focus on multiples

To find the greatest multiple of 7 that can be subtracted from 31,
1.   List the multiples of 7.
      7, 14, 21, 28, 35, 42, etc.
2.   Select the largest multiple that is less than 31.
      28 is less than 31.

Division Symbols

The division of 8 by 3 is indicated by

$$8 \div 3 \quad \text{or} \quad 3\overline{)8}$$

To divide 8 by 3 it is necessary to ask:

**What is the greatest multiple of 3 that can
be subtracted from 8?**

The answer, 6 (6 = 3 x 2), means that 3 goes into 6 two times.
When 6 is subtracted from 8 it gives a remainder (R) of 2.

$$
\begin{array}{r}
2 \\
3\,\overline{)\,8} \\
-\underline{6} \\
2
\end{array}
\qquad\qquad 8 \div 3 = 2\ \text{R2}
$$

Steps in Division

To do the division problem 17 ÷ 3

Ask:	What is the greatest multiple of 3 that can be subtracted from 17?
Answer:	15 because 3 x 5 is less than 17 and 3 x 6 is greater than 17.
Place:	5 is placed directly above the 17 and 3 x 5 or 15 is placed directly below 17.

$$
\begin{array}{r}
5 \\
3\,\overline{)\,1\,7} \\
-\underline{1\,5}
\end{array}
$$

Subtract: Subtract the multiple of 3 (3 x 5 = 15) from 17 to find the remainder.

$$
\begin{array}{r}
5\ \text{R2} \\
3\,\overline{)\,1\,7} \\
-\underline{1\,5} \\
2
\end{array}
$$

To divide 8,302 ÷ 6 the following steps are used:

1. Begin as if the problem
    were: 8 ÷ 6

```
      1
6 ) 8 3 0 2
   -6
    2
```

2. Bring down the 3 of 8,302
    and divide: 23 ÷ 6

```
     1 3
6 ) 8 3 0 2
   -6
    2 3
   -1 8
      5
```

3. Bring down the 0 of 8,302
    and divide: 50 ÷ 6

```
     1 3 8
6 ) 8 3 0 2
   -6
    2 3
   -1 8
      5 0
     -4 8
        2
```

*Focus on
division by a
single digit
divisor*

4. Bring down the 2 of 8,302
    and divide: 22 ÷ 6

```
     1 3 8 3
6 ) 8 3 0 2
   -6
    2 3
   -1 8
      5 0
     -4 8
        2 2
       -1 8
          4
```

5. The remainder is 4.

    8,302 ÷ 6 = 1,383 R4

In Division Move from Left to Right

To divide 4,395 by 6, each digit of 4,395 is divided by 6. This process moves from left to right.

The completed long division problem is shown at the right.

Notice that each step in the division requires that one more digit of 4,395 be "brought down." The division ends with a remainder when no more digits of 4,395 can be "brought down."

$$
\begin{array}{r}
0\,7\,3\,2 \text{ R3} \\
6\,)\overline{4\,3\,9\,5} \\
-\underline{0} \\
4\,3 \\
-\underline{4\,2} \\
1\,9 \\
-\underline{1\,8} \\
1\,5 \\
-\underline{1\,2} \\
3
\end{array}
$$

Focus on neatness in long division

In dividing 2,514 by 7, each digit of 2,514 has an "answer" digit directly above it.

Writing the problem neatly and maintaining the columns will, by itself, greatly improve your long division accuracy.

$$
\begin{array}{r}
0\,3\,5\,9 \text{ R1} \\
7\,)\overline{2\,5\,1\,4} \\
-\underline{0} \\
2\,5 \\
-\underline{2\,1} \\
4\,1 \\
-\underline{3\,5} \\
6\,4 \\
-\underline{6\,3} \\
1
\end{array}
$$

Maintain Vertical Columns Carefully

In dividing 2,741 by 6, it is important to remember that one digit of the answer goes directly above each digit of 2,741.

In practice, the first digit of the answer that is written is not zero.

This common practice is shown in the example at the right.

$$
\begin{array}{r}
4\,5\,6 \text{ R5} \\
6\,)\overline{2\,7\,4\,1} \\
-\underline{2\,4} \\
3\,4 \\
-\underline{3\,0} \\
4\,1 \\
-\underline{3\,6} \\
5
\end{array}
$$

Unit 3 Exercise

This exercise reviews the preceding unit. The exercise is divided into three parts. Part A reviews the foci of Unit 3. Part B offers opportunities to practice the skills and concepts of Unit 3. Part C contains problems that review your previous work in this text. All answers for Parts A and B are at the back of the book. Each problem of Part C is accompanied by a notation «CU» that refers to the Chapter (C) and Unit (U) in which that type of problem is studied.

Part A: Reviewing the foci of Unit 3.

1. What number can be multiplied by 4 to give 20?

2. What number can be multiplied by 9 to give 36?

3. 24 divided by 6 is a division problem. The answer is 4 because $24 = 6 \times$ _____.

4. 32 divided by 4 is a division problem. The answer is 8 because $32 = 4 \times$ _____.

5. $40 \div 8$ is a division problem. The answer is 5 because $40 = 8 \times$ _____.

6. $0 \div 7 = 0$ because $0 = 7 \times$ _____.

7. $12 \div 4 = 3$ because $12 = 4 \times$ _____.

8. $64 \div 8 = 8$ because $64 = 8 \times$ _____.

9. What is the greatest multiple of 3 that can be subtracted from 23?

10. What is the greatest multiple of 3 that can be subtracted from 2?

11. What is the greatest multiple of 4 that is less than 27?

12. What is the greatest multiple of 5 that is less than 32?

13. The division problem $12 \div 5$ is written at the right using the long division symbol. $5\,\overline{)\,1\,2}$ Write $6 \div 7$ using the long division symbol.

14. Write $13 \div 4$ using the long division symbol.

15. Write $9 \div 6$ using the long division symbol.

16. Show all your work for $9 \div 5$

17. Show all your work for $4 \div 6$

18. Show all your work for $6 \div 3$

19. Show all your work for $6 \div 9$

20. To divide $37 \div 5$ first divide $3 \div 5$. Show all your work for $37 \div 5$.

Part B: Drill and Practice

Divide.

1. $18 \div 9$

2. $12 \div 6$

3. $35 \div 7$

4. $54 \div 9$

5. $6 \div 6$

6. $0 \div 8$

7. $9 \div 3$

8. $12 \div 4$

9. $63 \div 7$

10. $0 \div 6$

11. $18 \div 3$

12. $8 \div 8$

13. $36 \div 9$

14. $24 \div 4$

15. $5 \div 4$

16. $4 \div 8$

17. $1 \div 2$

18. $8 \div 4$

19. $9 \div 8$

20. $6 \overline{)387}$

21. $4 \overline{)597}$

22. $9 \overline{)3845}$

23. $6 \overline{)1973}$

24. $3 \overline{)1826}$

25. $5 \overline{)6537}$

26. $7 \overline{)1815}$

Part C: Review. Answers for the problems of Part C are not given. However, the notation «C,U» refers to the Chapter (C) and Unit (U) in which problems of the same type were presented.

1. What is the hundreds digit in 47,936? «1,1»

2. What is the thousands digit in 980,509? «1,1»

3. Write the numeral with 5 as its hundreds digit, 2 as its ones digit, and 4 as its tens digit. «1,1»

4. 297 means   2 hundreds, 9 tens, 7 ones or
              1 hundred, _____ tens, 7 ones.
        «1,1»

5. 41,816 + 356 + 2,587   «1,2»

6. 68,413 + 83,592 + 5,927   «1,2»

7. 6,800 − 936   «1,3»

8. 30,907 − 19,238   «1,3»

9. 57,289 − 25,947   «1,3»

10. 27,882 − 34,158   «1,3»

11. 6708   «2,1»
    x   4

12. 5821   «2,1»
    x   8

13. 7328   «2,2»
    x   6

14. 4583   «2,2»
    x   7

15. Numbers can be added only if they have _____ (the same, different) labels.   «1,4»

16. Eric weighs 52 more pounds than John. If John weighs 139 pounds, then how much does Eric weigh?   «1,4»

17. A grocer received 28 cartons with 24 cans of corn in each carton. How many cans of corn were there?   «2,2»

18. Jim receives an allowance of $145 each month for his school costs. How much allowance does he receive in 9 months?   «2,2»

Unit 4:   LONG DIVISION

Using a Divisor with More Than One Digit

The long division examples below follow exactly the same process as that used in Unit 3.

```
        1 5 6  R4                          1 2 6 1  R61
   24)3 7 4 8                         347)4 3 7 6 2 8
     -2 4                                -3 4 7
      1 3 4                                9 0 6
     -1 2 0                               -6 9 4
        1 4 8                              2 1 2 2
       -1 4 4                             -2 0 8 2
            4                                 4 0 8
                                             -3 4 7
                                               6 1
```

Maintain all the vertical columns carefully.

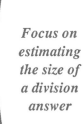

Focus on estimating the size of a division answer

In dividing whole numbers two important facts need to be noted before any computation is attempted.

1. Determine the position of the first non-zero digit in the answer.

2. Determine the number of digits in the completed answer.

In the problem at the right, the position of the first non-zero digit in the answer is indicated by the letter x. Other digits in the answer have been indicated by the letters a, b and c. The answer to the division will have four digits.

```
           0 0 x a b c
   463)8 1 4 3 7 6
     -0
      8 1
     -0 0
      8 1 4
```

Division Nomenclature And Applications

In a division problem each of the
numbers has a special name.

In the problem at the right,
628 is the **divisor**, 85 is the
quotient, 53,672 is the **dividend**,
and 292 is the **remainder**.

$$
\begin{array}{r}
85 \;\; R292 \\
628\,\overline{)\,53\,672} \\
-5\,0\,24 \\
\hline
3\,4\,3\,2 \\
-3\,1\,4\,0 \\
\hline
2\,9\,2
\end{array}
$$

Focus on checking division

To check the accuracy of a division answer multiply
the quotient and the divisor. This product is added
to the remainder. If the sum equals the dividend
the division is correct.

To check:

$$
\begin{array}{r}
14 \;\; R3 \\
5\,\overline{)\,73}
\end{array}
$$

Multiply: 14 x 5 = 70
Add: 70 + 3 = 73

Finding the Average

To find the **average** of the numbers 59, 68, and 62, add the numbers and divide
the sum by 3, the number of addends used to determine the sum.

$$
\begin{array}{r}
59 \\
68 \\
+62 \\
\hline
189
\end{array}
\qquad
\begin{array}{r}
63 \\
3\,\overline{)\,189} \\
-18 \\
\hline
09 \\
-\;9 \\
\hline
0
\end{array}
$$

The **average** of 59, 68, and 62 is 63.

Clue Words for Solving Problems Involving Division

The word "quotient" means the answer to a division problem and the words "divisor" and "dividend" mean numbers in a division problem. The presence of any of these words in a word problem is a good indication that division is needed to solve the problem.

Another good clue word indicating division is "each." When the word "each" appears in a word problem it usually means that either multiplication or division is needed.

Focus on the word "each" when solving word problems

Word Problem: Laura earns $52 each week. How many weeks does Laura need to work to earn $1,000?

The word **each** in the problem indicates that the problem is solved by division.

$$
\begin{array}{r}
19 \text{ R12} \\
52 \overline{)1000} \\
-52 \\
\hline
480 \\
-468 \\
\hline
12
\end{array}
$$

The quotient is 19 R12 and this means that after 19 weeks Laura will need $12 more to reach her $1,000 goal. The answer to the problem is best stated as 20 weeks.

In 19 weeks Laura will earn 19 x $52 or $988. To reach her goal of $1,000 she will have to earn $12 more in her twentieth week, ($1,000 − $988 = $12).

Unit 4 Exercise

This exercise reviews the preceding unit. The exercise is divided into three parts. Part A reviews the foci of Unit 4. Part B offers opportunities to practice the skills and concepts of Unit 4. Part C contains problems that review your previous work in this text. All answers for Parts A and B are at the back of the book. Each problem of Part C is accompanied by a notation «CU» that refers to the Chapter (C) and Unit (U) in which that type of problem is studied.

Part A: Reviewing the foci of Unit 4.

1. Use an X to mark the position of the first non-zero digit in the quotient of the problem shown at the right.

$$92 \overline{)43871}$$

2. How many digits will be in the quotient of the problem shown at the right.

$$92 \overline{)43871}$$

3. Use an X to mark the position of the first non-zero digit in the answer of the problem shown at the right.

$$52 \overline{)98372}$$

4. How many digits will be in the division answer of the problem shown at the right.

$$52 \overline{)98372}$$

5. Use an X to mark the position of the first non-zero digit in the answer of the problem shown at the right.

$$83 \overline{)72516}$$

6. How many digits will be in the division answer of the problem shown at the right.

$$83 \overline{)72516}$$

7. The example at the right shows 3 as the first non-zero digit of the answer. Multiply 3 x 576 and write the product in its correct position.

$$576 \overline{)16739} \quad {}^{3}$$

8. The work shown at the right clearly shows that 3 is not the first non-zero digit of the answer. Why?

$$576 \overline{)16739} \quad {}^{3}$$
$$- 1728$$

9. 3 is not the correct first non-zero digit of the answer. The correct first digit will be _____ (less, greater) than 3.

$$576 \overline{)16739} \quad {}^{3}$$
$$- 1728$$

10. Multiply 2 x 576 and place the product in its correct position in the problem.

$$576 \overline{)16739} \quad {}^{2}$$

11. Is 2 the correct first digit
    of the answer for the problem
    shown at the right?
    Why?

$$\begin{array}{r} 2 \\ 576\overline{)16739} \\ -\underline{1152} \\ 521 \end{array}$$

12. Complete the divison
    shown at the right.

$$\begin{array}{r} 2 \\ 576\overline{)16739} \\ -\underline{1152} \\ 521 \end{array}$$

13. Find the first digit of
    the answer for:

$$35\overline{)9362}$$

14. Find the first digit of
    the answer for:

$$78\overline{)57362}$$

15. Find the first digit of the
    answer for:

$$57\overline{)25738}$$

16. Find the first digit of the
    answer for:

$$83\overline{)47362}$$

17. Find the first digit of the
    answer for:

$$65\overline{)19836}$$

18. Complete the following.

$$\begin{array}{r} 26 \\ 35\overline{)9362} \\ -\underline{70} \\ 236 \\ -\underline{210} \\ 262 \end{array}$$

19. Complete the following.
    Caution: There is a zero
    in the answer. Be certain
    to place it correctly.

$$58\overline{)17635}$$

20. Complete the following.

$$92\overline{)40635}$$

21. Find the divisor of

$$\begin{array}{r} 15\ \text{R3} \\ 6\overline{)93} \end{array}$$

22. Find the dividend of

$$\begin{array}{r} 54\ \text{R4} \\ 8\overline{)436} \end{array}$$

23. Find the quotient of

$$\begin{array}{r} 62 \\ 7\overline{)434} \end{array}$$

24. In a division problem with remainder
    zero, the product of the _____ and
    _____ must equal the _____.

25. Find the average of 257, 364, 295, and 308
    by adding the numbers and dividing by 4.

26. Find the average of 19, 25, 30, 36, 25, and 27
    by counting the numbers and dividing their sum
    by that count.

27. A word problem with the word "each" usually
    indicates multiplication or _____.

28. Solve the problem: If 6 men earn a total of $864, what is the average earned by each man?

30. If 5 nurses have a total of 65 vacation days, what is the average vacation due each nurse?

29. Solve the problem: If a chemist has 845 milligrams of medicine in 5 test tubes, what is the average in each tube?

Part B: Drill and Practice

Divide.

1. $65 \overline{)8143}$

5. $17 \overline{)5736}$

9. $15 \overline{)9145}$

2. $59 \overline{)29367}$

6. $85 \overline{)9437}$

10. $271 \overline{)63083}$

3. $43 \overline{)9214}$

7. $72 \overline{)14314}$

11. $486 \overline{)90504}$

4. $28 \overline{)14632}$

8. $96 \overline{)25438}$

12. $763 \overline{)140987}$

13. Find the average of 261 and 313.

14. Find the average of 679, 513, and 521.

15. If 8624 fluid ounces is to be equally distributed into 7 containers, what amount should be put in each container?

16. Each tire in a sale is priced at $37. The total value of the tires is $2,775. Find the number of tires.

17. A total of 7,101 grams of gold is to be made into ornaments that weigh 9 grams each. How many ornaments can be made?

18. If 893 pearls are to be made into necklaces containing 19 pearls each, how many necklaces can be obtained?

19. How many statues weighing 23 tons each can be poured with 2,116 tons of concrete?

20. In the problem at the right, the divisor is _____, the dividend is _____, the quotient is _____, and the remainder is _____.

$$\begin{array}{r} 19 \ \text{R}38 \\ 46\overline{)912} \\ -46 \\ \hline 452 \\ -414 \\ \hline 38 \end{array}$$

21. Show how to check the division problem at the right.

$$\begin{array}{r} 507 \ \text{R}17 \\ 18\overline{)9143} \end{array}$$

22. Find the average of 57, 46, 52, and 49.

23. If 16 employees have total wages of $5,712, find the average pay of each employee.

24. 9,378 cards are to be separated into 6 equal piles. How many cards will be in each pile?

Part C: Review. Answers for the problems of Part C are not given. However, the notation «C,U» refers to the Chapter (C) and Unit (U) in which problems of the same type were presented.

1. Write the numeral with 6 as its hundreds digit, 0 as its ones digit, and 7 as its tens digit. «1,1»

2. 508 means   5 hundreds, 0 tens, 8 ones or
             4 hundreds, _____ tens, 8 ones.
        «1,1»

3. 90,016 + 347 + 2,568   «1,2»

4. 65,813 + 83,413 + 5,007   «1,2»

5. 7,200 – 716   «1,3»

6. 49,007 – 15,238   «1,3»

7. 5179   «2,1»
    x   8

8. 2761   «2,1»
    x   7

9. 3468   «2,2»
    x 46

10. 2173   «2,2»
     x 97

11. 8) 1923   «2,3»

12. 5) 1856   «2,3»

13. Numbers with the same label _____ (may, may not) be added.   «1,4»

14. Rick has 47 more baseball cards than Paul. If Paul has 125 cards, then how many does Rick have?   «1,4»

15. A produce store received 15 crates with 12 heads of cabbage in each crate. How many heads of cabbage were there?   «2,2»

16. Jane has 32 tutoring sessions each week. In a 16-week semester, how many tutoring sessions does she have?   «2,2»

CHAPTER 2 TEST

«2,U» shows the unit in which the problem is found in the chapter.

1. The numbers of a multiplication problem are the multiplier and multiplicand, but both may be called _____.   «2,2»

2. The answer to a multiplication problem is its _____.   «2,2»

3. 65 x 8   «2,1»

4. 3,175 x 7   «2,1»

5. 4,076 x 9   «2,1»

6. 234 x 18   «2,2»

7. 205 x 73   «2,2»

8. 3,852 x 602   «2,2»

9. 3,417 x 450   «2,2»

10. A druggist received 25 cartons with 48 jars of vitamins in each carton. How many vitamin jars are there?   «2,2»

11. Bob is paid $228 each week. How much does he earn in 26 weeks.   «2,2»

12. In the problem at the right, $46 \overline{) 2990}$ $^{65}$
    the divisor is ____
    the dividend is _____
    the quotient is _____   «2,4»

13. $4 \overline{) 3615}$ «2,3»

14. $7 \overline{) 6213}$ «2,3»

15. $8 \overline{) 1946}$ «2,3»

16. $28 \overline{) 5213}$ «2,4»

17. $37 \overline{) 36541}$ «2,4»

18. $58 \overline{) 91326}$ «2,4»

19. Find the average of 52, 60, 67, and 61.   «2,4»

20. 640 customers are to be equally distributed to 16 salespersons. How many customers will each salesperson receive?   «2,4»

3
Fractions

Unit 1:   INTRODUCTION TO FRACTIONS

The Meaning of a Fraction

Every fraction is an indicated division problem.

The fraction $\frac{1}{5}$ means 1 ÷ 5 or 1(one) divided into 5 (five) equal parts.

The fraction $\frac{2}{3}$ means 2 ÷ 3 or 2(two) divided into 3 (three) equal parts.

The fraction $\frac{7}{4}$ means 7 ÷ 4 or 7(seven) divided into 4 (four) equal parts.

Definitions   Numerator and Denominator

A fraction consists of two numbers, one above the fraction bar and the other beneath it. The number above the fraction bar is called the **numerator** of the fraction. The number below the fraction bar is called the **denominator**.

In the fraction $\frac{5}{7}$ the numerator is 5 and the denominator is 7.

Focus on the numerator and denominator of a fraction

The **numerator** of $\frac{5}{9}$ is 5 and the **denominator** is 9.

The **numerator** of $\frac{2}{3}$ is 2 and the **denominator** is 3.

Visualizing Fractions

In the figure at the right, the large rectangle has been divided into four smaller rectangles that are equal in size. One of those smaller rectangles has been shaded.

Because the large rectangle has been divided into four equal parts, each of the smaller rectangles represents $\frac{1}{4}$ of the large rectangle.

The shaded part of the figure is $\frac{1}{4}$ of the large rectangle.

The unshaded part of the figure is $\frac{3}{4}$ of the large rectangle.

In the figure at the right, one of the smaller rectangles is shaded. The three smaller rectangles are the same size.

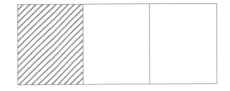

The fraction of the figure that is shaded is $\frac{1}{3}$.

Visualizing Fractions on a Line Segment

In the figure below, the line segment has its endpoints labeled 0 and 1. Four points labeled A,B,C,D separate the line segment into five equal parts. The point A marks the position of $\frac{1}{5}$; B marks $\frac{2}{5}$; C marks $\frac{3}{5}$; and D marks $\frac{4}{5}$.

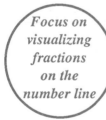

Focus on
visualizing
fractions
on the
number line

The line segment below has been separated into 12 equal parts. Point C represents $\frac{3}{12}$ or $\frac{1}{4}$ of the line.

Fractions Equal to One

One of the most important and useful properties of fractions is the fact that when the numerator and denominator are equal, the fraction is equal to one.

$$\frac{2}{2} = 1 \qquad \frac{8}{8} = 1 \qquad \frac{13}{13} = 1 \qquad \frac{27}{27} = 1 \qquad \frac{903}{903} = 1$$

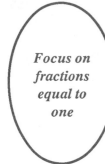

Focus on fractions equal to one

To write 1 as a fraction with denominator 12 means to select a replacement for the question mark in

$$1 = \frac{?}{12}$$

12 is the only number that can replace the question mark and make the equality true.

$$1 = \frac{12}{12}$$

Finding Common Factors (Divisors)

To simplify (reduce) a fraction like $\frac{8}{12}$ it is necessary to find a number that divides both 8 and 12. This process is called "finding a **common factor** for 8 and 12."

Definition   **Factors (Divisors)**

The counting   when D is divided
number f is a   **if and**   by f the remainder
factor (divisor) **only if** is zero.
of D

For example, the **factors** (divisors) of 20 are:

1, 2, 4, 5, 10, 20

Each of the **factors** divides 20 evenly so that the remainder is zero.

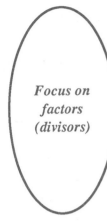

5 is a **factor** (divisor) of 20 because when 20 is divided by 5 there is a remainder of 0.

$$20 \div 5 = 4 \text{ R0}$$

4 is also a **factor** (divisor) of 20 because when 20 is divided by 4 there is a remainder of 0.

$$20 \div 4 = 5 \text{ R0}$$

6 is not a **factor** (divisor) of 20. When 20 is divided by 6 the remainder is not 0.

$$20 \div 6 = 3 \text{ R2}$$

Focus on factors (divisors)

Finding Factors by Long Division

To determine if 13 is a factor of 91, divide:

```
        7
13 ) 9 1
   − 9 1
        0
```

Since the division has a remainder of zero, we know both that 13 is a factor of 91 and the quotient, 7, is also a factor of 91.

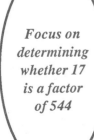

It is not obvious whether 17 is a factor of 544. To make this determination divide 544 by 17.

```
          3 2
17 ) 5 4 4
   − 5 1
        3 4
      − 3 4
           0
```

Focus on determining whether 17 is a factor of 544

The division has a remainder of zero and this means that both 17 and 32 are factors of 544.

Simplifying (Reducing) Fractions

The fraction $\frac{8}{12}$ is simplified (reduced) by dividing both
the numerator 8 and the denominator 12 by their common factor, 4.

$$\frac{8}{12} = \frac{8 \div 4}{12 \div 4} = \frac{2}{3}$$

Notice that $\frac{8}{12}$ is equal to $\frac{2}{3}$ in the figure below.

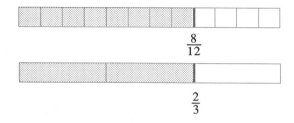

$$\frac{8}{12}$$

$$\frac{2}{3}$$

*Focus on
simplifying a
fraction*

$\frac{36}{42}$ is simplified by dividing both the numerator
and denominator by their common factor 6.

$$\frac{36}{42} = \frac{36 \div 6}{42 \div 6} = \frac{6}{7}$$

$\frac{36}{42}$ and $\frac{6}{7}$ are equal, but the simplest name for the fraction
is $\frac{6}{7}$.

Definition   Common Factor

The counting number c is a **common factor** for x and y	**if and only if**	c is a factor for x and c is also a factor for y.

For example, 3 is a **common factor** for 21 and 33 because
3 is a factor of 21 and 3 is also a factor of 33.

7 is not a **common factor** for 21 and 33 because 7 is a
factor of 21, but 7 is not a factor of 33.

*Focus on
the use of
a common
factor in
simplifying a
fraction*

To simplify $\frac{14}{16}$ first find a common factor of 14 and 16.
The only common factors of 14 and 16 are 1 and 2. Since
2 is the largest common factor it is used to simplify $\frac{14}{16}$.

$$\frac{14}{16} = \frac{14 \div 2}{16 \div 2} = \frac{7}{8}$$

$\frac{14}{16}$ and $\frac{7}{8}$ name the same fraction, but $\frac{7}{8}$ is the simplest
name for the number.

Definition		Highest Common Factor
The counting number h is the **highest common factor** of x and y	**if and only if**	h is a common factor of x and y and h is greater than any other common factor of x and y.

*Focus on
the highest
common factor
of 32 and 40*

The numbers 32 and 40 have more than one common factor.

Factors of 32: 1, 2, 4, 8, 16, 32
Factors of 40: 1, 2, 4, 5, 8, 10, 20, 40

The common factors of 32 and 40 are: 1, 2, 4, 8

The highest common factor (hcf) of 32 and 40 is 8.

Definition		Simplest Form of a Fraction
The fraction $\frac{x}{y}$ is in its **simplest form**	**if and only if**	1 is the highest common factor of x and y.

After Simplifying a Fraction, Try Again

1, 2, 4, and 8 are all common factors of 32 and 40, but 8 is
the highest common factor (hcf). If the hcf is not used in
simplifying $\frac{32}{40}$ the resulting fraction must be reduced again.

$$\frac{32}{40} = \frac{32 \div 2}{40 \div 2} = \frac{16}{20}$$

$$\frac{32}{40} = \frac{32 \div 4}{40 \div 4} = \frac{8}{10}$$

$$\frac{32}{40} = \frac{32 \div 8}{40 \div 8} = \frac{4}{5}$$

The fractions $\frac{32}{40}$, $\frac{16}{20}$, $\frac{8}{10}$, and $\frac{4}{5}$ are all equal, but $\frac{4}{5}$ is the **simplest**
name for the number.

Before ending any simplification, always check to determine that
the hcf of the numerator and denominator is 1.

*Focus on
simplifying
fractions*

To simplify $\frac{51}{85}$ the hcf for the numerator and denominator needs to be found. To
simplify this task, begin by finding only the factors of 51.

By long division it is fairly easy to
show that 3 is a factor of 51. The
division also shows that 17 is a factor
of 51.

$$\begin{array}{r} 17 \\ 3\overline{)51} \\ -3 \\ \hline 21 \\ -21 \\ \hline 0 \end{array}$$

Since both 3 and 17 are prime numbers
the list of all factors of 51 is: 1, 3, 17, 51.

Any common factor of 51 and 85 must be one of the numbers: 1, 3, 17, 51

Try any of these factors (from largest to smallest)
as a divisor of 85 to see if it is a common factor.
That process discloses the fact that 17 is a factor
of 85 and is the hcf of 51 and 85.

$$\begin{array}{r} 5 \\ 17\overline{)85} \\ -85 \\ \hline 0 \end{array}$$

The simplification of $\frac{51}{85}$ is accomplished
using 17 as the hcf.

$$\frac{51}{85} = \frac{51 \div 17}{85 \div 17} = \frac{3}{5}$$

Unit 1 Exercise

This exercise reviews the preceding unit. The exercise is divided into three parts. Part A reviews the foci of Unit 1. Part B offers opportunities to practice the skills and concepts of Unit 1. Part C contains problems that review your previous work in this text. All answers for Parts A and B are at the back of the book. Each problem of Part C is accompanied by a notation «CU» that refers to the Chapter (C) and Unit (U) in which that type of problem is studied.

Part A: Reviewing the foci of Unit 1.

1. The line segment below has been separated into two equal parts. What fraction represents Point A?

    0   A   1

2. The line segment below has been separated into 12 equal parts. Put a capital A at the point that is halfway from 0 to 1.

    0   1

3. The line segment below has been separated into 6 equal parts. Put a capital A at the point representing $\frac{1}{2}$.

    0   1

4. The line segment below has been separated into 3 equal parts by the Points A and B. Which point represents $\frac{1}{3}$?

    0   A   B   1

5. The line segment below has been separated into 6 equal parts. Place a capital letter A at the point representing $\frac{1}{3}$.

    0   1

6. In the figure below, the line segment has been separated into 12 equal parts. Place a capital letter A at the point representing $\frac{1}{3}$.

    0   1

7. The line segment below has been separated into 4 equal parts by the points A, B, and C. Which point represents $\frac{1}{4}$?

    0   A   B   C   1

8. The line segment below has been separated into 8 equal parts. Mark the point which represents $\frac{1}{4}$?

    0   1

9. The line segment below has been divided into 10 equal parts. Mark the point representing $\frac{1}{5}$ with a capital A.

    0   1

10. The line segment below has been divided into 6 equal parts by the points A, B, C, D, and E. What point represents $\frac{1}{6}$?

    0   A   B   C   D   E   1

11. The line segment below has been divided into 12 equal parts. Mark with a capital A the point representing $\frac{1}{6}$.

0 •••••••••••• 1

12. Below are shown two line segments; one is divided into two parts and the other into three parts. $\frac{1}{2}$ is greater than $\frac{1}{3}$ because $\frac{1}{2}$ is further to the right. Is $\frac{1}{2}$ greater than $\frac{2}{3}$?

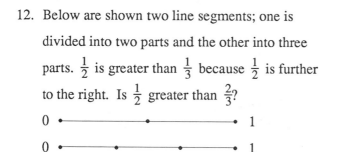

13. Below are shown two line segments. $\frac{1}{3}$ is greater than $\frac{1}{4}$. Which is greater, $\frac{2}{3}$ or $\frac{3}{4}$?

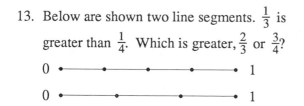

14. Below are shown two line segments. $\frac{1}{4}$ is greater than $\frac{1}{5}$. Is $\frac{3}{4}$ greater than $\frac{4}{5}$?

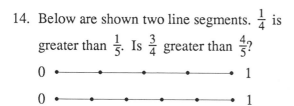

15. Below are shown two line segments. Is $\frac{2}{4}$ greater than $\frac{1}{2}$?

0 •——————•——————• 1

0 •————•————•————• 1

16. Below are shown two line segments. Is $\frac{1}{3}$ greater than $\frac{2}{6}$?

0 •——•——•——•——• 1

0 •————•————•————• 1

17. If a hospital has 250 beds and 233 patients, the fraction $\frac{233}{250}$ indicates the portion of the hospital beds that are occupied. Write a fraction indicating a hospital's occupancy if it has 140 beds and 73 patients.

18. A manufacturing plant with 392 employees has 13 absent employees. What fraction of the employees are absent?

19. A classroom has 35 seats and 24 students. Write a fraction indicating the number of occupied seats.

20. A fraction equals one when the numerator and denominator are _____.

21. Is 7 a factor of 546?

22. Is 8 a factor of 912?

23. Is 2 a factor of 436?

24. Is 3 a factor of 435?

25. Is 13 a factor of 52?

26. Is 6 a factor of 436?

27. To simplify $\frac{26}{39}$ first find all the factors of 26. The factors of 26 are _____, _____, _____, and _____.

28. 1, 2, 13, and 26 are the factors of 26. Try each of the factors and find the hcf of 26 and 39.

29. Simplify $\frac{51}{68}$

30. Simplify $\frac{38}{57}$

Part B: Drill and Practice

1. Write $5 \div 7$ as a fraction.

2. What is the numerator of $\frac{5}{9}$?

3. What is the denominator of $\frac{7}{10}$?

4. Use the figure below and determine
    the fraction of the rectangle that is shaded.

5. Use the two number lines below. Which
    of the following fractions is the largest?
    $$\frac{3}{4} \qquad \text{or} \qquad \frac{5}{6}$$

    0 •—•——•——•——•——•——• 1
    0 •————•———•———•——• 1

6. A hospital staff has 8 doctors and 45 nurses.
    Write a fraction indicating the relative number
    of doctors to nurses.

7. Replace the ? to make each equality true.
    $$? = \frac{9}{9} \qquad 1 = \frac{5}{?} \qquad 1 = \frac{14}{?} \qquad 1 = \frac{?}{653}$$

Simplify the following.

8. $\frac{33}{88}$

9. $\frac{41}{50}$

10. $\frac{63}{66}$

11. $\frac{16}{24}$

12. $\frac{24}{30}$

13. $\frac{18}{45}$

14. $\frac{15}{50}$

15. $\frac{18}{39}$

16. $\frac{7}{35}$

17. $\frac{26}{65}$

18. $\frac{13}{15}$

19. $\frac{40}{70}$

20. $\frac{24}{44}$

21. $\frac{45}{81}$

22. $\frac{57}{76}$

23. $\frac{42}{56}$

24. Is 2 a factor of 604?

25. Is 3 a factor of 814?

26. Is 4 a factor of 538?

27. Is 5 a factor of 436?

28. Is 6 a factor of 525?

29. Is 7 a factor of 475?

30. Is 8 a factor of 864?

31. Is 9 a factor of 618?

32. Is 10 a factor of 903?

Part C: Review. Answers for the problems of Part C are not given. However, the notation «C,U» refers to the Chapter (C) and Unit (U) in which problems of the same type were presented.

1. Write the numeral with 4 as its hundreds digit, 9 as its ones digit, and 6 as its tens digit. «1,1»

2. 560 means   5 hundreds, 6 tens, 0 ones or
               4 hundreds, _____ tens, 0 ones.
         «1,1»

3. 62,516 + 837 + 5,268   «1,2»

4. 8,613 + 29,413 + 5,447   «1,2»

5. 7,200 − 7,916   «1,3»

6. 94,007 − 43,638   «1,3»

7. 4379   «2,1»
    x   7

8. 2961   «2,1»
    x   8

9.   348   «2,2»
    x 57

10. 7213   «2,2»
    x   5

11. 8) 6323   «2,3»

12. 5) 5756   «2,3»

13. 48) 41604   «2,4»

14. 53) 10298   «2,4»

15. Is it possible to add 8 chairs and 5 tables.   «1,4»

16. Find the average of 261, 413, and 313.   «2,4»

17. Each aisle in a department store requires 13 display cases. How many display cases are there in 8 aisles?   «2,2»

18. Martha works 30 hours each week. In 19 weeks, how many hours will she have worked?   «2,2»

Unit 2:   EQUIVALENT FRACTIONS

Definition   Equal Fractions

$\frac{a}{b}$ and $\frac{c}{d}$ are   **if and** $a \times d = b \times c$
equal fractions   **only if**

For example, to determine if $\frac{3}{4}$ is equal to $\frac{6}{8}$
1. the first numerator 3 is multiplied by
    the second denominator 8.
    $3 \times 8 = 24$
2. the first denominator 4 is multiplied by
    the second numerator 6. $4 \times 6 = 24$
3. Since 3 x 8 equals 24 and 4 x 6 also
    equals 24 the fractions are equal.
    $$\frac{3}{4} = \frac{6}{8}$$

Finding Equivalent Fractions With Larger Denominators

A fraction can be written with a greater denominator by multiplying
both numerator and denominator by the same number.

$$\frac{5}{6} = \frac{5 \times 4}{6 \times 4} = \frac{20}{24}$$

This process is justified by the fact that $\frac{4}{4} = 1$.

Focus on changing to equivalent fractions

The fractions $\frac{7}{8}$ and $\frac{21}{24}$ are equal. To change $\frac{7}{8}$ to $\frac{21}{24}$ both numerator and denominator of $\frac{7}{8}$ are multiplied by 3.

$$\frac{7}{8} \times 1 = \frac{7 \times 3}{8 \times 3} = \frac{21}{24}$$

Definition   Multiples of a Counting Number

m is a **multiple** of the counting number c	**if and only if**	there is a counting number k such that k x c = m.

The multiples of 7 are

$$7, 14, 21, 28, \ldots$$

where the three dots mean "and so forth."

Focus on the meaning of multiple

Each of the multiples of 7 is found by multiplying a counting number by 7.

$$1 \text{ x } 7 = 7$$
$$2 \text{ x } 7 = 14$$
$$3 \text{ x } 7 = 21$$
$$4 \text{ x } 7 = 28$$

Definition   Least Common Multiple (LCM)

d is the **least common multiple** (lcm) of two counting numbers x and y	**if and only if**	d is a multiple of both x and y and d is less than any other number that is a multiple of both x and y.

For example,
Multiples of 4 are: 4, 8, 12, 16, 20, 24, 28, 32, 36, 40, etc.

Multiples of 6 are: 6, 12, 18, 24, 30, 36, 42, etc.

The numbers 12, 24, and 36 appear in both lists of multiples and are **common multiples** of 4 and 6.

The smallest number that appears in both lists of multiples is 12; therefore, 12 is the **least common multiple (lcm)** of 4 and 6.

Writing Two Fractions With a Common Denominator

A major use of the least common multiple of two numbers is the finding of a common denominator.

To write the fractions $\frac{3}{8}$ and $\frac{5}{6}$ with a common denominator, the following steps are used:

1. Find the lcm of the denominators of $\frac{3}{8}$ and $\frac{5}{6}$.

   The lcm of 8 and 6 is 24.

2. Write both fractions with the lcm, 24, as denominator.

$$\frac{3}{8} = \frac{3 \times 3}{8 \times 3} = \frac{9}{24}$$

$$\frac{5}{6} = \frac{5 \times 4}{6 \times 4} = \frac{20}{24}$$

Focus on writing two fractions with the same denominator

To write $\frac{5}{7}$ and $\frac{3}{8}$ as fractions with the same denominator we must first find the least common multiple (lcm) of 7 and 8.

The lcm of 7 and 8 is 56

$\frac{5}{7}$ is written with a denominator of 56 by multiplying both the numerator and denominator by 8.

$$\frac{5}{7} = \frac{5 \times 8}{7 \times 8} = \frac{40}{56}$$

$\frac{3}{8}$ is written with a denominator of 56 by multiplying both the numerator and denominator by 7.

$$\frac{3}{8} = \frac{3 \times 7}{8 \times 7} = \frac{21}{56}$$

Comparing Sizes of Two Fractions

To compare the sizes of $\frac{5}{8}$ and $\frac{2}{3}$ both fractions are written with a common denominator.

$$\frac{5}{8} = \frac{5 \times 3}{8 \times 3} = \frac{15}{24}$$

$$\frac{2}{3} = \frac{2 \times 8}{3 \times 8} = \frac{16}{24}$$

$\frac{5}{8}$ is less than $\frac{2}{3}$ because $\frac{15}{24}$ is less than $\frac{16}{24}$.

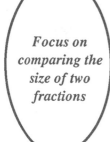

Focus on comparing the size of two fractions

To compare the sizes of $\frac{5}{8}$ and $\frac{4}{7}$,

1. Write both fractions with a common denominator.

$$\frac{5}{8} = \frac{5 \times 7}{8 \times 7} = \frac{35}{56}$$
$$\frac{4}{7} = \frac{4 \times 8}{7 \times 8} = \frac{32}{56}$$

2. The smaller fraction will have the smaller numerator.

$$\frac{32}{56} < \frac{35}{56}$$
$$\frac{4}{7} < \frac{5}{8}$$

$\frac{4}{7}$ is less than $\frac{5}{8}$ because $\frac{32}{56}$ is less than $\frac{35}{36}$.

Simplifying Fractions To Whole Or Mixed Numbers

Fractions like $\frac{5}{7}, \frac{11}{12}$, and $\frac{14}{17}$ are all less than 1 because their denominators are greater than their numerators. Such fractions are **proper fractions**.

Fractions like $\frac{4}{4}, \frac{11}{11}$, and $\frac{19}{19}$ are equal to 1 because their denominators are equal to their numerators.

Fractions like $\frac{7}{5}, \frac{48}{7}$, and $\frac{32}{3}$ are all greater than 1 because their denominators are less than their numerators. Such fractions are **improper fractions** and can be written as **mixed numbers**.

Whenever the numerator and denominator of a fraction
are equal the fraction can be simplified to 1.

$$\frac{4}{4} = 1 \quad \text{and} \quad \frac{6}{6} = 1 \quad \text{and} \quad \frac{37}{37} = 1$$

In general, for any counting number y

$$\frac{y}{y} = 1$$

*Focus on
fractions
equal to 1*

Definition   **Proper Fractions**

A fraction $\frac{x}{y}$ is   **if and**   x is less than y
a **proper**   **only if** $(x < y)$.
fraction

$\frac{3}{8}$ is a **proper fraction** because 3 is less than 8.

$\frac{6}{7}$ is a **proper fraction** because 6 is less than 7.

Notice that $\frac{6}{7}$ is less than 1.

$$\frac{6}{7} < \frac{7}{7} \quad \text{and} \quad \frac{7}{7} = 1$$

Every proper fraction is less than 1 (one) because the definition
requires that the numerator be smaller than the denominator.

$$\text{If } x < y, \text{ then } \frac{x}{y} < \frac{y}{y} \text{ and } \frac{y}{y} = 1$$

*Focus on
the meaning
of proper
fraction*

Definition   **Improper Fractions**

A fraction $\frac{x}{y}$ is   **if and**   x is greater than y
an **improper**   **only if** $(x > y)$.
fraction

$\frac{9}{7}$ is an **improper fraction** because 9 is greater than 7.

$\frac{8}{5}$ is an **improper fraction** because 8 is larger than 5.

Notice that $\frac{8}{5}$ is greater than 1.

Focus on the meaning of improper fraction

$\frac{8}{5} > \frac{5}{5}$ and $\frac{5}{5} = 1$

Every improper fraction is greater than 1 (one) because the definition requires that the numerator be larger than the denominator.

If $x > y$, then $\frac{x}{y} > \frac{y}{y}$ and $\frac{y}{y} = 1$

Changing An Improper Fraction to a Mixed Number

The improper fraction $\frac{18}{7}$ is equal to the mixed number $2\frac{4}{7}$.

To change $\frac{18}{7}$ to its mixed number equivalent:

1. Divide 18 by 7.

$$\begin{array}{r} 2 \ \ R4 \\ 7\overline{)18} \end{array}$$

2. The quotient, 2 R4, is used to write the mixed number.

2 R4 becomes $2\frac{4}{7}$

Converting Mixed Numbers To Improper Fractions

The mixed number $2\frac{5}{7}$ may be changed to its improper fraction using the fact that the whole number 2 is equal to $\frac{14}{7}$.

$2\frac{5}{7}$ means $2 + \frac{5}{7}$.

But $2 = \frac{2}{1} = \frac{2 \times 7}{1 \times 7} = \frac{14}{7}$

Therefore, $2\frac{5}{7}$ means $\frac{14}{7} + \frac{5}{7}$ or $\frac{19}{7}$.

Focus on writing a mixed number as an improper fraction

The mixed number $5\frac{3}{8}$ may be changed to its improper fraction using the fact that the whole number 5 is equal to $\frac{40}{8}$.

$5\frac{3}{8}$ means $5 + \frac{3}{8}$.

But $5 = \frac{5}{1} = \frac{5 \times 8}{1 \times 8} = \frac{40}{8}$.

Therefore, $5\frac{3}{8}$ means $\frac{40}{8} + \frac{3}{8}$ or $\frac{43}{8}$.

A Shortcut for Changing a Mixed Number to an Improper Fraction

To write a mixed number as an improper fraction,

1. Multiply the whole number, 4, by the denominator, 5.

   $4\frac{3}{5}$
   $4 \times 5 = 20$

2. Add the numerator, 3, of the proper fraction to the product found in step 1.

   $(4 \times 5) + 3 = 23$

3. Keep the same denominator.

   $4\frac{3}{5} = \frac{23}{5}$

Focus on a shortcut for writing a mixed number

To write $7\frac{1}{4}$ as an improper fraction

1. Multiply the whole number, 7, by the denominator, 4.

   $7 \times 4 = 28$

2. Add the original numerator, 1, and this sum is the new numerator

   $(7 \times 4) + 1 = 29$

3. Use the original denominator for the new improper fraction.

   $7\frac{1}{4} = \frac{29}{4}$

Unit 2 Exercise

This exercise reviews the preceding unit. The exercise is divided into three parts. Part A reviews the foci of Unit 2. Part B offers opportunities to practice the skills and concepts of Unit 2. Part C contains problems that review your previous work in this text. All answers for Parts A and B are at the back of the book. Each problem of Part C is accompanied by a notation «CU» that refers to the Chapter (C) and Unit (U) in which that type of problem is studied.

Part A: Reviewing the foci of Unit 2.

1. Multiply both numerator and denominator of $\frac{3}{4}$ by 9.

2. Are $\frac{3}{4}$ and $\frac{27}{36}$ equal?

3. To change $\frac{3}{4}$ to $\frac{27}{36}$, both numerator and denominator of $\frac{3}{4}$ are multiplied by _____.

4. Multiply both numerator and denominator of $\frac{1}{2}$ by 5.

5. Are $\frac{1}{2}$ and $\frac{5}{10}$ equal fractions?

6. To change $\frac{1}{2}$ to $\frac{3}{6}$, both the numerator and denominator of $\frac{1}{2}$ must be multiplied by _____.

7. What number must be multiplied by both the numerator and denominator of $\frac{5}{6}$ to give a fraction with denominator 12?

8. Find the numerator of the fraction below.
$\frac{1}{4} = \frac{?}{12}$

9. Find the numerator of the fraction below.
$\frac{5}{6} = \frac{?}{24}$

10. Write $\frac{3}{4}$ with 20 as the denominator.

11. Write $\frac{5}{6}$ with 30 as the denominator.

12. The numbers 18 and 36 are common multiples of 6 and 9. Is 6 x 9 or 54 also a common multiple of 6 and 9?

13. The numbers 18, 36, and 54 are common multiples of 6 and 9. Is 24 a common multiple of 6 and 9?

14. The product of two whole numbers is always a common multiple for them. Is 45 a common multiple for 9 and 5?

15. Is 48 a common multiple for 4 and 11?

16. The product of two whole numbers ____ (is, isn't) always a common multiple for them.

17. A common multiple of 6 and 9 is 54. Is 54 the smallest number that is a common multiple for 6 and 9?

18. A common multiple of 6 and 5 is 30 (6 x 5). Is 30 the smallest number that is a common multiple for 6 and 5?

19. The product of two whole numbers is always a common multiple for them. The product is _____ (always, sometimes) the smallest common multiple.

20. List the multiples of 2 and 3. Find the smallest common multiple for them.

21. List the multiples of 6 and 10 until you find a number in both lists. What is the smallest number in both lists?

22. The number 30 is called the least common multiple of 6 and 10 because it is the smallest whole number that is a multiple of both 6 and _____.

23. Find the least common multiple (lcm) of 4 and 5 by finding the smallest number that is a multiple of both of them.

24. List the multiples of 12 and 8 until you find their least common multiple (lcm).

25. List multiples of 3 and 5 and then select their lcm.

26. List multiples of 3 and 6 and then find their lcm.

27. List multiples of 7 and 5. What is the lcm of 7 and 5?

28. The fraction $\frac{8}{11}$ is less than $\frac{9}{11}$ because the denominators are the same and 8 is less than 9. Is $\frac{15}{17}$ less than $\frac{13}{17}$?

29. When fractions have the same denominator, compare the numerators. Is $\frac{14}{20}$ less than $\frac{17}{20}$?

30. If two fractions have the same denominator, the smaller fraction has the _____ (smaller, larger) numerator.

31. Write $\frac{2}{7}$ and $\frac{1}{4}$ with a common denominator. Is $\frac{2}{7}$ less than $\frac{1}{4}$?

32. Write $\frac{3}{5}$ and $\frac{7}{10}$ with a common denominator. Is $\frac{3}{5}$ less than $\frac{7}{10}$?

33. Write $\frac{5}{11}$ and $\frac{4}{9}$ with a common denominator. Is $\frac{5}{11}$ less than $\frac{4}{9}$?

34. Simplify $\frac{37}{37}$.

35. The fraction $\frac{15}{5}$ is indicated division, $15 \div 5$. $15 \div 5 =$ _____.

36. Is $\frac{18}{7}$ greater than 1?

37. Is $\frac{5}{6}$ greater than 1?

38. Change $\frac{10}{7}$ to a mixed number by dividing 10 by 7.

39. Change $\frac{11}{4}$ to a mixed number by dividing 11 by 4.

40. Change $\frac{15}{7}$ to a mixed number.

41. In the mixed number $2\frac{5}{7}$, the whole number 2 represents $\frac{14}{7}$ because $14 \div 7 = 2$. In the mixed number $3\frac{1}{8}$ the whole number 3 represents _____ because $24 \div 8 = 3$.

42. The whole number 5 is equal to $\frac{10}{2}, \frac{15}{3}, \frac{20}{4},$ etc. To make a fraction equal to the whole number 5 the numerator must be _____ times the denominator.

43. The whole number 7 is equal to $\frac{14}{2}, \frac{21}{3}, \frac{28}{4},$ etc. To make a fraction equal to the whole number 7 the numerator must be _____ times the denominator.

44. In the mixed number $7\frac{1}{4}$, the whole number 7 equals $\frac{28}{4}$. In the mixed number $5\frac{2}{3}$, the whole number 5 equals $\frac{?}{3}$.

45. The numerator of the improper fraction for $3\frac{5}{9}$ is $3 \times 9 + 5 = 32$. The denominator is still 9. Write the improper fraction for $3\frac{5}{9}$.

46. Find the numerator of the improper fraction for $2\frac{1}{7}$ by multiplying 2 times 7 and adding 1.

47. Find the numerator of the improper fraction for $3\frac{2}{7}$ by multiplying 3 times 7 and adding 2.

Part B: Drill and Practice

Find the numerator of the fractions
in problems 1-8.

1. $\frac{4}{7} = \frac{?}{35}$

2. $\frac{2}{3} = \frac{?}{12}$

3. $\frac{8}{9} = \frac{?}{36}$

4. $\frac{1}{4} = \frac{?}{44}$

5. $\frac{5}{6} = \frac{?}{24}$

6. $\frac{2}{3} = \frac{?}{15}$

7. $\frac{5}{8} = \frac{?}{16}$

8. $\frac{4}{7} = \frac{?}{21}$

Find the lcm for each of the pairs
of numbers in problems 9-14

9.   7, 9

10. 9, 6

11. 12, 4

12. 10, 15

13. 5, 9

14. 12, 6

Select the smaller fraction from each
pair in problems 15-22.

15. $\frac{7}{8}, \frac{9}{11}$

16. $\frac{3}{4}, \frac{8}{13}$

17. $\frac{3}{10}, \frac{4}{15}$

18. $\frac{7}{9}, \frac{5}{6}$

19. $\frac{5}{7}, \frac{11}{16}$

20. $\frac{1}{3}, \frac{2}{9}$

21. $\frac{3}{5}, \frac{2}{3}$

22. $\frac{8}{13}, \frac{2}{3}$

Label each of the fractions in problems
23-26 as a proper or improper fraction.

23. $\frac{14}{19}$

24. $\frac{15}{13}$

25. $\frac{87}{93}$

26. $\frac{57}{48}$

Write each fraction in problems 27-37 as a whole number or a mixed number.

Change the mixed numbers in problems 38-54 to improper fractions.

27. $\frac{32}{5}$

28. $\frac{47}{8}$

29. $\frac{19}{4}$

30. $\frac{28}{9}$

31. $\frac{35}{6}$

32. $\frac{41}{8}$

33. $\frac{65}{12}$

34. $\frac{42}{7}$

35. $\frac{39}{4}$

36. $\frac{47}{5}$

37. $\frac{8}{8}$

38. $4\frac{3}{8}$

39. $6\frac{5}{6}$

40. $8\frac{3}{5}$

41. $7\frac{3}{8}$

42. $9\frac{1}{4}$

43. $5\frac{7}{11}$

44. $9\frac{3}{8}$

45. $2\frac{5}{8}$

46. $5\frac{9}{11}$

47. $4\frac{3}{4}$

48. $8\frac{1}{5}$

49. $3\frac{2}{9}$

50. $6\frac{3}{7}$

51. $5\frac{4}{5}$

52. $1\frac{3}{8}$

53. $2\frac{4}{7}$

54. $1\frac{1}{3}$

Part C: Review. Answers for the problems of Part C are not given. However, the notation «C,U» refers to the Chapter (C) and Unit (U) in which problems of the same type were presented.

1. What is the thousands digit of 4,327,908?   «1,1»

10. $35 \overline{)51498}$ «2,4»

2. 56,216 + 397 + 5,628   «1,2»

11. Simplify $\frac{12}{18}$ «3,1»

3. 9,800 − 2,516   «1,3»

12. Simplify $\frac{25}{40}$ «3,1»

4. 94,007 − 43,638   «1,3»

13. Is 2 a factor of 547?   «3,1»

5. 345
    x 27   «2,2»

14. Is 7 a factor of 727?   «3,1»

6. 9917
    x   6   «2,1»

15. Find the average of 513, 286, 616, and 353.   «2,4»

7. $6 \overline{)6153}$ «2,3»

16. Each row in an auditorium has 58 seats. If there are 92 rows in the auditorium, how many seats are there?   «2,2»

8. $8 \overline{)6826}$ «2,3»

9. $83 \overline{)61404}$ «2,4»

Unit 3:   ADDING FRACTIONS AND MIXED NUMBERS

Common Labels—A Necessity for Addition

It is possible to add 5 apples and 3 apples because the numbers have the same label — apples.

It makes no sense to add 4 apples and 7 books because the numbers have different labels — apples and books.

Addition and subtraction can only be performed when the numbers have the same label.

Focus on adding numbers with the same label

It is possible to add 4 sevenths and 2 sevenths because the numbers 4 and 2 have the same label — sevenths.

$$4 \text{ sevenths } + 2 \text{ sevenths } = 6 \text{ sevenths}$$

It is not possible to add 5 eighths and 2 ninths by adding the numbers 5 and 2 because the numbers have different labels.

Adding Fractions With the Same Denominator

When two fractions have the same denominator (label), they can be added as shown by the following examples.

1. Add the numerators. $\frac{5}{13} + \frac{2}{13} = \frac{7}{13}$

2. Keep the same denominator (label). $\frac{4}{11} + \frac{1}{11} = \frac{5}{11}$

Two examples of correctly adding fractions with the same denominator are shown above.

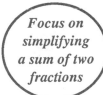

Focus on the addition of fractions with the same denominator

The fraction $\frac{5}{11}$ is read as 5 elevenths. The numerator, 5, tells us "how many" and the denominator, 11, is the label, "elevenths."

To add $\frac{5}{11}$ and $\frac{3}{11}$ first note that the fractions are:

5 elevenths   and   3 elevenths

The fractions $\frac{5}{11}$ and $\frac{3}{11}$ have the same label. They are added by adding their numerators and keeping their common denominator, elevenths.

$$\frac{5}{11} + \frac{3}{11} = \frac{8}{11}$$ (5 elevenths plus 3 elevenths is 8 elevenths)

Simplifying Sums of Addition Problems

In adding $\frac{5}{8} + \frac{1}{8}$ the answer (sum) can be simplified

$$\frac{5}{8} + \frac{1}{8} = \frac{6}{8}$$

The fraction $\frac{6}{8}$ should be simplified to $\frac{3}{4}$. $(\frac{6 \div 2}{8 \div 2} = \frac{3}{4})$

The sum of $\frac{5}{8} + \frac{1}{8}$ is $\frac{3}{4}$.

Focus on simplifying a sum of two fractions

The sum of $\frac{11}{12}$ and $\frac{5}{12}$ is a mixed number.

$$\frac{11}{12} + \frac{5}{12} = \frac{16}{12} \quad \text{and} \quad \frac{16}{12} = 1\frac{4}{12} = 1\frac{1}{3}$$

Notice: The proper fraction of $1\frac{4}{12}$ was simplified to give a final sum of $1\frac{1}{3}$.

Adding Fractions With Different Denominators

The fractions $\frac{3}{4}$ and $\frac{1}{6}$ cannot be added in their present forms because they have different denominators.

The first step in adding $\frac{3}{4}$ and $\frac{1}{6}$ is to write each fraction with the same denominator. After the denominators are made equal, the addition is completed by adding the new numerators.

To add $\frac{3}{4}$ and $\frac{1}{6}$,

1. Find the lcm of the denominators 4 and 6.

   The lcm of 4 and 6 is 12

2. Write both fractions with denominator 12.

$$\frac{3}{4} = \frac{9}{12}$$
$$+\frac{1}{6} = \frac{2}{12}$$
$$\overline{\quad\quad \frac{11}{12}}$$

3. Add the numerators of $\frac{9}{12}$ and $\frac{2}{12}$.

Focus on writing fractions with common denominators

Each of the fractions in the problem shown at the right can be written with their common denominator of 63, the lcm of 7 and 9.

$$\frac{3}{7} = \frac{}{63}$$
$$+\frac{4}{9} = \frac{}{63}$$

The addition is completed by adding the new numerators, 27 and 28.

$$\frac{3}{7} = \frac{27}{63}$$
$$+\frac{4}{9} = \frac{28}{63}$$
$$\overline{\quad\quad \frac{55}{63}}$$

The sum of $\frac{3}{7}$ and $\frac{4}{9}$ is $\frac{55}{63}$.

Simplifying a Sum Greater than One

The sum of $\frac{5}{8}$ and $\frac{1}{2}$ is greater than one and is therefore a mixed number.

$$\frac{5}{8} + \frac{1}{2} = \frac{5}{8} + \frac{4}{8} = \frac{9}{8} \quad \text{and} \quad \frac{9}{8} = 1\frac{1}{8}$$

The sum of $\frac{5}{8}$ and $\frac{1}{2}$ is $1\frac{1}{8}$.

Focus on changing an improper fraction sum to a mixed number

In the addition problem at the right the addition resulted in an improper fraction, $\frac{29}{24}$, which was then changed to a mixed number, $1\frac{5}{24}$.

$$\frac{3}{8} + \frac{5}{6}$$
$$\frac{9}{24} + \frac{20}{24}$$
$$\frac{29}{24}$$
$$1\frac{5}{24}$$

Adding Mixed Numbers

To add two mixed numbers, like $3\frac{1}{5}$ and $5\frac{2}{3}$,

1. Add the two proper fractions.

2. Add the two whole numbers.

$$3\frac{1}{5} = 3\frac{3}{15}$$
$$5\frac{2}{3} = 5\frac{10}{15}$$
$$8\frac{13}{15}$$

Focus on adding the whole numbers and fractions separately

Add $7\frac{3}{11}$ and $2\frac{4}{11}$ by:

1. Adding the fractions.

2. Adding the whole numbers.

$$7\frac{3}{11}$$
$$+ 2\frac{4}{11}$$
$$9\frac{7}{11}$$

Adding a Whole Number and an Improper Fraction

The example below shows how to add a whole number, 9, and an improper fraction, $\frac{5}{3}$.

$$9 + \frac{5}{3}$$
$$9 + 1\frac{2}{3}$$
$$10\frac{2}{3}$$

Focus on simplifying the sum of two mixed numbers

In adding $6\frac{4}{9}$ and $4\frac{5}{6}$, $6\frac{4}{9} + 4\frac{5}{6}$

1. Write each fraction with the common denominator. $6\frac{8}{18} + 4\frac{15}{18}$

2. Add the whole numbers and then add the fractions. $10\frac{23}{18}$

3. Change the improper fraction to a mixed number. $10 + 1\frac{5}{18}$

$$11\frac{5}{18}$$

Unit 3 Exercise

This exercise reviews the preceding unit. The exercise is divided into three parts. Part A reviews the foci of Unit 3. Part B offers opportunities to practice the skills and concepts of Unit 3. Part C contains problems that review your previous work in this text. All answers for Parts A and B are at the back of the book. Each problem of Part C is accompanied by a notation «CU» that refers to the Chapter (C) and Unit (U) in which that type of problem is studied.

Part A: Reviewing the foci of Unit 3.

1. It is possible to add 3 chairs and 5 chairs because the numbers have the same label— chairs. What is the sum of 3 chairs and 5 chairs?

2. It makes no sense to add 3 chairs and 2 desks because the numbers have different labels — chairs and desks. Does it make sense to add 4 pills and 7 beds?

3. Numbers can only be added when they have the same labels. Can 4 fifths be added to 7 thirds without changing them to fifteenths?

4. The denominator of a fraction is its label. The label of $\frac{2}{3}$ is thirds. The label of $\frac{3}{4}$ is _____

5. The label of $\frac{1}{9}$ is _____

6. Do the fractions $\frac{3}{8}$ and $\frac{5}{8}$ have the same label?

7. Do the fractions $\frac{3}{10}$ and $\frac{1}{10}$ have the same label?

8. Do the fractions $\frac{3}{17}$ and $\frac{5}{17}$ have the same denominator?

9. Add $\frac{3}{17}$ and $\frac{2}{17}$ by adding their numerators and keeping the same denominator.

10. Do the fractions $\frac{1}{5}$ and $\frac{3}{5}$ have the same denominator?

11. Add $\frac{1}{5} + \frac{3}{5}$ by adding the numerators and keeping the same denominator.

12. Do the fractions $\frac{4}{7}$ and $\frac{3}{4}$ have the same denominator?

13. Can $\frac{4}{7} + \frac{3}{4}$ be added by adding the numerators, 4 and 3?

14. Can $\frac{5}{11}$ and $\frac{1}{11}$ be added by adding the numerators, 5 and 1?

15. Sometimes the answer to an addition problem is a fraction that should be simplified. Add $\frac{3}{10} + \frac{1}{10}$ and simplify the result.

16. Add and simplify. $\frac{2}{9} + \frac{1}{9}$

17. Add and simplify. $\frac{5}{6} + \frac{1}{6}$

18. Sometimes the answer to an addition problem is greater than one (1) and needs to be changed to a mixed number.
    Add and simplify: $\frac{4}{5} + \frac{2}{5}$

19. Add and simplify: $\frac{4}{7} + \frac{5}{7}$

20. Add and simplify: $\frac{4}{5} + \frac{4}{5}$

21. Can the fraction problem shown at the right be completed by adding the numerators, 3 and 2? $\frac{3}{7} + \frac{2}{7}$

22. Can the fraction problem shown at the right be completed by adding the numerators, 4 and 2? $\frac{4}{11} + \frac{2}{13}$

23. Which of the following is the correct first step in adding $\frac{1}{4}$ and $\frac{2}{5}$?
    a) Add the numerators 1 and 2.
    b) Write each fraction with the same denominator.

24. Which of the following is the correct first step in adding $\frac{3}{8}$ and $\frac{1}{6}$?
    a) Add the numerators, 3 and 1.
    b) Write each fraction with the same denominator.

25. Which of the following is the correct first step in adding $\frac{7}{11}$ and $\frac{2}{11}$?
    a) Add the numerators, 7 and 2.
    b) Write the fractions with a new denominator.

26. Which of the following is the correct first step in adding $\frac{7}{10}$ and $\frac{1}{4}$?
    a) Add the numerators, 7 and 1.
    b) Write the fractions with a new denominator.

27. Add $\frac{3}{8} + \frac{5}{12}$ by first finding a common denominator.

28. Add $2\frac{1}{7} + 4\frac{4}{7}$ by separately adding the proper fractions and the whole numbers.

29. Add $5\frac{3}{17} + 8\frac{10}{17}$ by separately adding the proper fractions and the whole numbers.

30. Always check the sum of two fractions to see if it can be simplified.
    $4\frac{3}{8} + 9\frac{1}{8} = $ _____

31. Add $3\frac{2}{5} + 11\frac{1}{2}$ by first finding a common denominator for the fractions.

32. Add $3\frac{1}{3} + 5\frac{1}{9}$ by first finding a common denominator for the fractions.

33. Add $1\frac{5}{8} + 4\frac{1}{6}$ by first finding a common denominator for the fractions.

Part B: Drill and Practice

Add.

1. $\frac{5}{11} + \frac{1}{11}$

2. $\frac{4}{7} + \frac{2}{7}$

3. $\frac{1}{3} + \frac{1}{3}$

4. $\frac{5}{17} + \frac{8}{17}$

5. $\frac{1}{8} + \frac{3}{8}$

6. $\frac{2}{7} + \frac{3}{7}$

7. $\frac{5}{8} + \frac{1}{8}$

8. $\frac{9}{20} + \frac{3}{20}$

9. $\frac{9}{14} + \frac{3}{14}$

10. $\frac{5}{7} + \frac{6}{7}$

11. $\frac{10}{13} + \frac{7}{13}$

12. $\frac{7}{16} + \frac{3}{16}$

13. $\frac{5}{8} + \frac{7}{8}$

14. $\frac{11}{15} + \frac{7}{15}$

15. $\frac{5}{6} + \frac{5}{6}$

16. $\frac{7}{10} + \frac{9}{10}$

17. $\frac{7}{12} + \frac{11}{12}$

18. $\frac{9}{13} + \frac{6}{13}$

19. $\frac{5}{7} + \frac{1}{7}$

20. $\frac{8}{15} + \frac{4}{15}$

21. $\frac{3}{8} + \frac{7}{8}$

22. $\frac{5}{9} + \frac{7}{9}$

23. $\frac{4}{15} + \frac{7}{15}$

24. $\frac{14}{15} + \frac{4}{15}$

25. $\frac{3}{10} + \frac{4}{10}$

26. $\frac{5}{13} + \frac{2}{13}$

27. $\frac{3}{4} + \frac{4}{5}$

28. $\frac{8}{9} + \frac{5}{9}$

29. $\frac{5}{8} + \frac{7}{8}$

30. $\frac{9}{16} + \frac{11}{16}$

31. $\frac{6}{13} + \frac{7}{13}$

32. $\frac{5}{6} + \frac{5}{6}$

33. $\frac{8}{15} + \frac{11}{15}$

34. $\frac{9}{20} + \frac{13}{20}$

35. $\frac{7}{18} + \frac{5}{18}$

36. $\frac{8}{9} + \frac{8}{9}$

37. $\frac{3}{10} + \frac{3}{8}$

38. $\frac{2}{3} + \frac{1}{7}$

39. $\frac{5}{9} + \frac{1}{3}$

40. $\frac{6}{11} + \frac{1}{4}$

41. $\frac{9}{13} + \frac{1}{5}$

42. $\frac{7}{8} + \frac{1}{2}$

43. $\frac{5}{8} + \frac{3}{4}$

44. $\frac{5}{9} + \frac{4}{5}$

45. $\frac{3}{8} + \frac{11}{12}$

46. $\frac{5}{6} + \frac{1}{4}$

47. $\frac{2}{5} + \frac{7}{15}$

48. $\frac{7}{10} + \frac{1}{3}$

49. $\frac{5}{8} + \frac{1}{4}$

50. $\frac{3}{7} + \frac{2}{5}$

51. $\frac{4}{9} + \frac{5}{6}$

52. $\frac{8}{11} + \frac{3}{4}$

53. $\frac{2}{9} + \frac{2}{3}$

54. $\frac{3}{4} + \frac{3}{8}$

55. $\frac{5}{9} + \frac{7}{10}$

56. $\frac{5}{8} + \frac{3}{7}$

57. $\frac{4}{15} + \frac{1}{4}$

58. $\frac{7}{8} + \frac{5}{6}$

59. $\frac{3}{5} + \frac{8}{35}$

60. $\frac{15}{16} + \frac{19}{20}$

61. $7\frac{2}{13} + 12\frac{5}{13}$

62. $6\frac{3}{11} + \frac{5}{11}$

63. $3\frac{3}{16} + 6\frac{9}{16}$

64. $26\frac{5}{8} + 9\frac{1}{8}$

65. $2\frac{5}{7} + 8\frac{3}{7}$

66. $4\frac{5}{11} + 1\frac{8}{11}$

67. $6\frac{3}{5} + 7\frac{4}{5}$

68. $5\frac{3}{8} + 9\frac{7}{8}$

69. $4\frac{7}{10} + 1\frac{3}{10}$

70. $6\frac{7}{12} + 5\frac{1}{3}$

71. $2\frac{5}{8} + 1\frac{1}{3}$

72. $3\frac{5}{6} + 1\frac{2}{3}$

73. $7\frac{3}{4} + 2\frac{5}{8}$

74. $4\frac{6}{7} + 3\frac{1}{2}$

75. $4\frac{1}{3} + 3\frac{1}{3}$

76. $6\frac{5}{8} + 7$

77. $\frac{4}{7} + 3\frac{2}{7}$

78. $2\frac{7}{11} + 8\frac{8}{11}$

79. $2\frac{5}{16} + 5\frac{3}{16}$

80. $6\frac{3}{8} + 4\frac{3}{4}$

Part C: Review. Answers for the problems of Part C are not given. However, the notation «C,U» refers to the Chapter (C) and Unit (U) in which problems of the same type were presented.

1. What is the hundreds digit of 4,327,908?   «1,1»

2. 89,463 + 295 + 5,628   «1,2»

3. 93,607 − 43,638   «1,3»

4. 3,479 x 8   «2,1»

5. $6\overline{)4185}$ «2,3»

6. $23\overline{)51498}$ «2,4»

7. Simplify $\frac{24}{28}$ «3,1»

8. Write $\frac{38}{5}$ as a mixed number.   «3,2»

9. Find the numerator of $\frac{5}{8} = \frac{?}{48}$ «3,2»

10. Find the lcm for 20 and 15.   «3,2»

11. Select the smaller fraction from: $\frac{2}{3}$, $\frac{7}{11}$ «3,2»

12. Change $3\frac{3}{8}$ to an improper fraction.   «3,2»

13. Margaret has three bags to check for an airplane trip. They weigh 78, 43, and 28 pounds. What is the total weight of her bags?   «1,4»

14. Walter earns $4,934 on one investment and $2,438 on another. What is the difference in these amounts?   «1,4»

15. June scored 35, 78, 56, 84, and 92 on five math tests. What was her average score?   «2,4»

16. Each row in an auditorium has 58 seats. If there are 92 rows in the auditorium, how many seats are there?   «2,2»

Unit 4:   SUBTRACTING FRACTIONS AND MIXED NUMBERS

Subtracting Fractions

To subtract fractions they must have the same denominator.

The subtraction of $\frac{3}{8} - \frac{1}{5}$ is completed by:

1. Writing the fractions with the same denominator

2. Subtracting the numerators.

$$\frac{3}{8} = \frac{15}{40}$$
$$-\frac{1}{5} = \frac{8}{40}$$
$$\overline{\phantom{-\frac{1}{5} =}\;\frac{7}{40}}$$

Focus on the subtraction of fractions with the same denominator

The addition or subtraction of numbers is sensible only when the numbers have the same label.

The subtraction of $\frac{4}{17}$ from $\frac{15}{17}$ is possible because the fractions have the same label (denominator).

15 seventeenths − 4 seventeenths = 11 seventeenths

$$\frac{15}{17} - \frac{4}{17} = \frac{11}{17}$$

Subtracting Fractions With Different Denominators

To subtract $\frac{5}{8}$ from $\frac{11}{12}$

1. Write the problem with the subtrahend, $\frac{5}{8}$, and minuend, $\frac{11}{12}$, in the correct order.

$$\frac{11}{12} - \frac{5}{8}$$

2. Find a common denominator, 24, and write each fraction with it.

$$\frac{22}{24} - \frac{15}{24}$$

3. Subtract the numerators of the new fractions.

$$\frac{7}{24}$$

Focus on completing subtraction after finding a common denominator

The problem $\frac{9}{10} - \frac{1}{8}$ is completed by first changing each fraction to a common denominator.

$$\frac{9}{10} = \frac{36}{40}$$
$$-\ \frac{1}{8} = \frac{5}{40}$$
$$\overline{\phantom{-\ \frac{1}{8} =}\ \frac{31}{40}}$$

Subtracting Mixed Numbers

To subtract two mixed numbers,

1. Subtract the fractions, and

2. Subtract the whole numbers.

For the subtraction problem at the right, work from the right to the left.

$$4\frac{5}{7}$$
$$-\ 3\frac{2}{7}$$
$$1\frac{3}{7}$$

$\frac{5}{7} - \frac{2}{7} = \frac{3}{7}$ and $4 - 3 = 1$

Focus on the subtraction of mixed numbers

The subtraction of $2\frac{5}{8}$ from $8\frac{7}{8}$ is shown below. Parentheses, ( ), have been used to show that the fractions and whole numbers are separately subtracted.

$$8\frac{7}{8} - 2\frac{5}{8}$$
$$(8 - 2) \quad (\tfrac{7}{8} - \tfrac{5}{8})$$
$$(6) \quad (\tfrac{2}{8})$$
$$6\frac{2}{8}$$

The fraction $\frac{2}{8}$ should be simplified to give a final answer of $6\frac{1}{4}$.

Borrowing in the Subtraction of Mixed Numbers

To subtract $3\frac{8}{9}$ from $8\frac{4}{9}$ borrowing is necessary because $\frac{8}{9}$ is greater than $\frac{4}{9}$.

$$8\frac{4}{9} - 3\frac{8}{9}$$

1. Borrow one (1) from the whole number, 8, and put the one (1) with the proper fraction.

$$8\frac{4}{9} = 7 + 1\frac{4}{9}$$
$$-3\frac{8}{9} = -3 \quad \frac{8}{9}$$

2. Write $1\frac{4}{9}$ as an improper fraction.

$$7 + 1\frac{4}{9} = 7\frac{13}{9}$$
$$-3 \quad \frac{8}{9} = -3\frac{8}{9}$$
$$\rule{3cm}{0.4pt}$$
$$4\frac{5}{9}$$

3. Now $\frac{13}{9}$ is greater than $\frac{8}{9}$ so the subtraction can be completed.

Focus on borrowing in subtraction

The subtraction problem $6\frac{3}{7} - 3\frac{5}{7}$ requires borrowing because $\frac{5}{7}$ is greater than $\frac{3}{7}$.

$6\frac{3}{7}$ is equal to $5 + 1\frac{3}{7}$ and $1\frac{3}{7} = \frac{10}{7}$.

$$6\frac{3}{7} = 5\frac{10}{7}$$
$$-3\frac{5}{7} = -3\frac{5}{7}$$
$$\rule{3cm}{0.4pt}$$
$$2\frac{5}{7}$$

Subtracting Mixed Numbers With Different Denominators

To subtract two mixed numbers which have different denominators, a common denominator must first be found.

The steps in subtracting $5\frac{3}{5}$ from $8\frac{1}{4}$ are shown below.

$$
\begin{array}{rclclc}
8\frac{1}{4} &=& 8\frac{5}{20} &=& 7\frac{25}{20} \\
-\ 5\frac{3}{5} &=& -\ 5\frac{12}{20} &=& -\ 5\frac{12}{20} \\
\hline
& & & & 2\frac{13}{20}
\end{array}
$$

Focus on finding a common denominator before subtracting

Subtracting $2\frac{5}{6}$ from $7\frac{3}{4}$ requires finding a common denominator.

Borrowing is also needed.

$$
\begin{array}{rclclc}
7\frac{3}{4} &=& 7\frac{9}{12} &=& 6\frac{21}{12} \\
-\ 2\frac{5}{6} &=& -\ 2\frac{10}{12} &=& -\ 2\frac{10}{12} \\
\hline
& & & & 4\frac{11}{12}
\end{array}
$$

Focus on subtracting a whole number and a mixed number

Fractions such as $\frac{0}{7}$, $\frac{0}{3}$, and $\frac{0}{11}$ are equal to zero. Such fractions are useful in the two examples shown below.

$$
8\frac{4}{7} - 5 \qquad
\begin{array}{rclcl}
8\frac{4}{7} &=& 8\frac{4}{7} \\
-\ 5 &=& -\ 5\frac{0}{7} \\
\hline
& & 3\frac{4}{7}
\end{array}
$$

$$
6 - 2\frac{1}{3} \qquad
\begin{array}{rclclc}
6 &=& 6\frac{0}{3} &=& 5\frac{3}{3} \\
-\ 2\frac{1}{3} &=& -\ 2\frac{1}{3} &=& -\ 2\frac{1}{3} \\
\hline
& & & & 3\frac{2}{3}
\end{array}
$$

Unit 4 Exercise

This exercise reviews the preceding unit. The exercise is divided into three parts. Part A reviews the foci of Unit 4. Part B offers opportunities to practice the skills and concepts of Unit 4. Part C contains problems that review your previous work in this text. All answers for Parts A and B are at the back of the book. Each problem of Part C is accompanied by a notation «CU» that refers to the Chapter (C) and Unit (U) in which that type of problem is studied.

Part A: Reviewing the foci of Unit 4.

1. Addition and subtraction only make sense when the numbers have the same label. Does it make sense to subtract 3 doctors from 7 nurses.

2. The denominator of a fraction is its "label." $\frac{4}{5}$ can be read as 4 fifths. $\frac{5}{7}$ can be read as 5 sevenths. Do $\frac{4}{5}$ and $\frac{5}{7}$ have the same "label?"

3. The subtraction problem
$$2 - 5 = \underline{\hphantom{xx}}$$
has no whole number answer because the subtrahend, 5, is greater than the minuend, 2. Does the problem below have a whole number answer?
$$8 - 5 = \underline{\hphantom{xxx}}$$

4. The problem $\frac{2}{7} - \frac{5}{7} = X$ has its answer indicated by the capital letter X because $\frac{5}{7}$ is greater than $\frac{2}{7}$.
$$\frac{5}{12} - \frac{7}{12} = \underline{\hphantom{xxx}}$$

5. Subtract $2\frac{3}{11}$ from $7\frac{5}{11}$ by subtracting the fractions and then subtracting the whole numbers.

6. Subtract $1\frac{4}{5}$ from $3\frac{2}{5}$ by first borrowing from $3\frac{2}{5}$.

7. Use the borrowing process to subtract:
$$5\frac{3}{11} - 2\frac{5}{8} = \underline{\hphantom{xxx}}$$

8. The whole number 5 is equal to the mixed number $4\frac{7}{7}$.
$$5 - 1\frac{5}{7} = \underline{\hphantom{xxx}}$$

9. Subtract $4\frac{1}{2}$ from $6\frac{3}{4}$ by first finding a common denominator.

10. Subtract $5\frac{1}{16}$ from $5\frac{3}{8}$ by first finding a common denominator.

11. Subtract $2\frac{1}{2}$ from $7\frac{3}{5}$ by first finding a common denominator.

12. The fraction $\frac{0}{5}$ equals zero (0). Use this fact to subtract:
$$7\frac{3}{5} - 2 = 7\frac{3}{5} - 2\frac{0}{5}$$

Part B: Drill and Practice

Solve.

1. $\frac{4}{5} - \frac{1}{5}$

2. $\frac{6}{7} - \frac{3}{7}$

3. $\frac{6}{13} - \frac{8}{13}$

4. $\frac{8}{23} - \frac{15}{23}$

5. $\frac{8}{15} - \frac{4}{15}$

6. $\frac{1}{3} - \frac{1}{4}$

7. $\frac{9}{10} - \frac{1}{5}$

8. $\frac{3}{4} - \frac{1}{2}$

9. $\frac{5}{6} - \frac{7}{10}$

10. $\frac{5}{8} - \frac{2}{7}$

11. $\frac{9}{10} - \frac{3}{8}$

12. $\frac{7}{15} - \frac{3}{5}$

13. $\frac{2}{3} - \frac{1}{4}$

14. $\frac{3}{8} - \frac{2}{3}$

15. $\frac{5}{6} - \frac{5}{9}$

16. $\frac{9}{13} - \frac{2}{13}$

17. $\frac{7}{10} - \frac{9}{10}$

18. $\frac{4}{5} - \frac{3}{10}$

19. $\frac{2}{7} - \frac{1}{8}$

20. $\frac{5}{8} - \frac{1}{6}$

21. $\frac{5}{8} - \frac{1}{4}$

22. $\frac{4}{11} - \frac{1}{2}$

23. $\frac{5}{7} - \frac{3}{8}$

24. $\frac{5}{6} - \frac{3}{4}$

25. $\frac{9}{16} - \frac{3}{4}$

26. $\frac{4}{5} - \frac{2}{3}$

27. $\frac{5}{7} - \frac{3}{10}$

28. $5\frac{8}{9} - 3\frac{7}{9}$

29. $8\frac{2}{3} - 1\frac{1}{3}$

30. $9\frac{8}{11} - 5\frac{5}{11}$

31. $7\frac{9}{16} - 2\frac{10}{16}$

32. $5\frac{3}{10} - 2\frac{9}{10}$

33. $9\frac{1}{8} - 5\frac{7}{8}$

34. $4\frac{2}{8} - 3\frac{7}{8}$

35. $6\frac{3}{7} - 4\frac{1}{2}$

36. $5\frac{3}{10} - \frac{4}{5}$

37. $6\frac{1}{4} - 2\frac{2}{3}$

38. $6\frac{3}{8} - 5\frac{6}{7}$

39. $8\frac{7}{9} - 5$

40. $9 - 2\frac{3}{11}$

41. $5\frac{1}{5} - 2\frac{1}{8}$

42. $3\frac{5}{16} - 1\frac{1}{2}$

43. $3\frac{5}{8} - 1\frac{1}{4}$

44. $9\frac{3}{16} - 2\frac{1}{8}$

45. $4\frac{3}{7} - 2\frac{3}{8}$

46. $5\frac{9}{10} - 2\frac{1}{2}$

47. $8\frac{1}{3} - 3\frac{3}{5}$

48. $5\frac{1}{7} - 2\frac{1}{9}$

49. $4 - 3\frac{1}{2}$

50. $6\frac{1}{4} - 3\frac{1}{2}$

51. $8\frac{5}{9} - 3\frac{5}{6}$

52. $9\frac{5}{8} - 2\frac{1}{6}$

Part C: Review. Answers for the problems of Part C are not given. However, the notation «C,U» refers to the Chapter (C) and Unit (U) in which problems of the same type were presented.

1. What is the tens digit of 4,327,908?   «1,1»

2. 54,963 + 217 + 4,928   «1,2»

3. 80,907 − 43,638   «1,3»

4. 5,879 x 4   «2,1»

5. $6\overline{)7325}$ «2,3»

6. $74\overline{)59648}$ «2,4»

7. Simplify $\frac{26}{39}$ «3,1»

8. Write $\frac{34}{6}$ as a mixed number.   «3,2»

9. Find the numerator of $\frac{1}{4} = \frac{?}{48}$ «3,2»

10. Find the lcm for 12 and 15.   «3,2»

11. Select the smaller fraction from: $\frac{3}{4}$, $\frac{9}{13}$ «3,2»

12. Change $3\frac{3}{5}$ to an improper fraction.   «3,2»

13. $6\frac{3}{8} + 4\frac{3}{4}$ «3,3»

14. Katherine has money in two accounts. The amounts are $1,943 and $543. What is the sum of the money in her acounts?

15. The linemen on a football team weighted 239, 178, 256, 184, 189, 245, and 193 pounds. What was the average weight of the linemen?   «2,4»

16. A passenger train with fourteen cars can seat 64 people in each car. How many passenger seats are there on the train?   «2,2»

Unit 5:   MULTIPLYING FRACTIONS AND MIXED NUMBERS

Multiplying Fractions

Any two fractions can be multiplied by the following process:

1. Multiply the numerators to find the numerator of the product.

2. Multiply the denominators to find the denominator of the product.

$$\frac{5}{8} \times \frac{3}{4} = \frac{5 \times 3}{8 \times 4} = \frac{15}{32}$$

Focus on multiplication of fractions

The problem at the right is an example of the multiplication of two fractions. Multiplication is accomplished by separately multiplying the numerators and denominators.

$$\frac{3}{4} \times \frac{5}{7} = \frac{15}{28}$$

Cancelling — A Valuable Shortcut In Multiplying Fractions

One of the most valuable, useful shortcuts in arithmetic is the cancelling that is often possible when multiplying fractions.

Cancelling is possible when a numerator and a denominator have a common divisor greater than one. In the example at the right, 5 is a common factor. When the 5's are cancelled the multiplication problem becomes simpler, $\frac{3}{1} \times \frac{1}{7}$.

$$\frac{3}{5} \times \frac{5}{7} = \frac{3}{\underset{1}{\cancel{5}}} \times \frac{\cancel{5}^{1}}{7}$$

$$\frac{3}{1} \times \frac{1}{7} = \frac{3}{7}$$

Focus on cancelling in multiplication

The cancelling example shown at the right is dependent upon the fact that $2 \div 2 = 1$.

$$\frac{2}{5} \times \frac{1}{2} = \frac{1\cancel{2}}{5} \times \frac{1}{\cancel{2}_1} = \frac{1}{5} \times \frac{1}{1} = \frac{1}{5}$$

Cancelling Makes Multiplication of Fractions Easier

The problem $\frac{14}{15} \times \frac{4}{7}$ is made far easier by use of the cancelling process.

$$\frac{14}{15} \times \frac{4}{7} = \frac{2\cancel{14}}{15} \times \frac{4}{\cancel{7}_1}$$

$$= \frac{2}{15} \times \frac{4}{1}$$

$$= \frac{8}{15}$$

Focus on using cancellation to simplify a problem

Cancelling has been used to simplify the problem below.

$$\frac{3}{8} \times \frac{12}{13} = \frac{3}{_2\cancel{8}} \times \frac{\cancel{12}^{\,3}}{13}$$

$$= \frac{3}{2} \times \frac{3}{13}$$

$$= \frac{9}{26}$$

Cancelling Twice in a Multiplication Problem

Sometimes a multiplication problem offers two pairs of cancellation opportunities. The problem shown below illustrates this.

First, 5 is used as a common divisor.

$$\frac{5}{9} \times \frac{6}{25} = \frac{\cancel{5}^{\,1}}{9} \times \frac{6}{\cancel{25}_5}$$

Second, 3 is used as a common divisor.

$$= \frac{1}{_3\cancel{9}} \times \frac{\cancel{6}^{\,2}}{5}$$

$$= \frac{1}{3} \times \frac{2}{5}$$

The problem becomes $\frac{1}{3} \times \frac{2}{5}$.

$$= \frac{2}{15}$$

The problem below offers two pairs of cancellation opportunities.

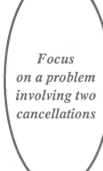

Focus on a problem involving two cancellations

$$\frac{4}{5} \times \frac{15}{16} = \frac{\overset{1}{\cancel{4}}}{5} \times \frac{15}{\cancel{16}_{4}}$$

$$= \frac{1}{\underset{1}{\cancel{5}}} \times \frac{\cancel{15}^{3}}{4}$$

$$= \frac{1}{1} \times \frac{3}{4}$$

$$= \frac{3}{4}$$

Multiplying Mixed Numbers

The simplest way to multiply two mixed numbers is to change them to improper fractions and then multiply using cancellation wherever possible.

$$2\frac{1}{4} \times 5\frac{1}{3}$$

$$\frac{9}{4} \times \frac{16}{3}$$

$$\frac{3}{1} \times \frac{4}{1}$$

$$12$$

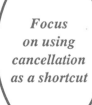

Focus on using cancellation as a shortcut

Cancellation is a valuable shortcut in multiplying mixed numbers. Cancellation is used to complete the problem shown below.

$$2\frac{1}{3} \times \frac{2}{7} = \frac{7}{3} \times \frac{2}{7}$$

$$= \frac{1}{3} \times \frac{2}{1}$$

$$= \frac{2}{3}$$

Unit 5 Exercise

This exercise reviews the preceding unit. The exercise is divided into three parts. Part A reviews the foci of Unit 5. Part B offers opportunities to practice the skills and concepts of Unit 5. Part C contains problems that review your previous work in this text. All answers for Parts A and B are at the back of the book. Each problem of Part C is accompanied by a notation «CU» that refers to the Chapter (C) and Unit (U) in which that type of problem is studied.

Part A: Reviewing the foci of Unit 5.

1. Multiply $\frac{7}{10}$ x $\frac{3}{8}$ by separately multiplying the numerators and denominators.

2. Multiply $\frac{5}{8}$ x $\frac{9}{11}$ by separately multiplying the numerators and denominators.

3. The whole number 6 can be written as $\frac{6}{1}$. Multiply the problem below.

    6 x $\frac{5}{7}$

4. In the multiplication problem shown at the right, what number is a divisor of both a numerator and a denominator? $\frac{2}{5}$ x $\frac{1}{2}$

Cancel whenever possible in problems 5-14.

5. $\frac{3}{4}$ x $\frac{1}{3}$

6. $\frac{6}{7}$ x $\frac{8}{9}$

7. $\frac{7}{10}$ x $\frac{5}{9}$

8. $\frac{6}{11}$ x $\frac{5}{6}$

9. $\frac{7}{11}$ x $\frac{11}{12}$

10. $\frac{4}{9}$ x $\frac{3}{8}$

11. $\frac{8}{11}$ x $\frac{11}{12}$

12. $\frac{15}{16}$ x $\frac{8}{25}$

13. $\frac{3}{7}$ x $\frac{5}{6}$

14. $\frac{10}{21}$ x $\frac{3}{20}$

In problems 15-22 change the mixed numbers to improper fractions. Then multiply using cancellation whenever possible.

15. $6\frac{1}{2}$ x $\frac{3}{4}$

16. $\frac{6}{7}$ x $1\frac{1}{2}$

17. $5\frac{1}{3}$ x $\frac{3}{4}$

18. $\frac{2}{3}$ x $2\frac{3}{4}$

19. $5\frac{1}{3}$ x $2\frac{1}{4}$

20. $\frac{5}{9}$ x $3\frac{3}{8}$

21. $5\frac{1}{3}$ x $\frac{3}{8}$

Part B: Drill and Practice

Multiply.

1. $\frac{2}{5} \times \frac{2}{7}$

2. $\frac{6}{7} \times \frac{5}{13}$

3. $\frac{4}{9} \times \frac{2}{3}$

4. $\frac{9}{16} \times \frac{1}{2}$

5. $4 \times \frac{3}{7}$

6. $\frac{3}{8} \times 3$

7. $\frac{3}{4} \times 7$

8. $\frac{2}{3} \times 5$

9. $\frac{6}{7} \times \frac{2}{5}$

10. $\frac{5}{8} \times 3$

11. $\frac{3}{7} \times 9$

12. $4 \times \frac{4}{11}$

13. $\frac{2}{5} \times \frac{4}{9}$

14. $\frac{5}{7} \times \frac{3}{11}$

15. $\frac{4}{5} \times \frac{1}{3}$

16. $\frac{6}{7} \times \frac{3}{5}$

17. $4 \times \frac{1}{5}$

18. $\frac{6}{11} \times 3$

19. $\frac{4}{9} \times \frac{2}{5}$

20. $9 \times \frac{3}{5}$

21. $\frac{8}{11} \times \frac{2}{3}$

22. $\frac{8}{9} \times 2$

23. $7 \times \frac{4}{9}$

24. $\frac{3}{4} \times 9$

25. $\frac{8}{9} \times \frac{5}{14}$

26. $\frac{5}{18} \times \frac{12}{17}$

27. $\frac{18}{23} \times \frac{5}{9}$

28. $\frac{7}{8} \times \frac{2}{5}$

29. $\frac{7}{18} \times \frac{27}{28}$

30. $\frac{5}{21} \times \frac{18}{25}$

31. $\frac{9}{26} \times \frac{13}{15}$

32. $\frac{12}{25} \times \frac{15}{16}$

33. $\frac{35}{38} \times \frac{20}{21}$

34. $\frac{12}{55} \times \frac{25}{18}$

35. $\frac{15}{16} \times \frac{12}{35}$

36. $\frac{3}{4} \times \frac{2}{5}$

37. $\frac{4}{9} \times \frac{6}{7}$

38. $\frac{15}{17} \times \frac{17}{19}$

39. $\frac{4}{9} \times \frac{5}{12}$

40. $\frac{5}{22} \times \frac{11}{12}$

41. $\frac{6}{7} \times \frac{14}{15}$

42. $\frac{8}{9} \times \frac{15}{16}$

43. $\frac{12}{13} \times \frac{13}{18}$

44. $\frac{33}{35} \times \frac{7}{22}$

45. $\frac{8}{15} \times \frac{3}{4}$

46. $\frac{5}{8} \times \frac{6}{7}$

47. $\frac{7}{10} \times \frac{3}{14}$

48. $\frac{3}{5} \times 4\frac{1}{2}$

49. $\frac{2}{5} \times 3\frac{1}{3}$

50. $5\frac{1}{2} \times \frac{7}{11}$

51. $2\frac{1}{2} \times \frac{3}{5}$

52. $6\frac{3}{4} \times \frac{1}{9}$

53. $\frac{2}{7} \times 2\frac{4}{5}$

54. $1\frac{2}{7} \times \frac{2}{3}$

55. $1\frac{3}{5} \times \frac{1}{2}$

56. $\frac{4}{5} \times 1\frac{1}{5}$

57. $1\frac{5}{6} \times \frac{6}{11}$

58. $\frac{9}{13} \times 3\frac{1}{4}$

59. $2\frac{2}{3} \times \frac{1}{4}$

60. $3\frac{1}{8} \times \frac{3}{10}$

Part C: Review. Answers for the problems of Part C are not given. However, the notation «C,U» refers to the Chapter (C) and Unit (U) in which problems of the same type were presented.

1. What is the ten thousands digit of 4,327,908?   «1,1»

2. 54,417 + 907 + 2,828   «1,2»

3. 78,007 − 43,638   «1,3»

4. 6,179 x 6   «2,1»

5. 8) 7325   «2,3»

6. 15) 59648   «2,4»

7. Simplify $\frac{30}{35}$ «3,1»

8. Write $\frac{19}{3}$ as a mixed number.   «3,2»

9. Find the numerator of $\frac{2}{3} = \frac{?}{48}$ «3,2»

10. Find the lcm for 11 and 15.   «3,2»

11. Select the smaller fraction from: $\frac{1}{4}, \frac{3}{11}$ «3,2»

12. Change $3\frac{2}{3}$ to an improper fraction.   «3,2»

13. $6\frac{5}{7} + 3\frac{3}{4}$ «3,3»

14. $8\frac{1}{3} - 3\frac{3}{5}$ «3,4»

15. Wendy read 394 pages from her book, while Art read 217 from his book. What is the difference in these amounts?   «1,4»

16. A mailman thinks he can deliver to 14 houses each hour. How many houses can he deliver to in an eight hour day?   «2,2»

Unit 6:   DIVIDING FRACTIONS AND MIXED NUMBERS

Definition   **The Reciprocal of a Fraction**

Two fractions are   **if and**   their product
reciprocals   **only if**   is 1 (one).

For example, $\frac{5}{7}$ and $\frac{7}{5}$ are reciprocals because

$$\frac{\overset{1}{\cancel{5}}}{\underset{1}{\cancel{7}}} \times \frac{\overset{1}{\cancel{7}}}{\underset{1}{\cancel{5}}} = \frac{1}{1} = 1$$

Focus on the meaning of reciprocal

The reciprocal of $\frac{1}{3}$ is found by interchanging its numerator and denominator.

The **reciprocal** of $\frac{1}{3}$ is $\frac{3}{1}$ (note the numerator and denominator were interchanged).

$$\frac{1}{\underset{1}{\cancel{3}}} \times \frac{\overset{1}{\cancel{3}}}{1} = \frac{1}{1} = 1$$

Focus on reciprocals

The reciprocal of $\frac{3}{4}$ is $\frac{4}{3}$ because the numerator and denominator have been interchanged.

$$\frac{\overset{1}{\cancel{3}}}{\underset{1}{\cancel{4}}} \times \frac{\overset{1}{\cancel{4}}}{\underset{1}{\cancel{3}}} = \frac{1}{1} = 1$$

The reciprocal of $\frac{5}{8}$ is $\frac{8}{5}$.

$$\frac{5}{8} \times \frac{8}{5} = \frac{1}{1} = 1$$

The reciprocal of $\frac{5}{6}$ is $\frac{6}{5}$.

$$\frac{5}{6} \times \frac{6}{5} = \frac{1}{1} = 1$$

The reciprocal of $\frac{9}{10}$ is $\frac{10}{9}$.

$$\frac{9}{10} \times \frac{10}{9} = \frac{1}{1} = 1$$

The reciprocal of 6 is $\frac{1}{6}$. $(6 = \frac{6}{1})$

$$\frac{6}{1} \times \frac{1}{6} = \frac{1}{1} = 1$$

Naming the Numbers in a Division Problem

In the problem shown at the right, the **dividend** is $\frac{3}{8}$ and the **divisor** is $\frac{2}{7}$. The symbol ÷ means "divided by" and the **divisor** is always to the right of the division symbol.

$$\frac{3}{8} \div \frac{2}{7} = \frac{21}{16}$$

(dividend ÷ divisor = quotient)

or

$$\left(\frac{\text{dividend}}{\text{divisor}} = \text{quotient}\right)$$

The answer, $\frac{21}{16}$, to the division problem is its **quotient.**

Dividing Fractions

To divide $\frac{3}{8}$ by $\frac{2}{7}$, the dividend is multiplied by the reciprocal of the divisor. In the example at the right $\frac{3}{8}$ is multiplied by $\frac{7}{2}$.

$$\frac{3}{8} \div \frac{2}{7}$$

(dividend ÷ divisor)

$$\frac{3}{8} \times \frac{7}{2} = \frac{21}{16} = 1\frac{5}{16}$$

(dividend x divisor's reciprocal)

The answer, $\frac{21}{16}$ or $1\frac{5}{6}$, is the **quotient.**

The division of fractions is accomplished by:
1. Changing the division to multiplication, and
2. Changing the divisor to its reciprocal

Focus on changing division to multiplication

The two examples at the right show the division process. Note the two steps listed above and the fact that the multiplication allows for cancellation.

┌── change to multiplication
│ ┌─ use the divisor's reciprocal
$$\frac{2}{5} \div \frac{3}{5} = \frac{2}{5} \times \frac{5}{3} = \frac{2}{1} \times \frac{1}{3} = \frac{2}{3}$$
┌── change to multiplication
│ ┌─ use the divisor's reciprocal
$$\frac{3}{4} \div 2 = \frac{3}{4} \times \frac{1}{2} = \frac{3}{8}$$

Justifying the Division Process

The method for dividing fractions may seem like magic, but that is not the best way to learn mathematics. Understanding is necessary for learning mathematics.

Recall that both numerator and denominator of a fraction can be multiplied by the same number and give a new, equivalent fraction. This principle is illustrated by:

$$\frac{5}{8} = \frac{5 \times 4}{8 \times 4} = \frac{20}{32}$$

This same principle is applied below to a division problem that is first written as a fraction.

$$\frac{2}{3} \div \frac{3}{4} = \frac{\frac{2}{3}}{\frac{3}{4}} = \frac{\frac{2}{3} \times \frac{4}{3}}{\frac{3}{4} \times \frac{4}{3}} = \frac{\frac{2}{3} \times \frac{4}{3}}{1} = \frac{2}{3} \times \frac{4}{3}$$

the divisor's reciprocal

Dividing Mixed Numbers

To divide two mixed numbers, first change the mixed numbers to improper fractions.

$$3\frac{1}{5} \div 2\frac{1}{3}$$

$$\frac{16}{5} \div \frac{7}{3}$$

Then complete the problem by multiplying the dividend by the divisor's reciprocal.

$$\frac{16}{5} \times \frac{3}{7}$$

$$\frac{48}{35}$$

The example at the right shows the steps in dividing two mixed numbers.

$$1\frac{13}{35}$$

Focus on the division of two mixed numbers

To divide two mixed numbers, $5\frac{2}{5} \div \frac{9}{10}$

1. Change the mixed numbers to improper fractions $\frac{27}{5} \div \frac{9}{10}$

2. Change the division to multiplication. $\frac{27}{5}$ x ??

3. Change the divisor to its reciprocal. $\frac{27}{5}$ x $\frac{10}{9}$

4. Multiply (cancel where possible). $\frac{\overset{3}{\cancel{27}}}{\underset{1}{\cancel{5}}}$ x $\frac{\overset{2}{\cancel{10}}}{\underset{1}{\cancel{9}}} = \frac{6}{1} = 6$

Checking Division

The long division problem shown at the right is checked by multiplying the divisor and the quotient.

$$
\begin{array}{r}
575 \\
7\overline{)4025}
\end{array}
\qquad
\begin{array}{r}
\text{Check} \\
575 \\
\times \quad 7 \\
\hline
4025 = 4{,}025
\end{array}
$$

The check shows the division to have been done correctly because the product of the divisor and quotient is the dividend.

Division of fractions is checked in exactly the same way. When the product of the divisor and quotient is the dividend, the division checks.

Focus on the process for checking division

$$2\frac{5}{8} \div 1\frac{1}{2} = \frac{21}{8} \div \frac{3}{2} = \frac{21}{8} \times \frac{2}{3} = \frac{7}{4} \times \frac{1}{1} = \frac{7}{4} = 1\frac{3}{4}$$

To check this division, multiply the quotient, $1\frac{3}{4}$, and the divisor, $1\frac{1}{2}$, to show that the product is the dividend, $2\frac{5}{8}$.

$$1\frac{3}{4} \times 1\frac{1}{2} = \frac{7}{4} \times \frac{3}{2} = \frac{21}{8} = 2\frac{5}{8}$$

Since $2\frac{5}{8}$ is the original dividend the check shows the division was correct.

Unit 6 Exercise

This exercise reviews the preceding unit. The exercise is divided into three parts. Part A reviews the foci of Unit 6. Part B offers opportunities to practice the skills and concepts of Unit 6. Part C contains problems that review your previous work in this text. All answers for Parts A and B are at the back of the book. Each problem of Part C is accompanied by a notation «CU» that refers to the Chapter (C) and Unit (U) in which that type of problem is studied.

Part A: Reviewing the foci of Unit 6.

1. Find the reciprocal of $\frac{7}{3}$.

2. The reciprocal of $\frac{5}{9}$ is _____ .

3. Multiply $\frac{7}{8}$ by its reciprocal.

4. Multiply $\frac{17}{5}$ by its reciprocal.

5. Multiply 5 by its reciprocal.

6. Complete the division problem shown below.

    $$\frac{4}{9} \div \frac{5}{8} = \frac{4}{9} \times \frac{8}{5} = \underline{\hspace{1cm}}$$

7. Complete the division problem shown below.

    $$2 \div \frac{5}{7} = \frac{2}{1} \times \frac{7}{5} = \underline{\hspace{1cm}}$$

8. Complete the division problem shown below.

    $$\frac{7}{8} \div 5 = \frac{7}{8} \div \frac{5}{1} = \frac{7}{8} \times \frac{1}{5} = \underline{\hspace{1cm}}$$

9. Divide $3\frac{1}{5} \div \frac{2}{3}$ by first changing $3\frac{1}{5}$ to a mixed number.

10. Divide $\frac{3}{4} \div 2\frac{1}{3}$ by first changing $2\frac{1}{3}$ to a mixed number.

11. Divide and check. $\frac{8}{9} \div 6$

12. Divide and check. $\frac{3}{8} \div \frac{3}{4}$

13. Divide and check. $4 \div \frac{3}{5}$

14. Divide and check. $\frac{4}{9} \div 2\frac{1}{3}$

15. Divide and check. $1\frac{1}{4} \div 2\frac{3}{5}$

Part B: Drill and Practice

Divide and check.

1. $\frac{9}{10} \div \frac{3}{7}$

2. $\frac{7}{8} \div \frac{7}{5}$

3. $6\frac{1}{4} \div \frac{5}{8}$

4. $3\frac{1}{3} \div \frac{4}{5}$

5. $2\frac{2}{5} \div \frac{3}{4}$

6. $\frac{5}{6} \div 1\frac{1}{3}$

7. $\frac{3}{4} \div 5\frac{3}{4}$

8. $\frac{3}{8} \div \frac{1}{3}$

9. $\frac{4}{7} \div \frac{2}{9}$

10. $\frac{4}{5} \div 2$

11. $5 \div \frac{3}{8}$

12. $\frac{6}{7} \div 2$

13. $2\frac{1}{7} \div \frac{5}{8}$

14. $3\frac{1}{9} \div \frac{7}{9}$

15. $5\frac{1}{5} \div 1\frac{1}{3}$

16. $2\frac{1}{4} \div 1\frac{1}{8}$

17. $3\frac{2}{5} \div 1\frac{2}{15}$

18. $1\frac{2}{5} \div 1\frac{5}{9}$

19. $5\frac{1}{4} \div 2\frac{1}{3}$

20. $6\frac{1}{4} \div 1\frac{3}{8}$

Part C: Review. Answers for the problems of Part C are not given. However, the notation «C,U» refers to the Chapter (C) and Unit (U) in which problems of the same type were presented.

1. What is the thousands digit of 6,258,403?
    «1,1»

2. 62,217 + 437 + 3,528   «1,2»

3. 58,430 − 43,638   «1,3»

4. 3,129 x 14   «2,2»

5. 12 $\overline{)41748}$ «2,4»

6. Simplify $\frac{12}{32}$ «3,1»

7. Write $\frac{17}{7}$ as a mixed number.   «3,2»

8. Find the numerator of $\frac{2}{3} = \frac{?}{42}$ «3,2»

9. $6\frac{1}{2} + 5\frac{3}{4}$ «3,3»

10. $9\frac{1}{2} - 3\frac{3}{5}$ «3,4»

11. $\frac{9}{13}$ x $3\frac{1}{4}$ «3,5»

12. Central High School has 2,598 students. Superior High School has 1,865 students. What is the sum of their students?
    «1,4»

13. Last year Carl paid $784 in taxes, Mike paid $1,408 in taxes, and Kate paid $3,856 in taxes?. What was the average taxes paid by the three?   «2,4»

Unit 7:   WORD PROBLEMS

Word Problems Requiring Addition or Subtraction of Fractions

The numbers in an addition
problem are called addends.
The answer to an addition
problem is its sum.

$$4\tfrac{3}{8} \qquad \textbf{addend}$$
$$+\ 1\tfrac{1}{8} \qquad \textbf{addend}$$
$$5\tfrac{4}{8}\ =\ 5\tfrac{1}{2}\ \ \textbf{sum}$$

The numbers in a subtraction
problem are the minuend and
subtrahend. The answer to
a subtraction problem is the
difference.

$$6\tfrac{7}{12} \qquad \textbf{minuend}$$
$$-\ 5\tfrac{4}{12} \qquad \textbf{subtrahend}$$
$$1\tfrac{3}{12}\ =\ 1\tfrac{1}{4}\ \ \textbf{difference}$$

Clue Words for Addition Problems

When arithmetic problems are stated in words, the student must
develop an ability to determine when the problem requires an
addition of the numbers.

Although there are no rules with word problems that should always
be followed, some words frequently are good clues that addition
is necessary.

The words "sum" and "total" are answers for addition problems.

The terms "more," "increased by," and "plus" also might indicate
that addition is the operation to be used.

Word Problem: One jar contains $3\frac{1}{8}$ ounces of syrup and a second jar contains $2\frac{9}{10}$ ounces of syrup. What is the total amount of syrup in the two jars?

Focus on a word problem solved by addition

The word "total" is the

clue that the numbers

should be added.

$$3\frac{1}{8} = 3\frac{5}{40}$$
$$+2\frac{9}{10} = 2\frac{36}{40}$$
$$5\frac{41}{40} = 6\frac{1}{40}$$

$6\frac{1}{40}$ ounces of syrup is the total in the two jars.

Solving Word Problems Requiring Subtraction

There are clue words which often indicate subtraction is the way to solve a word problem.

The word "difference" is the answer to a subtraction problem.

Subtraction is also normally used to solve problems which ask "how much more" or "how many more."

The terms "decreased by," "less than," "diminished by," or "minus" often indicate a problem that will be solved using subtraction.

Focus on finding the difference

The answer to a subtraction problem is its difference. To find the difference between $6\frac{7}{11}$ and $3\frac{2}{11}$ subtract as shown below.

$$6\frac{7}{11}$$
$$-3\frac{2}{11}$$
$$3\frac{5}{11}$$

Word Problem: How much more is $2\frac{1}{2}$ yards than $1\frac{5}{8}$ yards?

Focus on solving a word problem requiring subtraction

The words "how much more" indicate subtraction.

$$2\frac{1}{2} = 2\frac{4}{8} = 1\frac{12}{8}$$
$$-1\frac{5}{8} = -1\frac{5}{8} = -1\frac{5}{8}$$
$$\overline{\qquad\qquad\qquad\frac{7}{8}}$$

$2\frac{1}{2}$ yards is $\frac{7}{8}$ yards more than $1\frac{5}{8}$ yards.

A Fraction "Of" A Number Indicates Multiplication

One of the best clue words in mathematics is the use of the word "of" in a phrase such as $\frac{3}{4}$ **of** 28.

The phrase "**a fraction of a number**" always can be interpreted as indicating multiplication of the fraction and the number.

$\frac{3}{4}$ **of** 28 means $\frac{3}{4}$ x 28.

$\frac{7}{5}$ **of** 46 means $\frac{7}{5}$ x 46.

Focus on a word problem solved using multiplication

Word Problem: Two-thirds of the students in a school are males. If the school has 495 students, how many are males?

The phrase "Two-thirds of the students" is the clue that $\frac{2}{3}$ should be multiplied by 495 (the number of students).

$$\frac{2}{3} \text{ x } 495 = \frac{2}{{}_1 3} \text{ x } \frac{\overset{165}{\cancel{495}}}{1} = \frac{2}{1} \text{ x } \frac{165}{1} = \frac{330}{1} = 330$$

330 of the 495 students are males.

Division Is Indicated By The Question "How Many Are In?"

The problem:

"How many 5 cent pencils can be purchased for 85 cents?"

can be re-phrased as

"How many 5's are in 85?"

The question is answered by dividing 85 by 5.

Whenever a problem asks:

"How many of a first number are in a second number?"

the answer is found by dividing the second number by the first.

Focus on dividing in word problems requiring division

Word Problem: How many $\frac{1}{2}$ inches are in 8 inches?

The phrase "How many are in" indicates this problem is solved by division. 8 is divided by $\frac{1}{2}$.

$$8 \div \frac{1}{2} = \frac{8}{1} \div \frac{1}{2} = \frac{8}{1} \times \frac{2}{1} = \frac{16}{1} = 16$$

There are 16 one-half inches in 8 inches.

Unit 7 Exercise

This exercise reviews the preceding unit. The exercise is divided into three parts. Part A reviews the foci of Unit 7. Part B offers opportunities to practice the skills and concepts of Unit 7. Part C contains problems that review your previous work in this text. All answers for Parts A and B are at the back of the book. Each problem of Part C is accompanied by a notation «CU» that refers to the Chapter (C) and Unit (U) in which that type of problem is studied.

Part A: Reviewing the foci of Unit 7.

1. What is the sum of $\frac{3}{4}$ lbs. of sugar and $1\frac{1}{2}$ lbs. of sugar?

2. Does it make sense to add $4\frac{1}{2}$ lbs of flour and 3 lbs of sugar?

3. Does it make sense to add $4\frac{1}{2}$ ounces of flour and $3\frac{1}{4}$ lbs. of flour?

4. To find $\frac{9}{10}$ of 15, _____ (divide, multiply) $\frac{9}{10}$ and 15.

5. The problem: "How many $\frac{3}{5}$ gram rings can be made from 12 grams of gold?" can be re-phrased as: How many _____ are in _____?

6. How many $\frac{3}{5}$ gram rings can be made from 12 grams of gold?

7. The problem: "Find the number of $\frac{7}{10}$ grams pills that can be made from 21 grams of medicine" can be re-phrased to: How many _____ are in _____?.

8. Find the number of $\frac{7}{10}$ grams pills that can be made from 21 grams of medicine.

Part B: Drill and Practice

1. How much more is $7\frac{1}{2}$ days than $1\frac{2}{3}$ days?

2. One capsule holds $20\frac{3}{10}$ grams while another holds $15\frac{7}{10}$ grams. How much more is in the bigger capsule?

3. Ben bought two steaks. One weighed $2\frac{1}{4}$ lbs. and the other $1\frac{1}{2}$ lbs. How much steak did Ben buy?

4. One bottle held $4\frac{3}{8}$ liters of alcohol and another held $1\frac{5}{8}$ liters. How much more did the first bottle hold?

5. A specimen of iron ore weighed $4\frac{7}{10}$ kilograms and another weighed $3\frac{1}{10}$ kilograms. What was the sum of the weights of the two specimens?

6. Mary worked $2\frac{5}{6}$ hours and Sally worked $7\frac{1}{4}$ hours. How much more did Sally work?

7. Container A held $5\frac{1}{2}$ cups of milk while container B held only $2\frac{3}{4}$ cups. Find the difference in the amounts of milk the two containers can hold.

8. The price of a stock was $\$10\frac{1}{8}$ on Thursday and $\$8\frac{3}{4}$ on Friday. What was the difference in the price those two days?

9. Bill ran $1\frac{1}{4}$ hours in the morning and $2\frac{3}{5}$ hours in the afternoon. How many hours did he run altogether?

10. Find $\frac{9}{10}$ of 15.

11. Find $\frac{3}{4}$ of 8.

12. Find $\frac{5}{8}$ of 24.

13. Suppose $\frac{3}{4}$ of a class of 20 students were absent. How many students were absent?

14. If a carpenter has finished $\frac{2}{3}$ of a 30 hour job, how many hours has he worked?

15. If a student has read $\frac{5}{6}$ of his 120 page reading lesson, how many pages have been read?

16. Find how many $\frac{1}{4}$ years there are in 5 years.

17. Find how many $\frac{1}{3}$ acres there are in 12 acres.

18. Find how many $\frac{1}{2}$ ounces there are in 32 ounces.

19. How many $\frac{1}{4}$ inches are in 3 inches?

20. How many $\frac{1}{10}$ grams are in 12 grams?

21. How many $\frac{4}{5}$ liters are in 7 liters?

22. How many $3\frac{4}{5}$ kilogram bricks can be made from 950 kilograms of concrete?

23. Razor blades weighing $1\frac{3}{10}$ grams each are to be made from 91 grams of steel. How many can be made?

24. If capsules containing $3\frac{4}{5}$ milligrams of medicine are to be filled from a supply of 76 mg, how many can be made?

25. How many home mortgages are there if the average interest per year is $1 $\frac{1}{2}$ thousand for each mortgage and the total interest collected per year is $45 thousand?

26. The answer to a subtraction problem is its

_____.

27. The answer to an addition problem is its

_____.

28. How much more is $9\frac{1}{4}$ inches than $4\frac{7}{8}$ inches?

29. Find $\frac{4}{7}$ of 21.

30. Sarah bought two pieces of silk. One was $1\frac{1}{2}$ yards long and the other was $1\frac{3}{4}$ yards long. What was the total length of the silk?

31. Find how many $\frac{1}{3}$ years there are in 6 years.

32. One stock sells for 11\frac{3}{8}$ while another sells for 9\frac{1}{4}$. Find the sum of the prices of the stocks.

33. Find how many $\frac{1}{2}$ quart bottles can be filled with $8\frac{1}{2}$ quarts of liquid.

34. A block of iron weighed $12\frac{7}{10}$ kilograms while a block of gold weighed $2\frac{9}{10}$ kilograms. How much more did the iron weigh than the gold?

35. A general estimates that $\frac{3}{8}$ of his 32 battalions are well-prepared. How many battalions are well-prepared?

36. One capsule weighed $5\frac{1}{10}$ grams. Another weighed $3\frac{3}{5}$ grams. Find the difference in their weights.

37. How many $\frac{1}{4}$ ounce coins can be made from 25 ounces of silver?

38. If $2\frac{73}{100}$ meters of rope is cut from a piece $6\frac{3}{5}$ meters long, how much will be left?

39. How many $\frac{3}{5}$ kilogram blocks of lead can be made from 63 kilograms of lead?

40. Bill has to put $\frac{3}{5}$ of a 7 ton load of sand on a vacant lot. How many tons of sand should he put on the lot?

41. John spends $86 to buy stock that costs 5\frac{3}{8}$ per share. How many shares does he buy?

Part C: Review. Answers for the problems of Part C are not given. However, the notation «C,U» refers to the Chapter (C) and Unit (U) in which problems of the same type were presented.

1. What is the hundreds digit of 6,258,403?   «1,1»

2. $57,917 + 463 + 3,417$ «1,2»

3. $58,430 - 35,938$ «1,3»

4. $8,709 \times 67$ «2,2»

5. $57\overline{)41748}$ «2,4»

6. Find the lcm for 18 and 12.   «3,2»

7. Select the smaller fraction from $\frac{2}{5}, \frac{5}{12}$ «3,2»

8. Change $3\frac{6}{7}$ to an improper fraction.   «3,2»

9. $4\frac{2}{3} + 5\frac{3}{8}$ «3,3»

10. $6\frac{1}{3} - 2\frac{3}{8}$ «3,4»

11. $3\frac{1}{5} \times \frac{1}{4}$ «3,5»

12. $5\frac{1}{2} \div 1\frac{1}{8}$ «3,6»

13. Carmen fixed dinner 17 times last month while her husband, Keith, fixed dinner 12 times. What is the difference in these amounts? «1,4»

14. Each month, Henry saves $240. How much will he save in twelve months?   «2,2»

Chapter 3 Test

«3,U» shows the unit in which the problem is found in the chapter.

1. Which of the following is smaller? $\frac{7}{10}$ or $\frac{2}{3}$
«3,2»

2. In the fraction equal to $5 \div 6$ the numerator is _____ .   «3,1»

3. In the fraction $\frac{10}{9}$ the numerator is _____ .   «3,1»

4. What fraction of the figure is shaded?   «3,1»

5. Is 3 a factor of 412?   «3,1»

6. Is 5 a factor of 655?   «3,1»

7. Simplify $\frac{8}{10}$ «3,1»

8. Simplify $\frac{45}{54}$ «3,1»

9. Write $\frac{12}{5}$ as a mixed number.   «3,2»

10. Write $\frac{75}{8}$ as a mixed number.   «3,2»

11. Write $5\frac{7}{8}$ as an improper fraction.   «3,2»

12. Write $3\frac{10}{13}$ as an improper fraction.   «3,2»

13. Find the least common multiple (lcm) for 12 and 15.   «3,2»

14. Write $\frac{7}{8}$ with denominator 24.   «3,2»

15. $\frac{2}{5} + 1\frac{1}{2}$ «3,3»

16. $\frac{1}{7} + \frac{2}{3}$ «3,3»

17. $\frac{4}{7} + 2\frac{3}{4}$ «3,3»

18. $\frac{5}{6} - \frac{1}{3}$ «3,4»

19. $2\frac{3}{8} - 1\frac{7}{10}$ «3,4»

20. $1\frac{1}{8} - \frac{3}{4}$ «3,4»

21. Two rocks, one weighing $2\frac{3}{4}$ kilograms and the other weighing $1\frac{1}{2}$ kilograms, were placed on a scale. What was their total weight?   «3,7»

22. A piece of rope $5\frac{3}{4}$ meters long was cut from a rope $10\frac{1}{2}$ meters long. How much rope was left?   «3,7»

23. How much more is $5\frac{1}{8}$ cups than $3\frac{3}{8}$ cups?   «3,7»

24. $\frac{15}{16}$ x $\frac{32}{35}$ «3,5»

25. $\frac{9}{10}$ x $\frac{22}{18}$ «3,5»

26. $1\frac{5}{8}$ x 3   «3,5»

27. $1\frac{2}{3}$ x $\frac{1}{2}$ «3,5»

28. $3\frac{1}{2}$ ÷ $1\frac{3}{4}$ «3,6»

29. $8 \div \frac{1}{5}$ «3,6»

30. $2\frac{3}{4} \div \frac{3}{7}$ «3,6»

31. In the problem $\frac{2}{3} \div \frac{5}{9} = \frac{6}{5}$, the quotient is ____, the dividend is ____, and the divisor is ____.   «3,6»

32. Find $\frac{4}{5}$ of 25.   «3,7»

33. Bill had $3\frac{2}{3}$ liters of medicine. One half of it was needed on a wound. How much was needed?   «3,7»

34. Clay cups weighing $\frac{3}{10}$ kilograms are to be made from $5\frac{7}{10}$ kilograms of clay. How many cups can be made?   «3,7»

35. The cost of a motorcycle is $\frac{5}{6}$ of the sales price. The sales price is $1800. Find the cost of the motorcycle.   «3,7»

4
Decimal
Fractions

Unit 1: CHANGING FRACTIONS TO DECIMALS

Writing Fractions as Decimal Fractions

Decimal notation provides an easy way to write fractions with
denominators of 10, 100, 1000, and so forth.

$\frac{1}{10}$ is written as 0.1 $\frac{1}{10} = 0.1$

$\frac{1}{100}$ is written as 0.01 $\frac{1}{100} = 0.01$

$\frac{1}{1000}$ is written as 0.001 $\frac{1}{1000} = 0.001$

Focus on the decimal notation for a fraction

The decimal notation for $\frac{1}{100}$ is 0.01; 100 has two zeros and the "1" in 0.01 is in the second place to the right of the decimal point. The decimal notation for $\frac{1}{10}$ is 0.1

The decimal notation for $\frac{3}{10000}$ is 0.0003; 10,000 has four zeros and the "3" in 0.0003 is in the fourth place to the right of the decimal point.

Decimal Fractions That Are Greater Than One

To find the decimal equal to $\frac{258}{10}$:

1. Count the number of zeros in 10.

   There is one zero in the 10 of $\frac{258}{10}$

2. Place the "8" in 258 one place right of the decimal point. The decimal point is moved one place to the left.

   There is one zero in the 10, place the "8" one place to the right of the decimal point.

$$\frac{258}{10} = 25.8$$

Focus on writing fractions as decimals

To write the decimal fraction equal to $\frac{3689}{100}$

1. Count the number of zeros in 100.

   Two zeros in 100.

2. Place the decimal in 3,689 so that the "9" is two places to the right.

$$\frac{3689}{100} = 36.89$$

Place Values of Decimal Fractions

The diagram below shows the names of the place values on both sides of the decimal.

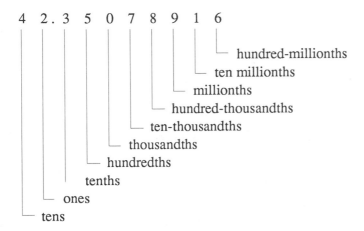

There are several important lessons to be learned about the place value names.

1. Each place is one-tenth as great as the place on the left.
    $\frac{1}{10}$ is one-tenth as great as 1.
    $\frac{1}{100}$ is one-tenth as great as $\frac{1}{10}$.
2. There is no one's place on the right of the decimal point.

Focus on writing decimals for fractions

To write a decimal numeral for 57 thousandths note that "thousandths" requires three decimal places.

Hence, 57 thousandths $= 0.057$

0.057 has **three** decimal places which makes it "thousandths." 1,000 has three zeros.

Focus on reading a decimal numeral

The decimal numeral 0.000081 has six decimal places and **six** decimal places means "millionths." 1,000,000 has six zeros.

Hence, 0.000081 $= 81$ millionths

Writing Decimal Numerals As Fractions

To write the decimal numeral 45.07 as a fraction,

1. Determine the place value of the last digit on the right.

    > 7 is the last digit on the right of 45.07 and the place value of 7 is hundredths.

2. Write the decimal numeral omitting the decimal point in the numerator and use the place value found in step 1 to write the denominator.

$$45.07 = \frac{4507}{100}$$

Notice that when the decimal number has a non-zero number to the left of the decimal point the equivalent fractional value is an improper fraction.

Comparing Sizes Of Two Decimal Fractions

0.56 is less than 0.63 because both numbers are hundredths and 56 is less than 63.

When decimal fractions have a different number of decimal places, like 0.95 and 0.461, their sizes may be compared as follows:

1. Write the fractions one directly above the other with their decimal points in the same vertical column.

    0.9 5
    0.4 6 1

2. Affix zeros to the "shorter" decimal fraction until the right-hand digits are aligned in the same vertical column.

    0.9 5 0
    0.4 6 1

3. Since $\frac{950}{1000}$ is greater than $\frac{461}{1000}$, 0.95 is greater than 0.461.

<table>
<tr><td>Focus on arranging decimals vertically to compare sizes</td><td>To determine whether 0.45 is greater than 0.169, arrange them vertically as shown at the right.</td><td>0.4 5
0.1 6 9</td></tr>
<tr><td></td><td>Affix a zero to 0.45.</td><td>0.4 5 0
0.1 6 9</td></tr>
</table>

It is clear that 0.45 is greater than 0.169.

"Greater Than" and "Less Than"

The symbol ">" means "greater than."
The symbol "<" means "less than."

0.05 > 0.0009 is true because:
0.0500 is greater than 0.0009.

$$\frac{500}{10000} > \frac{9}{10000}$$

Focus on the symbol for less than

The symbol "<" means "less than."

0.098 < 0.0981 is true because:
0.098 = 0.0980 and is less than 0.0981.

$$\frac{980}{10000} < \frac{981}{10000}$$

Focus on the symbol for greater than

The symbol ">" means "greater than."

0.041 > 0.0095 is true because:
0.041 = 0.0410 and is greater than 0.0095.

$$\frac{410}{10000} > \frac{95}{10000}$$

Approximating a Decimal Fraction

1.573 can be approximated by the two-place decimals
    1.57 and 1.58

1.57 can be found from 1.573 by simply omitting the last digit.

    1.573 is slightly more than 1.57

1.57 is 157 hundredths and the next largest two-place decimal is 158 hundredths or 1.58

    1.573 is slightly less than 1.58

1.573 is approximately equal to both 1.57 and 1.58

To round off 1.573 to the nearest hundredth, it is necessary to find the two-place decimal that is closest to 1.573.

Focus on approximating three-place decimals by two-place decimals

0.798 is a three-place decimal. It can be approximated by the two-place decimals

    0.79   or   0.80

0.79 is slightly less than 0.798 and is found by omitting the last digit of 0.798

0.80 is slightly more than 0.798 and is found by increasing 79 hundredths to 80 hundredths.

The three-place decimal 0.132 is greater than the two-place decimal 0.13 and less than the two place decimal 0.14

    $0.132 > 0.13$ and $0.132 < 0.14$

Rounding Off Decimal Fractions To Two Places

The **rounding off process** for decimal numerals is a method of approximation.

The rule for rounding off to two decimal places (nearest hundredth) has two parts.

1. If the digit in the third decimal place (thousandths) is less than 5 drop all the digits to the right of the hundredth's place without changing any other digits.

2. If the digit in the third decimal place (thousandths) is greater than or equal to 5, add one (1) to the digit in the hundredth's place and drop all digits to its right.

Focus on rounding off to two decimal places

To round off 8.9728 to two decimal places,
1.   Note that the digit in the third decimal place is a 2.
2.   Since 2 is less than 5, drop all digits to the right of the second decimal place.

      8.9728 is rounded off to 8.97

To round off 13.5863 to the nearest hundredth,
1.   Note that the digit in the thousandths place is a 6.
2.   Since 6 is more than 5, add one to the hundredths digit and drop all digits to the right of the hundredths place.

      13.5863 is rounded off to 13.59

Four examples of rounding to two decimal (hundredths) places are listed below:

Numeral	Thousandth's Digit	Rounded to Two Decimal Places
1.27351	3	1.27
0.095034	5	0.10
14.271	1	14.27
4.5069	6	4.51

Rounding Decimal Fractions

Decimal fractions can be rounded off to any number of decimal places or to any place value.

The "rounding" rule is:

To round a numeral to a particular place, look at the digit right of the place.

1. If it is greater than or equal to 5, add one to the digit in the place and drop all the digits to the right.

2. If it is less than 5, drop all the digits to the right of the particular place without changing the digit.

1.27 rounded to one decimal (tenths) place is 1.3

0.0948 rounded to three decimal (thousandths) places is 0.095

0.10293 rounded to four decimal (ten-thousandths) places is 0.1029

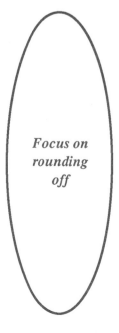

Focus on rounding off

To round 4.26739 to one decimal place, look at the digit in the second decimal place. The digit there is 6. Since 6 is 5 or greater, 1 is added to the first decimal place and the digits on the right are dropped.

4.26739

6 is in the second decimal place

4.26739 rounded to one decimal place is 4.3

To round 0.9263 to the thousandth's place, look at the first digit right of the thousandth's place. Since 3 is less than 5, the digits right of the thousandth's place are dropped.

0.9263

3 is the digit immediately to the right of the thousandth's place

0.9263 rounded to the nearest thousandth is 0.926

Unit 1 Exercise

This exercise reviews the preceding unit. The exercise is divided into three parts. Part A reviews the foci of Unit 1. Part B offers opportunities to practice the skills and concepts of Unit 1. Part C contains problems that review your previous work in this text. All answers for Parts A and B are at the back of the book. Each problem of Part C is accompanied by a notation «CU» that refers to the Chapter (C) and Unit (U) in which that type of problem is studied.

Part A: Reviewing the foci of Unit 1.

1. To write a decimal fraction for $\frac{12}{100000}$, place the "2" five places right of the decimal. $\frac{12}{100000} = 0.00012$
Write the decimal fraction for $\frac{13}{1000}$.

2. $\frac{125}{10000} = 0.0125$

    The "5" in 0.0125 is four places right of the

    decimal point and 10,000 has four zeros.

    Write $\frac{372}{100000}$ using a decimal numeral.

3. 0.8 is less than one.
    1.2 is more than one.
    0.92 is _____ (less, greater) than one.

4. $\frac{4}{10}$ = 0.4 and is less than one.
    $\frac{15}{10}$ = 1.5 and is greater than one.
    $\frac{17}{10}$ is _____ (greater, less) than one.

5. $\frac{34051}{1000}$ is equal to 34.051 because 1,000 has
    three zeros. 10,000 has four zeros. Write
    $\frac{102394}{10000}$ as a decimal fraction.

6. To find the greater of 0.2 or 0.13,   0.2
    arrange them vertically as shown   0.13
    at the right. Then affix a zero to
    0.2 so that is has the same number   0.20
    of decimal places as 0.13 and make   0.13
    the comparison. Which is greater,
    0.2 or 0.13?

7. To determine whether 0.0941 is
    greater then 0.12, arrange them   0.0941
    vertically as shown at the right.   0.12
    Affix two zeros to 0.12. Which
    is greater, 0.0941 or 0.12?

8. 0.352 rounded off to two decimal places is
    0.35 because "2" is in the third decimal place.
    Round off 0.584 to two decimal places.

9. Rounding off 1.283 to two decimal places is
    the same as rounding off 1.283 to the
    hundredth's place. Round off 1.283 to the
    hundredth's place.

10. 0.975 rounded off to the nearest hundredth is
    0.98. Round 0.389 to the nearest hundredth.

Part B: Drill and Practice

In problems 1-14 write the
fraction as a decimal numeral.

1. $\frac{28}{100}$

2. $\frac{3}{1000000}$

3. $\frac{1246}{100000}$

4. $\frac{152}{100}$

5. $\frac{4205}{100}$

6. $\frac{5428}{1000}$

7. $\frac{38}{1000}$

8. $\frac{45125}{1000}$

9. $\frac{1}{10}$

10. $\frac{37}{10}$

11. $\frac{37}{100}$

12. $\frac{125}{10}$

13. $\frac{7}{10000}$

14. $\frac{2300681}{10000}$

15. 1.07 equals the fraction $\frac{107}{100}$.
    Write 1.925 as a fraction.

In problems 16-23 write each
decimal as a fraction.

16. 0.0008

17. 0.00204

18. 28.2

19. 90.07

20. 0.0900

21. 0.028

22. 10.8

23. 0.514

In problems 24-27 select
the greater numeral.

24. 0.98 or 0.1

25. 0.11 or 0.9

26. 0.09 or 0.3

27. 51.6 or 5.16319

In problems 28-35 state whether
the relationship is true or false.

28. 0.08 > 0.009

29. 0.07 > 0.056

30. 0.00125 > 0.125

31. 0.12 < 0.13

32. 0.05 < 0.006

33. 0.07 > 0.004

34. 0.0092 < 0.00925

35. 3.6 > 0.0928

In problems 36-57 round off as indicated.

36. 5.3172 to two decimal places

37. 0.326 to the hundredth's place

38. 1.892 to two decimal places

39. 0.0939 to the nearest hundredth

40. 1.995 to two decimal places

41. 0.473 to two decimal places

42. 1.049 to two decimal places

43. 1.495 to two decimal places

44. 12.301 to the nearest hundredth

45. 1.894 to two decimal places

46. 13.999 to the hundredth's place

47. 0.455 to the second decimal place

48. 103.007 to two decimal places

49. 301.993 to the nearest hundredth

50. 0.505 to two decimal places

51. 13.50293 to the nearest one

52. 2.91256 to the ten thousandth's place

53. 1.09052 to 1 decimal place

54. 13.5 to the nearest one

55. 0.49 to one decimal place

56. 243.53 to the nearest tenth

57. 4.04938 to four decimal places

Part C: Review. Answers for the problems of Part C are not given. However, the notation «C,U» refers to the Chapter (C) and Unit (U) in which problems of the same type were presented.

1. What is the tens digit of 6,258,403?   «1,1»

2. 43,917 + 663 + 2,417   «1,2»

3. 25,430 − 13,918   «1,3»

4. 8,359 x 37   «2,2»

5. $32 \overline{)41748}$ «2,4»

6. Simplify $\frac{15}{24}$ «3,1»

7. Write $\frac{18}{5}$ as a mixed number.   «3,2»

8. Find the numerator of $\frac{5}{7} = \frac{?}{42}$ «3,2»

9. $6\frac{3}{4} + 5\frac{7}{10}$ «3,3»

10. $5\frac{1}{2} - 4\frac{3}{4}$ «3,4»

11. $7\frac{1}{5} \times \frac{5}{9}$ «3,5»

12. $4\frac{1}{2} \div 1\frac{5}{6}$ «3,6»

13. A recipe calls for $1\frac{1}{2}$ cups of flour and $\frac{1}{3}$ cup of sugar. What is the sum of these ingredients? «3,7»

14. A printer did $\frac{2}{3}$ of a job in 30 hours. How many hours are needed to do the full job? «3,7»

Unit 2:   ADDING AND SUBTRACTING DECIMALS

Adding Decimal Fractions

The addition of decimal fractions
is accomplished vertically by
arranging digits with the same
place value in columns.

6.5 3	addend
+ 0.2 4 6	addend
6.7 7 6	sum

The addition of 6.53 and 0.246 is
shown at the right.

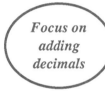

*Focus on
adding
decimals*

To add 0.5 and 3.17 the decimal
points are arranged in a
vertical column.

The addition is shown at the right.

0.5
+ 3.1 7
3.6 7

Adding "Ragged Decimals"

The decimal numerals 0.2 and 0.35 are sometimes called
"ragged decimals" because they have a different number
of decimal places. The addition of "ragged" decimals is
aided by affixing zeros so the numerals have the same
number of decimal places.

Two ways of doing 0.2 + 0.35 are shown below.

0.2	0.2 0	addend
+ 0.3 5	+ 0.3 5	addend
0.5 5	0.5 5	sum

The example on the left shows the "ragged" decimal addition.
In the example on the right the decimals are not "ragged".

In either case, the decimal points must be aligned in a vertical column.

Focus on adding ragged decimals

The addition of 7.83, 91.5, 4.812, and 0.26 is shown at the right.

Notice that all decimal points were aligned vertically. Digits of the same place value are in the same column.

Carrying was necessary because the sum exceeded ten in several columns.

```
  7.83          7.830
 91.5          91.500
  4.812         4.812
+ 0.26        + 0.260
              104.402
```

The sum is 104.402

Subtraction of Decimal Fractions

The addition of decimal fractions depends on placing the decimal points in a vertical column so that the digits of the same place value will be in the same column. The subtraction of decimal fractions also depends upon placing the decimal points in a vertical column. Also, both numbers should be written with the same number of decimal places. Sometimes affixing of zeros will be required. The subtraction is completed by subtracting the digits in each column.

To subtract 0.03 from 0.09

1. The word "from" indicates the problem $0.09 - 0.03$ which is written vertically as shown at the right.

    ```
      0.0 9    minuend
    - 0.0 3    subtrahend
    ```

2. Complete the subtraction by placing the decimal point in the answer directly below the decimals in the problem.

    ```
      0.0 9
    - 0.0 3
      0.0 6    difference
    ```

Focus on subtracting decimals

To subtract 0.28 from 6:

1. Write 6 with a decimal and two zeros so it will have the same number of decimal places as 0.28

$6 = 6.00$

2. Place 0.28 in the second row and arrange the decimal points in a vertical column.

$$\begin{array}{r} 6.0\,0 \\ -\ 0.2\,8 \end{array}$$

3. Subtract in columns bringing the decimal point straight down.

$$\begin{array}{r} 6.0\,0 \\ -\ 0.2\,8 \\ \hline 5.7\,2 \end{array}$$

The subtraction of 0.28 from 6 gives 5.72

Borrowing in the Subtraction of Decimals

Subtracting decimal fractions vertically is accomplished as though subtracting whole numbers. This includes the borrowing process

$$\begin{array}{rl} 0.2\,3\,0 & \text{minuend} \\ -\ 0.1\,3\,7 & \text{subtrahend} \\ \hline 0.0\,9\,3 & \text{difference} \end{array}$$

Notice that "borrowing" was needed in this subtraction and it was accomplished in the same way as it would have been with whole numbers.

Focus on the subtraction of decimals requiring borrowing

To subtract 9.35 from 12.8,

1. First note that "subtract 9.35 from 12.8" means $12.8 - 9.35$ and is written vertically as shown at the right.

$$\begin{array}{r} 1\,2.8 \\ -\,9.3\,5 \end{array}$$

2. Affix a zero to 12.8 so that it will have the same number of decimal places as 9.35

$$\begin{array}{r} 1\,2.8\,0 \\ -\,9.3\,5 \end{array}$$

3. Subtract in columns, borrowing where necessary exactly as would be done with whole numbers.

$$\begin{array}{r} 1\,2.8\,0 \\ -\,9.3\,5 \\ \hline 3.4\,5 \end{array}$$

Focus on a subtraction problem that cannot be completed

To subtract 57.83 from 42.184
begin by recognizing this means

42.184 – 57.83

and is written vertically as shown at
the right.

$$\begin{array}{r} 42.184 \\ -\ 57.83 \\ \hline X \end{array}$$

Because 57.83 is greater than 42.184
the problem cannot be completed here.

Checking Subtraction

In the subtraction example below:

0.50 is the minuend
0.12 is the subtrahend
and 0.38 is the difference.

$$\begin{array}{r} 0.50 \\ -\ 0.12 \\ \hline 0.38 \end{array}$$

The subtraction is checked by the
addition shown at the right. The
sum of the subtrahend and the
difference must be the minuend.

$$\begin{array}{rl} 0.12 & \text{subtrahend} \\ +\ 0.38 & \text{difference} \\ \hline 0.50 & \text{minuend} \end{array}$$

Focus on checking subtraction

In the subtraction example on the right,
the minuend is 0.03, the subtrahend is
0.007, and the difference is 0.023.

$$\begin{array}{r} 0.03 \\ -\ 0.007 \\ \hline 0.023 \end{array}$$

To check the subtraction, add the
subtrahend and difference to get a
sum that is equal to the minuend.

Check

$$\begin{array}{r} 0.007 \\ +\ 0.023 \\ \hline 0.030 \end{array}$$

Clue Words for Addition Problems

When arithmetic problems are stated in words, certain clue words are valuable in indicating that the problem will be solved using addition.

The words "sum" and "total" are answers for addition problems.

The terms "more," "increased by," and "plus" also might indicate that addition is the operation to be used.

Focus on solving a word problem requiring addition

Word Problem: George has 1.05 kilograms of aspirin in one bottle and 0.98 kilograms of aspirin in another. What is the total amount of aspirin he has?

The word "total" indicates the answer to the question is found by addition.

The total (sum) of the aspirin is found by adding 1.05 and 0.98 as shown at the right.

$$\begin{array}{r} 1.05 \\ + \ 0.98 \\ \hline 2.03 \ \text{kilograms} \end{array}$$

Solving Word Problems Requiring Subtraction

There are clue words which often indicate subtraction is the way to solve a word problem.

The word "difference" is the answer to a subtraction problem.

Subtraction is also normally used to solve problems which ask "how much more" or "how many more."

The terms "decreased by," "less than," "diminished by," or "minus" often indicate a problem that will be solved using subtraction.

The word "difference" is the name of the answer to a subtraction problem. It indicates subtraction is needed to solve the problem.

Focus on the word "difference" to answer a subtraction problem

Word Problem: If one book weighs 28.3 kilograms and another book weighs 16.2 kilograms (kgm), what is the difference of the weights?

The word "difference" tells you to subtract to solve this word problem.

$$\begin{array}{r} 28.3 \\ -\ 16.2 \\ \hline 12.1 \end{array}$$

The answer is 12.1 kilograms.

Unit 2 Exercise

This exercise reviews the preceding unit. The exercise is divided into three parts. Part A reviews the foci of Unit 2. Part B offers opportunities to practice the skills and concepts of Unit 2. Part C contains problems that review your previous work in this text. All answers for Parts A and B are at the back of the book. Each problem of Part C is accompanied by a notation «CU» that refers to the Chapter (C) and Unit (U) in which that type of problem is studied.

Part A: Reviewing the foci of Unit 2.

1. Add 0.05 + 0.028 affixing a zero to eliminate ragged decimals.

2. Add 0.007 + 0.1 + 0.0006 affixing zeros where necessary.

3. Add 0.8 and 0.9 and show the carrying required.

4. Add 1.3 and 9.7 and show the carrying.

5. Add 1.05 and 2.37 and show the carrying.

6. Subtract 0.08 from 0.12 and show all your
    work.

7. Subtract 0.8 from 4 by first writing 4 with a
    decimal point.

8. In the subtraction $0.008 - 0.001 = 0.007$,
    the minuend is._____,
    the subtrahend is _____,
    and the difference is _____.

9. In the subtraction, "take 0.02 from 0.08"
    the minuend is _____,
    the subtrahend is _____,
    and the difference is _____.

10. The phrase "how much more" implies a
    subtraction situation. Jose has $1,268.09 in
    his savings account and Maria has $932.78
    in hers. How much more is in Jose's
    account than in Maria's?

11. The phrase "how much was left" indicates
    a subtraction situation. $15,600 was with-
    drawn from a bank account of $92,500.
    How much was left?

12. The word "less" implies subtraction.
    4 less than 9 means 9   4.
    Does the sum of 8 and 2 mean 8 − 2?

Part B: Drill and Practice

Add.

1. 13.081, 2.5, 4.62, and 0.4

2. $0.027 + 0.99 + 0.038$

3. 3.098, 0.409, and 1.047

4. 0.12, 0.378, 0.9, 0.8073

5. $\quad 0.3$
    $+\, 0.05$

6. $\quad 0.08$
    $+\, 0.09$

7. $\quad 0.27$
    $+\, 0.8$

8. $\quad 1.047$
    $+\, 9.128$

9. $\quad 0.037$
    $\quad 0.46$
    $+\, 0.1282$

10. $\quad 1.5$
    $\quad 7.05$
    $\quad 2.49$
    $+\, 0.21$

11. 0.006
     0.013
   + 0.49

12. 0.0025
     0.1037
   + 0.056

13. 2.3, 1.09, 4.35, and 0.08

14. 0.026, 0.055, and 1.9203

15. 28.026 and 37.082

16. 1,350.5 and 2,259.21

Subtract.

17. 1.5 from 2.5

18. 0.2 from 5

19. 0.12 from 0.5

20. 0.025 from 0.8

21. 0.039 − 0.018

22. 5 − 0.08

23. 14 − 0.3

24. 0.0427 − 0.0116

25. 15.715 − 0.8

26. 3.65 − 1.025

27. 0.00287 − 0.00092

28. 0.2904 − 0.0915

29.   0.48
      − 0.27

30.   0.038
      − 0.02

31.   10.4
    − 4.079

32.   0.6
      − 0.16

33. 1.049 from 9.1052

34. 0.105 from 0.82

35. 10.4 − 5.7

36. 0.0493 − 0.0486

37. 0.1027 − 0.0923

38. 182.04 − 29.82

39. An architect's fees were 3.975% and 4.1092%. Find the sum of the fees.

40. The ingredients in a medicine tablet weighed 2.03 mg, 14 mg, 8.785 mg and 0.573 mg. Find the sum of the weights of the ingredients.

41. The amounts of three solutions to be mixed are 4.5 liters, 9.236 liters, and 2 liters. What is the sum of the three amounts?

42. The unit prices of 3 sizes of screws were found to be $.0397, $.125, and $.38275. Find the sum of the prices.

43.   1.27
      – 0.16

44.   0.0056
     – 0.0019

45.   0.5
    – 0.13

46.   128.57
    – 75.2

47. 4.75 – 2.92

48. 0.406 – 0.01932

49. 0.0125 – 0.0118

50. Take 0.7 from 12.2

51. Find the sum of 5.6 milligrams (mg) and 23.92 mg.

52. What is the sum of 0.45 liters and 1.9 liters?

53. 1.4 kilograms and 0.5 feet cannot be added because the numbers have different labels. Can the sum of 483.25 grams and 37.099 grams (gm) be found?

54. Can 57.3 pounds be added to 0.49 feet?

55. Nurse Reynolds needed to give dosages of 6 minims, 28 minims, and 14 minims. Find the sum of the dosages.

56. Cotes Sporting Goods store received two shipments costing $1,468.05 and $82,025.36. Find the sum of the costs of the shipments.

57. Find the sum of forces on an object if they are 42.5 kilograms (kgm), 1.07 kgm, 0.78 kgm, and 38 kgm.

58. A pharmacy received the following amounts of alcohol: 4.05 liters of alcohol, 0.561 liters, and 0.26 liters. Find the sum of the amounts of alcohol.

59. Bill ran a mile race with the following quarter mile split times: 58.3 sec, 61.1 sec., 58 sec., 59.28 sec. Find the sum of the split times.

60. Find the sum of the commission fees for a job when the four fees were 1.936%, 2.05%, 2.19%, and 2.335%.

61. Sheerif has $92.56 and Abdullah has 48 kopeks. How much more money does Sheerif have than Abdullah?

62. A supply of 38.5 liters of solvent was decreased by 10.6 liters. How much was left?

63. A first tablet weighed 3.9 milligrams (mg.) less than another tablet that weighed 12.1 milligrams. Find the weight of the first tablet.

64. The interest rate on Iris's savings account is 1.42% per quarter. Percy's account pays 1.53% per quarter. How much more is Percy's interest rate than Iris'?

65. If Vito's salary is raised to $6.85 per hour and Luigi makes $5.87 per hour, how much more does Vito make each hour than Luigi?

66. Sammy makes $1.25 less per hour than Johnny who makes $8.19 per hour. How much does Sammy make per hour?

67. A virus traveled 0.0068 millimeters (mm) through a human cell which had a thickness of 0.0092 millimeters. How much farther does the virus need to go to travel completely through the cell?

68. A specimen of rock feldspar weighs 106.2 milligrams and a specimen of lignite weighs 28.3 milligrams. Find the difference in their weights.

69. A supply of 42 cubic centimeters (cc) of morphine was decreased by 9.7 cubic centimeters. How much was left?

70. Find the difference between 4.0798 and 3.526.

71. If a subtraction problem has a difference of 12.05 and a subtrahend of 10.39, what is the minuend?

Part C: Review. Answers for the problems of Part C are not given. However, the notation «C,U» refers to the Chapter (C) and Unit (U) in which problems of the same type were presented.

1. What is the ten thousands digit of 6,258,403?   «1,1»

2. 612,517 + 4,663 + 25,417   «1,2»

3. 5,430 − 2,918   «1,3»

4. 4,159 x 78   «2,2»

5. 28) 41748   «2,4»

6. Simplify $\frac{8}{14}$ «3,1»

7. Write $\frac{65}{8}$ as a mixed number.   «3,2»

8. Find the numerator of $\frac{5}{6} = \frac{?}{42}$ «3,2»

9. $6\frac{3}{8} + 4\frac{7}{10}$ «3,3»

10. $5\frac{2}{3} - 3\frac{5}{6}$ «3,4»

11. $4\frac{1}{5} \times \frac{4}{7}$ «3,5»

12. $6\frac{1}{3} \div 1\frac{1}{6}$ «3,6»

13. Write $\frac{2300681}{10000}$ as a decimal numeral.   «4,1»

14. Write 0.0007 as a fraction.   «4,1»

15. Karen weighed $112\frac{1}{2}$ pounds before she played tennis and only $111\frac{3}{4}$ pounds after her exercise. How much weight did she lose?   «3,7»

16. Each month, Gayle spends $\frac{1}{3}$ of her income on food. In a month when she earned $660 how much did she spend on food?   «3,7»

Unit 3:   MULTIPLYING DECIMALS

Multiplying Whole Numbers

The multiplication of decimal fractions is almost identical to the
multiplication of whole numbers. The only new task that is required
is the placement of the decimal point in the answer (product). If
you are uncertain of your ability to correctly multiply whole numbers,
it would be wise to review the work of Chapter 2.

The problem at the right	3 6	**factor (multiplicand)**
offers a quick review of	x 2 8	**factor (multiplier)**
the words and processes	2 8 8	= 36 x 8 **partial product**
used in the multiplication	7 2	= 36 x 2 **partial product**
of whole numbers.	1 0 0 8	= 36 x 28 **product**

36 x 28 = 1,008

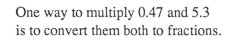

*Focus on
the number of
decimal places
in a product*

One way to multiply 0.47 and 5.3
is to convert them both to fractions. $0.47 = \frac{47}{100}$ and $5.3 = \frac{53}{10}$

The multiplication of 0.47 and 5.3, $\frac{47}{100} \times \frac{53}{10} = \frac{47 \times 53}{100 \times 10} = \frac{47 \times 53}{1,000}$
as fractions, is shown at the right.

Notice that the numerator is the multiplication of the whole numbers, 47 and 53.
Notice also that the denominator, 1,000, has the same number of zeros as the
total zeros in 100 and 10.

(2 zeros in 100) + (1 zero in 10) = 3 zeros in 1,000

Since, 47 x 53 = 2,491 $\frac{47 \times 53}{1,000} = \frac{2,491}{1,000}$

Decimal Places in the Product of Decimal Fractions

When multiplying decimal fractions, the location of the decimal point in the product is determined by the total number of digits to the right of the decimal point in the two factors of the problem.

To determine the number of digits right of the decimal point in the product of 0.12 x 13.005:

1. Count the digits to the right of the decimal point in each factor.

    0.12 has two digits right of the decimal point.
    13.005 has three digits right of the decimal point.

2. Find the sum of 2 and 3 to find the number of digits right of the decimal point in the product of 0.12 and 13.005.

    2 + 3 = 5 digits right of the decimal point.

Focus on counting the decimal places in a product

In multiplying 8.915 and 0.0067, the total number of decimal places is needed.

8.915 has 3 decimal places
0.0067 has 4 decimal places

The product of 8.915 and 0.0067 will be found by multiplying the whole numbers 8915 and 67 and marking off 7 decimal places in the answer.

(3 decimal places) + (4 decimal places) = 7 decimal places

Placing the Decimal Point in the Product of Decimal Fractions

The number of digits to the right of the decimal point in the product of decimal fractions is:

The sum of the number of digits to the right of the decimal points in the two factors which is the total number of decimal places in the two factors.

To multiply 0.24 by 0.02:

1. Multiply the factors as though multiplying whole numbers. Ignore the decimal points.

$$
\begin{array}{r}
0.2\,4 \\
\times\ 0.0\,2 \\
\hline
4\,8 \\
0\,0 \\
0\,0 \\
\hline
0\,0\,4\,8
\end{array}
$$

2. Since there is a total of four digits right of the decimal points in the two factors, start at the right side of 0048 and count four digits to the left. After the fourth digit, place the decimal point in the product.

0.0 0 4 8   0.24 x 0.02 = 0.0048

Focus on multiplying decimals

To multiply 1.29 by 0.01:

1. Multiply and ignore the decimal points.

$$
\begin{array}{r}
1.2\,9 \\
\times\ 0.0\,1 \\
\hline
1\,2\,9 \\
0\,0\,0 \\
0\,0\,0 \\
\hline
0\,0\,1\,2\,9
\end{array}
$$

2. The multiplication is completed by marking off 4 decimal places.   .0 1 2 9

Multiplying Decimal Fractions With Two or More Digits

Multiplying decimal fractions with two or more non-zero digits
in the multiplier is done as it was in the preceding unit.

The decimal point is placed according to the total number of digits to the
right of the decimal point in the two factors of the problem.

The example at the right shows the
multiplication of 8.5 by 0.68.

$$\begin{array}{r} 8.5 \\ \times\ 0.6\,8 \\ \hline 6\,8\,0 \\ 5\,1\,0 \\ \underline{0\,0} \\ 5.7\,8\,0 \end{array}$$

Notice that there are three decimal
places in the product because the total
number of digits right of the decimal
points in the multiplier and the
multiplicand is three.

$$8.5 \ \times \ 0.68 \ = \ 5.780 \text{ or } 5.78$$

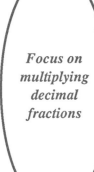

*Focus on
multiplying
decimal
fractions*

To multiply 2.2 by 0.34:

1. Multiply 22 by 34.

    $$22 \ \times \ 34 \ = \ 748$$

2. Count the total number of digits
    to the right of the decimal points
    in the factors 2.2 and 0.34

    The total is 3 digits

3. The multiplication is completed by
    placing the decimal point correctly
    in 748.

    0.7 4 8

Multiplying by Powers of Ten

"Powers of Ten" are numbers such as 10, 100, 1000, 10000, etc.

There is a simple shortcut for multiplying any decimal fraction by a power of ten. The shortcut is:

>   Move the decimal point in the decimal fraction to the
>   right the same number of places as there are zeros
>   in the power of ten.

For example, the multiplication of 5.78 by 1,000 is accomplished by

$$5.78 \ \times \ 1000 \ = \ 5.780 \ = \ 5,780$$

The decimal was moved 3 places right because there are 3 zeros in 1,000.

Focus on multiplying by powers of ten

To multiply 100,000 by 0.281 first count the number of zeros in 100,000 .

100,000 has 5 zeros

Then move the decimal point to the right that many places in 0.281.

$100,000 \ \times \ 0.281 \ = \ 28,100$

The Clue Word "Each" for Multiplication

In the problem, "What is the cost of 120 books if **each** book costs $4.95?" the word "**each**" gives the clue to multiply to find the answer.

$$120 \ \times \ \$4.95 \ = \ \$594.00 \text{ total cost}$$

Focus on the word "each" to indicate multiplication

Word Problem: If 250 test tubes hold 102.5 milliliters (ml) each, what is the total number of milliliters that the test tubes can hold?

The word **"each"** indicates a multiplication situation. The total number of milliliters that the test tubes can hold is the product of 250 and 102.5. Multiplication is required to solve this problem.

$$250 \ \times \ 102.5 \ = \ 25{,}625 \text{ milliliters}$$

"Of" as a Clue Word for Multiplication

To find "A fraction **of** a number" the fraction is always multiplied by the number.

Decimal numerals are fractions and any sentence of the form

"A decimal numeral of a number"

indicates that the decimal numeral should be multiplied by the number.

Focus on the word "of" to indicate multiplication of decimals

Word Problem: Find 0.8 of 53.

To find "0.8 of 53" the word **"of"** indicates that 0.8 should be multiplied by 53.

$$0.8 \textbf{ of } 53 \ = \ 0.8 \ \times \ 53 \ = \ 42.4$$

Unit 3 Exercise

This exercise reviews the preceding unit. The exercise is divided into three parts. Part A reviews the foci of Unit 3. Part B offers opportunities to practice the skills and concepts of Unit 3. Part C contains problems that review your previous work in this text. All answers for Parts A and B are at the back of the book. Each problem of Part C is accompanied by a notation «CU» that refers to the Chapter (C) and Unit (U) in which that type of problem is studied.

Part A: Reviewing the foci of Unit 3.

In problems 1-7 find the number of decimal places in the product.

1.   0.3 x 1.343

2.   0.025 x 394

3.   10.1 x 0.28387

4.   0.8 x 0.2

5.   0.03 x 0.9

6.   4 x 0.5

7.   0.28 x 0.03

8.   Multiply the problem shown at the right. Count the total number of decimal places in 0.23 and 0.009; then place the decimal point correctly in the product.

$$\begin{array}{r} 0.23 \\ \times\ 0.009 \\ \hline \end{array}$$

9.   Find 0.3 of 4.8 by multiplying the numbers.

10.   0.05 of a vat of beer is sediment. The vat holds 5,000 liters. The "of" phrase "0.05 of a vat of beer" is equivalent to the phrase 0.05 of 5,000 liters because the vat holds _____ liters.

11.   1.2 of the cost of a refrigerator is the sale price. The cost is $300. From the information in the two previous sentences the "of" phrase
            "1.2 of $300"
can be made. In the first sentence the words "the cost" were replaced by _____.

12.   0.6 of the people seeing a doctor have no physical problems. A doctor sees 60 people. The phrase "0.6 of the people" can be replaced by the "of" phrase: 0.6 of _____.

13.   0.875 of the volume of milk is water. A tank truck holds 20,000 liters of milk. The phrase "0.875 of the volume of milk" can be replaced by the "of" phrase _____.

14. Ariane has 300 ml of a medicine. 0.3 of the medicine is a solvent. Using the information in the two sentences above, the phrase "0.3 of the medicine" can be replaced by the "of" phrase _____.

15. 0.095 of the price of a car goes toward the administrative expense of the company. The price of a car is $5,900. $5,900 can be substituted for "price" in the phrase
0.095 of the price
to obtain the "of" phrase _____.

Part B: Drill and Practice

Multiply.

1.   1.98 x 0.6

2.   0.222 x 0.07

3.   0.4 x 0.3

4.   2.03 x 0.009

5.   0.045 x 1.6

6.   1.02
       x 0.3

7.   0.078
       x 0.001

8.   1.625
       x 0.03

9.   102.037
       x 0.008

10. 12 x 0.9

11. 5 x 0.01

12. 10.25 x 0.8

13. 7.1 x 7

14. 3.09 x 0.009

15. Find the product of 780 and 0.005.

16. 0.128 x 25

17. 1.032 x 3.5

18. 12.0897 x 68

19. 3.8 x 12.7

20. 0.483 x 1.11

21. 203 x 2.51

22. 0.462 x 248

23. 92.8 x 1.79

24. 921.7 x 856.9

25. 0.0971 x 100

26. 46.31 x 1000

27. 1.293 x 10000

28. 0.039 x 10

29. 100 x 2.0398

30. 100000 x 0.3

31. 300 x 2.5

32. 4.6 x 7.9

33. 0.97 x 0.08

34. 2,000 x 0.12

35. 1.03 x 7.9

36. 4,028 x 0.906

37. 0.056 x 0.97

38. 0.28 x 1.829

39. 0.28 x 10

40. 0.278 x 100

41. 3.2784 x 1000

42. 0.00829 x 10,000

43. If 4,000 razor blades weigh 0.129 grams each, what is the total weight of the blades?

44. If each theater ticket costs $3.50, how much will 2,000 tickets cost?

45. If the fee for each dollar value of a travelers check is $0.0075, what is the fee for $300 worth of travelers checks?

46. A bank charges $0.11 for each dollar loaned for one year. What will it charge for loaning $12,000 for one year?

47. Find 0.079 of 11,000.

48. Find 0.0037 of 4.6.

49. In the problem "John lost .34 of his $300. How much money did he lose?" the word "of" indicates that 0.34 and 300 should be

_____

50. A general estimates that .35 of his 2000 rifles are defective. How many rifles are defective?

51. An engineer has .42 of his 40 day project completed. How many days are completed?

52. 0.116 of the amount of a mortgage is the interest on the mortgage for one year. The mortgage is for $30,000. How much is the interest?

53. A microwave oven can cook a roast in 0.6 of the time needed by an electric oven. An electric oven cooks a roast in 3 hours. How much time does the microwave oven need to cook the roast?

54. 0.21 of the atmosphere is oxygen. A closet contains 3 cubic meters of air. How much of the air is oxygen?

55. The selling price of a book is $20.50. The printing cost is 0.12 of the selling price. Find the printing cost of the book.

56. A diver's tank will hold 203.8 liters of air. 0.79 of the volume of the air is nitrogen. How much of the air in the tank is nitrogen?

57. 0.8 of a bottle of medicine is codeine. The bottle contains 200 drams of medicine. How much codeine is in the bottle?

58. A loan payment is 1.3 of the principal part of the payment. The principal part of the payment is $80. Find the size of the loan payment.

59. If 25 radios cost $9.50 each, what is the cost of all the radios?

60. Each cubic meter of gas weighs 1.56 kgm. A tank holds 40 cubic meters of the gas. Find the weight of the gas in the tank. _____

61. A department store charges $0.015 a month for each dollar it loans. Janie borrows $300 to buy a set of silver. How much is the interest on the loan for one month?

62. Sam buys 24 peaches. They cost $0.12 each. How much will he pay for the peaches?

63. A processed meat product weighs 2.5 kgm. Each kgm of the product is allowed to have 0.0009 kgm of insect parts. How many kgm of insect parts could be in the product?

64. 0.3 of 8

65. 1.6 of 500

66. 0.2 of a tablet is caffeine. The tablet weighs 5 milligrams. How much caffeine is in the tablet?

67. The price of brand X deodorant is 1.13 of the price of Snickle deodorant. Snickle costs $2 per can. What is the price of brand X?

68. For a genetics experiment 200 brown rats were used. 0.87 of the rats had brown eyes. How many rats had brown eyes?

Part C: Review. Answers for the problems of Part C are not given. However, the notation «C,U» refers to the Chapter (C) and Unit (U) in which problems of the same type were presented.

1. What is the thousands digit of 4,825,369?   «1,1»

2. 52,547 + 4,283 + 28,217   «1,2»

3. 31,430 − 15,918   «1,3»

4. 6,942 x 85   «2,2»

5. $26\overline{)47248}$ «2,4»

6. Find the lcm for 14 and 20.   «3,2»

7. Select the smaller fraction from $\frac{3}{7}, \frac{5}{12}$ «3,2»

8. Change $2\frac{5}{8}$ to an improper fraction.   «3,2»

9. $6\frac{2}{3} + 3\frac{7}{10}$ «3,3»

10. $7\frac{1}{4} - 3\frac{5}{6}$ «3,4»

11. $5\frac{1}{4} \times \frac{4}{7}$ «3,5»

12. $7\frac{1}{4} \div 1\frac{1}{2}$ «3,6»

13. Write $\frac{56}{1000}$ as a decimal numeral.   «4,1»

14. Write 0.0203 as a fraction.   «4,1»

15. Add 4.715, 23.91, and 0.4189   «4,2»

16. Subtract 5.68 from 8.3   «4,2»

17. Allen bought two steaks weighing 2.37 pounds and 3.6 pounds. What was the total weight of the steaks?   «4,2»

18. If a carpenter has finished $\frac{3}{5}$ of a 30 hour job, how many hours work has he finished?   «3,7»

Unit 4:   DIVIDING DECIMALS

Dividing a Decimal Fraction by a Whole Number

To divide 1.32 (a decimal fraction) by 6 (a whole number):

1. Place the decimal point in the quotient space directly above the decimal point in 1.32.

$$6\overline{)\,1.3\,2}$$

2. Complete the division as if dividing 132 by 6.

$$
\begin{array}{r}
0.2\ 2 \\
6\overline{)\,1.3\,2} \\
-\underline{1\,2} \\
1\,2 \\
-\underline{1\,2} \\
0
\end{array}
$$

Focus on placing the decimal point in a division problem

When dividing by a whole number, the decimal in the quotient (answer) goes directly above the decimal in the dividend.

In the problem at the right, the decimal point in the quotient would go where the letter "c" now is.

$$57\overline{)\,6\,5.4\,7\,9}\ \ \ (c)$$

Always place the decimal point in the quotient as a first step to long division. Then divide as if the numbers were whole numbers.

The first step in dividing 65.479 by 57 is the correct placement of the decimal point in the quotient. That step is shown at the right.

$$57\overline{)\,6\,5.4\,7\,9}$$

Affixing Zeros to the Dividend

The decimal numeral 4.35 is equal to 4.350 or 4.3500.
Zeros may be affixed to decimal fractions like 4.35 without
changing their value.

In dividing a decimal fraction, zeros may be affixed to the
dividend to extend the number of decimal places in the quotient.

To divide 0.01 by 4, zeros can be
affixed.

$$0.01 \ = \ 0.0100$$

By affixing 2 zeros to 0.01 the
division can be continued until
a zero remainder is found.

```
         0.0 0 2 5
  4 ) 0.0 1 0 0
       - 8
         2 0
       - 2 0
           0
```

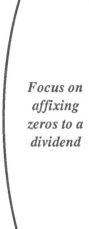

*Focus on
affixing
zeros to a
dividend*

In dividing 6.13 ÷ 5, the first three
digits of the quotient are shown at the
right.

Notice that a remainder of 3 exists.

```
        1.2 2
  5 ) 6.1 3
     - 5
       1 1
     - 1 0
         1 3
       - 1 0
           3
```

If one zero is affixed to 6.13 the
division can be continued for
four digits and ends with a zero
remainder.

```
        1.2 2 6
  5 ) 6.1 3 0
     - 5
       1 1
     - 1 0
         1 3
       - 1 0
           3 0
         - 3 0
             0
```

Changing a Fraction to a Decimal Fraction

The fraction $\frac{3}{16}$ can be changed to a decimal fraction using long division. The steps in the process are listed below.

1. Write $\frac{3}{16}$ as a long division problem.

    $16\overline{)3}$

2. Affix a decimal point and zeros to the right of the dividend, 3.

    $16\overline{)3.0\,0\,0\,0}$

3. Complete the long division

    $$\begin{array}{r} 0.1\,8\,7\,5 \\ 16\overline{)3.0\,0\,0\,0} \end{array}$$

$\frac{3}{16} = 0.1\,8\,7\,5$

Focus on changing a fraction to a decimal fraction

To change the fraction $\frac{5}{6}$ to a decimal fraction rounded off at two decimal places, begin by recognizing that $\frac{5}{6}$ means $5 \div 6$.

The division of 5 by 6 is carried to three decimal places so that it can be rounded off at two decimal places.

$$\begin{array}{r} 0.8\,3\,3 \\ 6\overline{)5.0\,0\,0} \\ -4\,8 \\ \hline 2\,0 \\ -1\,8 \\ \hline 2\,0 \\ -1\,8 \\ \hline 2 \end{array}$$

0.833 is rounded off to 0.83

The fraction $\frac{5}{6}$ is approximately equal to the decimal fraction 0.83

Equal Fractions

The fractions

$$\frac{53.76}{0.048} \quad \text{and} \quad \frac{53760}{48}$$

are equal. The decimal point in the denominator of $\frac{53.76}{0.048}$ was
moved three places to the right and, therefore, the decimal point
in the numerator was also moved three places to the right. This is
equivalent to multiplying both numerator and denominator of $\frac{53.76}{0.048}$
by 1,000.

$$\frac{53.76 \times 1000}{0.048 \times 1000} = \frac{53760}{48}$$

Focus on equal fractions

To write $\frac{1.576}{0.4}$ as a fraction with a whole number
denominator, the decimal point in 0.4 must be moved one
decimal place. This can be accomplished by multiplying both
numerator and denominator by 10.

$$\frac{1.576}{0.4} = \frac{1.576 \times 10}{0.4 \times 10} = \frac{15.76}{4}$$

Dividing with a Decimal Fraction Divisor

To place the decimal point in the quotient of
$53.76 \div 0.048$ the following steps are used.

$$.048 \overline{)\ 53.76}$$

1. Move the decimal point in the divisor
    to the right until the divisor is a whole
    number.

$$.048. \overline{)\ 53.76}$$

2. Move the decimal point in the dividend
    to the right the same number of places
    it was moved in the divisor. This may
    require the affixing of one or more zeros.

$$48 \overline{)\ 53.760.}$$

3. Since the divisor is now a whole
    number the decimal point is placed
    in the quotient directly above the new
    decimal point in the dividend.

$$48 \overline{)\ 53760.}$$

Focus on moving the decimal points to prepare a problem for division

In the problem at the right, change the divisor to a whole number by moving its decimal point two places to the right. Also, move the decimal point in the dividend two places to the right and place the decimal point in the quotient directly above the decimal point in the dividend.

$$.06 \overline{)\,5.316}$$

$$6 \overline{)\,531.6}$$

Division Problems with the Same Quotient

The fractions

$$\frac{2.\,172}{0.\,7} \quad \text{and} \quad \frac{21.\,72}{7}$$

are equal which means the division problems

$$0.7 \overline{)\,2.172} \quad \text{and} \quad 7 \overline{)\,21.72}$$

have the same answer.

To prepare 2.172 ÷ 0.7 for long division, move the decimal points in both the divisor and dividend the same number of places so that the divisor will be a whole number. Place the decimal point in the quotient.

$$.7 \overline{)\,2.172}$$

$$7 \overline{)\,21.72}$$

The problem is now 21.72 ÷ 7

Focus on placing the decimal point in the quotient

To prepare the problem at the right for long division move the decimal points in both the divisor and dividend. Place the decimal point in the quotient space.

$$.0025 \overline{)\,0.36}$$

$$25 \overline{)\,3600.}$$

The division can now be completed as 3600 ÷ 25.

Rounding Off Decimal Division Quotients

Previous examples in this unit had zero remainders when the division was carried to enough decimal places. For some problems the long division of decimals will never arrive at a zero remainder and rounding off is required.

Division answers (quotients) are often approximated by rounding off the answer at a particular number of decimal places. The example below illustrates the procedure for rounding off the answer of $1 \div 3$ at two decimal places:

1. Since the quotient is to be rounded off at two places, the division of 1 by 3 is carried to three decimal places.

2. The decimal 0.333 has the digit "3" in its third decimal place and is rounded off to 0.33

$$
\begin{array}{r}
.333 \\
3 \overline{)1.000} \\
-9 \\
\hline
10 \\
-9 \\
\hline
10 \\
-9 \\
\hline
1
\end{array}
$$

If the digit in the third place is less than 5 then it is dropped and the second digit remains unchanged.

   $0.333 \doteq 0.33$

The symbol "$\doteq$" means "approximately equal to."

Focus on rounding off

To round off the quotient of $4.28 \div 0.75$ to two decimal places first complete the division to three decimal places.

$$
\begin{array}{r}
5.706 \doteq 5.71 \\
75 \overline{)428.000}
\end{array}
$$

Solving Word Problems Using Division of Decimals

Word problems using the word "each" frequently are solved by division.

When numbers are given for the total and for each item that makes up the total, the word problem is solved by dividing the "total" number by the "each" number.

("total" number) ÷ ("each" number) = answer

Word Problem: Beer is sold by the case or by the can, but the cost is the same. Each can costs $.60 and a case costs $14.40. Find the number of cans in a case.

$$.60 \overline{)1\,4.4\,0}$$

Focus on the words "each" and "total"

The word "each" is a clue that division will be used in solving this problem. Since the total cost for a case is $14.40, and each can costs $.60, the division is

$$
\begin{array}{r}
24. \\
60 \overline{)1440.} \\
-120 \\
\hline
240 \\
-240 \\
\hline
0
\end{array}
$$

$14.40 ÷ $.60 = 24   There are 24 cans in a case.

Averages Are Found by Using Division

To find the average of a group of numbers, first add the numbers and then divide the sum by the number of addends.

(total or sum) ÷ (number of addends) = average

If the division does not end with a zero remainder then round off the quotient at two decimal places.

Focus on finding the average

Word Problem: Tom has taken four math tests and received scores of 92, 85, 67, and 82. What is his average?

To find his average add the numbers and divide by 4 as shown at the right.

```
  92
  85
  67
+ 82
 326
```

```
        8 1.5
  4 ) 3 2 6.0
     -3 2
        0 6
       - 4
        2 0
      - 2 0
          0
```

His average is 81.5 in math.

Unit 4 Exercise

This exercise reviews the preceding unit. The exercise is divided into three parts. Part A reviews the foci of Unit 4. Part B offers opportunities to practice the skills and concepts of Unit 4. Part C contains problems that review your previous work in this text. All answers for Parts A and B are at the back of the book. Each problem of Part C is accompanied by a notation «CU» that refers to the Chapter (C) and Unit (U) in which that type of problem is studied.

Part A: Reviewing the foci of Unit 4.

1. Divide $37.2 \div 2$ by first placing the decimal point in the quotient and then dividing as if the numbers were whole numbers.

2. Divide $3.12 \div 3$ by first placing the decimal point in the quotient and then dividing as if the numbers were whole numbers.

3. Divide 1.5 by 6. Affix zeros to 1.5 until a zero remainder is obtained.

4. Divide 1.8 by 16. Affix zeros to 1.8 until a zero remainder is obtained.

5. Change $\frac{5}{8}$ to a decimal fraction by dividing 5 by 8 until a zero remainder is obtained.

6. Change $\frac{7}{20}$ to a decimal fraction by dividing 7 by 20 until a zero remainder is obtained.

7. Divide 16.054 by 4.6 until a zero remainder is obtained.

8. Divide 10.816 by 0.52 until a zero remainder is obtained.

9. Divide 400 by 0.02 by first changing the divisor to a whole number.

10. Round off the quotient of $5 \div 7$ to two decimal places by first dividing 5 by 7 and carrying the division to three decimal places.

11. Divide 4.1 by 7 to three decimal places. Then round off the answer to two places.

12. Divide 6.2 by 6 to three decimal places. Then round off the answer to two places.

13. Divide 8.5 by 4.6 to three decimal places. Then round off the answer to two places.

14. Divide 0.56 by 0.015 to three decimal places. Then round off the answer to two places.

15. Change $\frac{3}{7}$ to a decimal. Round off the answer to two decimal places.

16. Change $\frac{5}{9}$ to a decimal. Round off the answer to two decimal places.

Part B: Drill and Practice

Divide until a zero remainder is obtained.

1. $47.24 \div 4$

2. $30.1 \div 7$

3. $121.5 \div 9$

4. $0.008 \div 4$

5. $1.82 \div 13$

6. $47.5 \div 19$

7. $1.638 \div 39$

8. $93.6 \div 78$

9. $2.065 \div 59$

10. 0.14 by 4

11. 0.02 by 125

12. $\frac{13}{4}$

13. $\frac{5}{16}$

14. $\frac{17}{4}$

15. $0.00765 \div 15$

16. $0.075 \div 3$

17. $0.8 \div 4$

18. $0.0625 \div 25$

19. $80.08 \div 16$

20. $4.07 \div 50$

21. $28.38 \div 48$

22. $24.84 \div 36$

23. $\frac{3}{8}$

24. $\frac{42}{16}$

25. 320 by 0.16

26. 4.256 by 0.2

27. 5.25 by 0.005

28. 1.5762 by 0.37

29. 17.34 by 0.085

30. $0.6 \div 0.5$

31. $8.2 \div 0.016$

32. $714 \div 1.25$

33. $4.1 \div 0.0082$

34. $0.04536 \div 7.2$

35. $2.8458 \div 0.93$

36. Round off $9.132 \div 0.081$ at one decimal place by first carrying the division to two decimal places.

37. Round off $8.57 \div 3$ at two decimal places.

38. Round off $47 \div 5.3$ at three decimal places.

39. Round off $4.126 \div 0.8$ at one decimal place.

40. Approximate $\frac{2}{9}$ with a three-place decimal by dividing 2 by 9.

41. Approximate $\frac{17}{14}$ with a two-place decimal.

Round off each of the following to two decimal places.

42. $1.3 \div 46$

43. $4 \div 0.6$

44. $0.0035 \div 0.14$

45. $8.945 \div 7.2$

46. $0.17 \div 5.6$

47. $0.0049 \div 0.13$

48. $4.57 \div 0.27$

49. $84.34 \div 0.09$

Approximate each fraction with a two place decimal.

50. $\frac{4}{17}$

51. $\frac{4}{7}$

52. $\frac{15}{23}$

53. $\frac{5}{32}$

54. Homer has a loan at the bank which costs him $.13 interest each day. When his total interest is $4.55 how many days have passed?

55. A pile of coins weighs 243.6 grams (gm). If each coin weighs 1.2 gm, how many coins are in the pile?

56. A carnival sold tickets for $.75 each. At the end of the performance the box office had a total of $51.00. How many tickets had been sold?

57. A manufacturing plant uses 4.2 cm of wire on each item. If 3549 cm of wire are used in a day, how many items were manufactured?

58. A new carpet is purchased for $9 a square yard. The total bill was $725.40. How many square yards were purchased?

59. The seven linemen on the football team weigh 182 lbs, 202 lbs, 175 lbs, 210 lbs, 194 lbs, 183 lbs, and 198 lbs. Find the average weight by adding the weights and dividing by 7.

60. Jake earned three pay checks over his Christmas vacation of $75.63, $81.48, and $95.13. What was his average pay check?

61. Scott took a 5-day trip and registered his miles traveled each day. His miles were 450, 292, 527, 139, and 387. What was his average miles traveled each day?

62. Six runners ran in a 3,000 meter race. Their times, in minutes, were 10.3, 12.2, 19.6, 15.4, 14.7, and 11.8. What was the average time of the runners?

63. The pollution in a stream was measured on five consecutive days. The measures, in grams per liter, were 14, 8, 25, 45, and 38. Find the average pollution each day.

64. Sara mailed 45 packages at a total cost of $37.52. What was the average mailing cost per package?

65. Seven passengers boarded a plane. Their total weight was 1085 pounds. What was the average weight per passenger?

66. Twelve packages weighing a total of 369 lbs were loaded on a plane. What was the average weight per package?

67. Sally earned $78, $93, $150, $225, and $392 on successive days as a waitress. What was her average earnings per day?

68. A small theater sold tickets for $.63 each, tax included, and collected a total of $569.52. Find the number of tickets sold.

69. The heights of the members of the basket-
ball team were, in meters, 1.84, 2.10, 1.92,
1.89, and 1.95. Find the average height of
a player.

71. Smith bought 6.5 square yards of carpet for
$54.60. What was the price for each square
yard of carpet?

70. The druggist had sales of $244.15 last
Monday from 95 separate purchases. What
was the average purchase price?

Part C: Review. Answers for the problems of Part C are not given. However, the notation «C,U» refers to the Chapter (C) and Unit (U) in which problems of the same type were presented.

1. 68,947 + 3,583 + 28,493   «1,2»

8. Write $\frac{569}{10000}$ as a decimal numeral.   «4,1»

2. 5386 x 6   «2,1»

9. Add 47.15, 2.391, and 0.4189   «4,2»

3. Simplify $\frac{16}{24}$ «3,1»

10. Find the sum of 285 and 418. «1,4»

4. Write $\frac{52}{7}$ as a mixed number.   «3,2»

11. Find the difference of 58.72 and 34.6   «4,2»

5. Find the numerator of $\frac{5}{8} = \frac{?}{32}$ «3,2»

12. Find the product of 8,432 and 19. «2,2»

6. $8\frac{3}{5} + 3\frac{7}{10}$ «3,3»

13. Find 0.08 of 4,700. «4,3»

7. $6\frac{1}{4} \times \frac{4}{5}$ «3,5»

14. How many $\frac{1}{2}$ years are there in 5 years?.
«3,7»

Chapter 4 Test

«4,U» shows the unit in which the problem is found in the chapter.

1. Write $\frac{5}{1000}$ as a decimal numeral.   «4,1»

2. Write 0.47 as a fraction.   «4,1»

3. True or false? $0.06 > 0.5$ «4,1»

4. Round off 1.273 to two decimal places.   «4,1»

5. Find the sum of 0.085 and 0.016   «4,2»

6. Find the sum of 0.025, 0.0962, 0.2, and 1.75
   «4,2»

7. Find the difference between 0.01 and 0.008
   «4,2»

8. Find the difference between 5 and 2.0025
   «4,2»

9. If Sandra has $2,049 and Tom has $3,150.75, how much more does Tom have than Sandra?
   «4,2»

10. The chemicals in a mixture weigh 0.25 kgm, 1.6 kgm, 0.097 kgm, and 4.97 kgm. Find the sum of the weights.   «4,2»

11. A small flask holds 4.51 liters less than a jar holding 8.49 liters. How much does the small flask hold?   «4,2»

12. Robert decreases his interest rate of 8.23% by 1.51%. What is his new interest rate after the decrease?   «4,2»

13. 9.643×0.78 «4,3»

14. Round off $38.46 \div 0.65$ to two decimal places.
    «4,4»

15. Write $\frac{5}{13}$ as a decimal rounded off to two decimal places.   «4,4»

16. Farmer Hank discovered that 0.43 of his grain had spoiled. Of his 60,000 bushels of wheat, how much had spoiled?   «4,3»

17. A pharmacist needs 25 does of phenobarbital. Each dose is to be 5.72 milligrams. How much of this medicine is needed to fill the order?
    «4,3»

18. The bank charges $0.115 per year for each dollar it loans. Burt needs a $30,000 loan. How much will the bank charge to borrow $30,000 for one year?   «4,3»

19. Greeting cards are on sale for $.18 each. How many can Pete buy with $16.92?   «4,4»

20. Max received four orders for $45.93, $87.50, $19.95, and $32.10. What was the average of these four orders?   «4,4»

5
Algebra of the Counting Numbers

Unit 1: COUNTING NUMBER EXPRESSIONS

Definition	Addition Expression	
x + y is an **addition expression**	**if and only if**	x and y are replaced by whole numbers.

Examples of addition expressions are:

$$5 + 7 \qquad 41 + 5 \qquad 813 + 512$$

Focus on the meaning of addition expressions

Whenever the addition of two whole numbers is indicated by symbols such as:

$$18 + 4,567 \qquad 0 + 0 \qquad 53 + 17$$

these are called **addition expressions**.

Definition	**Evaluation**	
The number e is the **evaluation** of an expression	**if and only if**	the simplification of the expression is e.

The **evaluation**:   of 5 + 8 is 13 and 8 + 5 is 13
of 6 + 19 is 25 and 19 + 6 is 25

Notice that the order in which the numbers are placed in an addition expression results in the same evaluation.

Evaluating Addition Expressions With Parentheses

To evaluate (4 + 3) + 5 it is necessary to understand the role of the parentheses ( ). Regardless of how many numbers are involved in an addition expression, only two counting numbers can be added at a time.

Parentheses or square brackets are used to indicate which two numbers are to be added in the first step.

$$(4 + 3) + 5 \qquad\qquad 4 + (3 + 5)$$
$$7 + 5 \qquad\qquad 4 + 8$$
$$12 \qquad\qquad 12$$

In the addition expressions above, the parentheses indicate the numbers to be added first. Then their sum is added to the other number to obtain the evaluations. Notice that the evaluations are the same.

Focus on the use of grouping symbols

In the expression (5 + 3) + [4 + 2], the parentheses indicate that the 3 is to be added to the 5, and the square brackets indicate that the 2 is to be added to the 4.

$$(5 + 3) + [4 + 2]$$

$$8 \quad + \quad [4 + 2]$$

Notice that each grouping symbol encloses only two numbers. This is because only two counting numbers can be added at one time.

$$8 \quad + \quad 6$$

$$14$$

Definition		Subtraction Expression
x − y is a subtraction expression	if and only if	x and y are replaced by numbers

Examples of subtraction expressions are:

$$15 - 7 \qquad 41 - 35 \qquad 813 - 512$$

Focus on the meaning of subtraction expressions

Whenever the subtraction of two whole numbers is indicated by symbols such as:

$$418 - 67 \qquad 0 - 0 \qquad 53 - 17$$

these are called **subtraction expressions**.

Evaluating Subtraction Expressions With Square Brackets

To evaluate [11 − 3] − 5 it is necessary to understand the role of the square brackets [ ]. Regardless of how many numbers are involved in a subtraction expression, only two counting numbers can be subtracted at a time.

Parentheses or square brackets are used to indicate which two numbers are to be subtracted in the first step.	[11 − 3] − 5
	8 − 5
	3

In the subtraction expression above, the square brackets indicate that 3 is to be subtracted from 11 first. Then 5 is subtracted from 8.

Focus on the order of subtraction with parentheses

The subtraction expression (10 − 3) − 2 involves the same counting numbers as 10 − (3 − 2). The difference in the two expressions is that the parentheses enclose different numbers indicating a different first step in the evaluations.

The evaluations of these two expressions are shown below.

(10 − 3) − 2	10 − (3 − 2)
7 − 2	10 − 1
5	9

Notice that the placement of parentheses in a subtraction expression results in different evaluations.

Definition   **Multiplication Expression**

x • y is a	if and	x and y are
multiplication	**only if**	replaced by
expression		numbers

Notice that this definition uses the symbol "•" for multiplication rather than "x." Parentheses may also be used to indicate multiplication. Examples of multiplication expressions are:

$$5 \times 7 \qquad 41 \times 5 \qquad 813 \times 512$$
$$5 \cdot 7 \qquad 41 \cdot 5 \qquad 813 \cdot 512$$
$$5(7) \qquad 41(5) \qquad 813(512)$$

Focus on the meaning of multiplication expressions

Whenever the multiplication of two whole numbers is indicated by symbols such as:

$$18 \cdot 4{,}567 \qquad 0 \cdot 0 \qquad 53 \cdot 17$$

these are called **multiplication expressions**.

Evaluating Multiplication Expressions With Parentheses

To evaluate $(4 \cdot 3) \cdot 5$ it is necessary to understand the role of the parentheses ( ). Regardless of how many counting numbers are involved in a multiplication expression, only two counting numbers can be multiplied at a time.

Parentheses or square brackets are used to indicate which two numbers are to be multiplied in the first step.

$$(4 \cdot 3) \cdot 5$$

$$12 \cdot 5$$

$$60$$

In the multiplication expression above, the parentheses indicate that 4 and 3 are to be multiplied first. Then their product is multiplied by 5.

Focus on square brackets when multiplying expressions

The multiplication expressions [6 • 2] • 4 and 6 • [2 • 4] involve the same three counting numbers, but the placement of the square brackets means that the expressions are evaluated in a different manner. The two evaluations are shown below.

$$[6 \bullet 2] \bullet 4 \qquad\qquad 6 \bullet [2 \bullet 4]$$
$$12 \bullet 4 \qquad\qquad\qquad 6 \bullet 8$$
$$48 \qquad\qquad\qquad\qquad 48$$

Notice that the movement of parentheses in a multiplication expression does not effect the evaluation.

Definition   **Division Expression**

$x \div y$ or $\frac{x}{y}$ is   **if and**   x and y are replaced
a **division**   **only if**   by numbers and
expression   y is not zero.

Notice that this definition uses the division symbol "÷" or a fraction bar. Examples of division expressions are:

$$5 \div 7 \qquad \frac{45}{5} \qquad \frac{813}{512}$$

Focus on the meaning of division expressions

Whenever the division of two whole numbers is indicated by symbols such as:

$$\frac{18}{567} \qquad 58 \div 58 \qquad \frac{53}{17}$$

these are called **division expressions**.

Evaluating Division Expressions With Parentheses

To evaluate $(40 \div 5) \div 2$ it is necessary to understand the role of the parentheses ( ). Regardless of how many numbers are involved in a division expression, only two counting numbers can be divided at a time.

Parentheses or square brackets are used to indicate which two numbers are to be divided in the first step.	$(40 \div 5) \div 2$
	$8 \div 2$
	4

In the division expression above, the parentheses indicate that 40 is to be divided by 5 first. Then their quotient is divided by 2.

Focus on order of division with square brackets

The division expression $[128 \div 8] \div 2$ involves the same numbers as $128 \div [8 \div 2]$. The difference in the two expressions is that the square brackets enclose different numbers indicating a different first step in the evaluations.

The evaluations of these two expressions are shown below.

$[128 \div 8] \div 2$	$128 \div [8 \div 2]$
$16 \div 2$	$128 \div 4$
8	32

Notice that the placement of square brackets in a division expression results in different evaluations.

Evaluating Expressions With "Nested" Grouping Symbols

In the expression [(4 + 3) + 5] + 7, the square brackets completely enclose the parentheses. Such a situation is called "nested" grouping symbols.

There are three numbers in the square brackets and two numbers in the parentheses. Since only two numbers can be added at one time, the numbers in the parentheses are added first.

$$[(4 + 3) + 5] + 7$$

$$[7 + 5] + 7$$

$$12 + 7$$

The steps in evaluating the addition expression [(4 + 3) + 5] + 7 are shown at the right.

$$19$$

Focus on evaluating with nested grouping symbols

When the expression 17 − ([58 − 45] − 3) is evaluated, the grouping symbols indicate that the steps are:

1. Subtract 45 from 58.

2. Subtract 3 from 13.

3. Subtract 10 from 17.

$$17 − ([58 − 45] − 3)$$

1. $17 − (13 − 3)$

2. $17 − 10$

3. 7

The expression 5 • [2 • (4 • 7)] has three counting numbers in the square brackets and two numbers in the parentheses. The numbers 4 and 7 are to be multiplied first in evaluating the expression. The evaluation of the expression 5 • [2 • (4 • 7)] is shown below.

$$5 \cdot [2 \cdot (4 \cdot 7)]$$
$$5 \cdot [2 \cdot 28]$$
$$5 \cdot 56$$
$$280$$

Unit 1 Exercise

This exercise reviews the preceding unit. The exercise is divided into three parts. Part A reviews the foci of Unit 1. Part B offers opportunities to practice the skills and concepts of Unit 1. Part C contains problems that review your previous work in this text. All answers for Parts A and B are at the back of the book. Each problem of Part C is accompanied by a notation «CU» that refers to the Chapter (C) and Unit (U) in which that type of problem is studied.

Part A: Reviewing the foci of Unit 1.

1. The parentheses in $(6 + 7) + 8$ indicate that 7 is to be added to _____.

2. What numbers are to be added first in the expression $(7 + 9) + 6$?

3. In the expression $(8 + 9) + [4 + 7]$, the 7 is to be added to _____.

4. In the expression $[9 + 3] + (8 + 7)$, the 3 is to be added to _____.

5. Do $3 + 9$ and $9 + 3$ have the same evaluation?

6. Do $9 - 7$ and $7 - 9$ have the same evaluation?

7. Do $(15 + 9) + 3$ and $15 + (9 + 3)$ have the same evaluation?

8. Do $(15 - 9) - 3$ and $15 - (9 - 3)$ have the same evaluation?

9. Do $3 \cdot 8$ and $8 \cdot 3$ have the same evaluation?

10. Do $8 \div 4$ and $4 \div 8$ have the same evaluation?

11. Do $(8 \cdot 5) \cdot 2$ and $8 \cdot (5 \cdot 2)$ have the same evaluation?

12. Do $(48 \div 4) \div 2$ and $48 \div (4 \div 2)$ have the same evaluation?

13. In the expression $(4 \cdot 7) \cdot (3 \cdot 5)$, which number is to be multiplied by the 3?

14. In the expression $[9 \cdot 6] \cdot (4 \cdot 7)$, which number is to be multiplied by the 9?

15. What is the first step in evaluating $[19 + (3 + 8)] + 6$?

16. What is the first step in evaluating $100 \div ([60 \div 2] \div 6)$?

Part B: Drill and Practice

Evaluate.

1. $5 + [9 + (3 + 6)]$

2. $([9 + 3] + 6) + 4$

3. $[9 + (8 + 6)] + 7$

4. $4 + ([7 + 9] + 5)$

5. $(7 + 3) + [2 + 5]$

6. $[7 + 9] + (3 + 8)$

7. $[7 + (3 + 2)] + 8$

8. $(37 - 12) - 8$

9. $19 - (8 - 4)$

10. $17 - (12 - 9)$

11. $(17 - 4) - 8$

12. $15 - (8 - 5)$

13. $(10 - 5) - 3$

14. $4 \cdot (5 \cdot 6)$

15. $[3 \cdot 2] \cdot 7$

16. $9 \cdot [1 \cdot (3 \cdot 2)]$

17. $(7 \cdot 8) \cdot 9$

18. $14 \cdot (5 \cdot 2)$

19. $100 \div (8 \div 2)$

20. $18 \div (36 \div 6)$

21. $[8 \div 4] \div 2$

22. $8 \div [4 \div 2]$

23. $16 \div (40 \div 5)$

24. $[90 \div 3] \div 5$

25. $17 - [4 - (25 - 23)]$

26. $48 \div [(18 \div 9) \div 2]$

Part C: Review. Answers for the problems of Part C are not given. However, the notation «C,U» refers to the Chapter (C) and Unit (U) in which problems of the same type were presented.

1. What is the hundreds digit of 4,825,369?   «1,1»

2. 45,308 − 27,614   «1,3»

3. $33 \overline{)\ 47248}$ «2,4»

4. Find the lcm for 15 and 25.   «3,2»

5. Select the smaller fraction from: $\frac{7}{8}, \frac{2}{3}$ «3,2»

6. Change $5\frac{4}{9}$ to an improper fraction.   «3,2»

7. $7\frac{2}{3} - 4\frac{5}{6}$ «3,4»

8. $7\frac{1}{8} \div 1\frac{3}{4}$ «3,6»

9. Write 0.413 as a fraction.   «4,1»

10. Subtract 4.37 from 17.4   «4,2»

11. Find the total of 423 and 68.   «1,4»

12. Find the sum of 8.345 and 13.42   «4,2»

13. Divide 8.32 by 0.6 and round off the answer to 2 decimal places.   «4,4»

14. Margaret has prepared 114 ounces of tomatoes for canning. How many 6 ounce jars are needed for these tomatoes?   «4,4»

15. The weatherman estimates 0.3 days in June will be rainy. In the 30 days of June, how many days does he estimate will be rainy?   «4,4»

Unit 2:   EQUIVALENT NUMERICAL EXPRESSIONS

Definition		Numerical Expression
An expression is a **numerical** **expression**	**if and** **only if**	the expression is an addition, subtraction, multiplication, and/or division expression, or a combination of such expressions.

Examples of numerical expressions are:

$$15 \cdot (5 + 7) \qquad (4 \div 5) \cdot 4 \qquad 8 + (5 + 512)$$

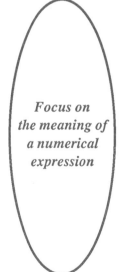

Focus on the meaning of a numerical expression

A numerical expression is generated whenever one or more whole number expressions are combined by the use of addition, subtraction, multiplication, and/or division.

$$(6{,}312 - 5{,}305) \div (18 + 4)$$

This is a combination of two expressions, one a subtraction and the other an addition, which are joined by the symbol $\div$.

$$0 \cdot (0 + 75)$$

This is a combination of 0 and an addition expression joined by the multiplication symbol $\cdot$.

$$19 + 41$$

This is a simple addition expression, but is also a numerical expression.

Notice that the term "numerical expression" encompasses addition, subtraction, multiplication, and/or division expressions, but also allows for more than one operation in an expression.

Definition   **Equivalent Numerical Expressions**

If K and L are numerical expressions, they are **equivalent** **if and only if** K and L have exactly the same evaluations.

For example $5 + 3$ and $4 \cdot 2$ are **equivalent** because both of the numerical expressions have an evaluation of 8.

Focus on the meaning of equivalent numerical expressions

To determine if $9 \cdot (5 + 2)$ is equivalent to $70 - (3 + 4)$ the numerical expressions are evaluated separately.

$$
\begin{array}{cc}
9 \cdot (5 + 2) & 70 - (3 + 4) \\
9 \cdot 7 & 70 - 7 \\
63 & 63
\end{array}
$$

Notice that the expressions in the parentheses are evaluated first.

Since $9 \cdot (5 + 2)$ and $70 - (3 + 4)$ have the same evaluation, 63, the expressions are equivalent.

Focus on numerical expressions that are not equivalent

To determine whether $8 - (5 - 3)$ is equivalent to $(8 - 5) - 3$, the two numerical expressions are evaluated separately.

$$
\begin{array}{cc}
8 - (5 - 3) & (8 - 5) - 3 \\
8 - 2 & 3 - 3 \\
6 & 0
\end{array}
$$

The two numerical expressions are not equivalent because they have different evaluations. $8 - (5 - 3)$ has an evaluation of 6. $(8 - 5) - 3$ has an evaluation of 0.

Addition Expressions With Identical Addends

Addition expressions which have the same addends will always be equivalent regardless of the order in which the addends are arranged or the placement of grouping symbols.

For example, if 5, 7, 8, and 2 are used as addends a large number of addition expressions can be written, but all will have the same evaluation and, therefore, all are equivalent. Four of these equivalent expressions are shown below.

$$5 + [(7 + 8) + 2]$$

$$(7 + 2) + [8 + 5]$$

$$([8 + 2] + 5) + 7$$

$$[2 + 5] + (8 + 7)$$

Each of the addition expressions has an evaluation of 22. All are equivalent.

Focus on equivalent addition expressions

When two numerical expressions only involve addition and contain exactly the same numbers, the expressions are equivalent.

For example,

$$(235 + 8,764) + [3,245 + 499]$$

is equivalent to

$$(3,245 + [8,764 + 499]) + 235$$

because both expressions are addition expressions which contain the addends 235, 499, 3,245, and 8,764.

Subtraction Expressions With Identical Numbers

Subtraction expressions which have the same numbers
will almost never be equivalent.

For example, if 9, 5, and 2 are used as numbers in two different
subtraction expressions the result will be expressions that are
not equivalent.

$(9 - 5) - 2$ has an evaluation of 2.

$9 - (5 - 2)$ has an evaluation of 6.

$(5 - 2) - 9$ has no evaluation because 9 is greater than 3.

Focus on subtraction expressions

When two numerical expressions only involve subtraction and contain
exactly the same numbers, the expressions will rarely be equivalent.

For example,

$$10 - 8 \text{ is not equivalent to } 8 - 10$$

because $10 - 8$ has an evaluation of 2 and $8 - 10$ has no evaluation
because 10 is greater than 8.

The expressions $(80 - 30) - 24$ and $80 - [30 - 24]$ have different
evaluations.

$$
\begin{array}{cc}
(80 - 30) - 24 & 80 - [30 - 24] \\
50 - 24 & 80 - 6 \\
26 & 74
\end{array}
$$

The expressions are not equivalent even when they contain the same numbers.

Multiplication Expressions With Identical Factors

Multiplication expressions which have the same factors
will always be equivalent regardless of the order in which the
factors are arranged or the placement of grouping symbols.

For example, if 5, 7, 8, and 2 are used as factors a large
number of multiplication expressions can be written, but all will have
the same evaluation and, therefore, all are equivalent. Four of
these equivalent expressions are shown below.

$$5 \cdot [(7 \cdot 8) \cdot 2]$$

$$(7 \cdot 2) \cdot [8 \cdot 5]$$

$$([8 \cdot 2] \cdot 5) \cdot 7$$

$$[2 \cdot 5] \cdot (8 \cdot 7)$$

Each of the addition expressions has an evaluation of 560. All
are equivalent.

*Focus on
equivalent
multiplication
expressions*

When two numerical expressions only involve multiplication and contain
the same numbers, the equivalency of the expressions is guaranteed.

For example,

$$(235 \cdot 8{,}764) \cdot [3{,}245 \cdot 499]$$

is equivalent to

$$(3{,}245 \cdot [8{,}764 \cdot 499]) \cdot 235$$

because both expressions are multiplication expressions which contain the
factors 235, 499, 3,245, and 8,764.

Division Expressions With Identical Numbers

Division expressions which have the same numbers
will almost never be equivalent.

For example, if 16, 8, and 2 are used as numbers in three different
division expressions the result will be three expressions that are
not equivalent.

$(16 \div 4) \div 2$ has an evaluation of 2.

$16 \div (4 \div 2)$ has an evaluation of 8.

$(4 \div 2) \div 16$ has an evaluation of $\frac{2}{16}$ or $\frac{1}{8}$ which is not a counting number.

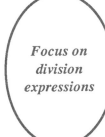

*Focus on
division
expressions*

When two numerical expressions only involve division and contain
the same numbers, the expressions will rarely be equivalent.

For example,

$$30 \div 5 \text{ is not equivalent to } 5 \div 30$$

because $30 \div 5$ has an evaluation of 6 and $5 \div 30$ has an evaluation
of $\frac{1}{6}$ which is not a counting number.

Expressions Containing Two Operations

The numerical expression $5 + [3 \cdot (4 + 2)]$ contains four
counting numbers and the operations of addition and
multiplication. Only two counting numbers may be operated
on at one time. Consequently, the first step in evaluating
$5 + [3 \cdot (4 + 2)]$ is to find the sum of 4 and 2. The
evaluation of this expression is shown below.

$$5 + [3 \cdot (4 + 2)]$$
$$5 + [3 \cdot 6]$$
$$5 + 18$$
$$23$$

*Focus on
the order
of steps in
evaluating
expressions*

The evaluation of the expression $7 + [(4 + 1) \cdot 6]$ requires that 4 and 1 be added first. The parentheses and square brackets are used to indicate the first and second steps in this evaluation. The evaluation of $7 + [(4 + 1) \cdot 6]$ is shown below.

$$7 + [(4 + 1) \cdot 6]$$
$$7 + [5 \cdot 6]$$
$$7 + 30$$
$$37$$

Notice that the expression in the parentheses is evaluated first, because the parentheses are inside the square brackets.

Evaluating Expressions With No Grouping Symbols

Whenever parentheses or square brackets do *not* appear in an expression, *multiplication is always performed before addition.*

For example, in the expression $3 + 4 \cdot 5$ the first step is to multiply 4 and 5.

There are no parentheses or square brackets to indicate the first step in the evaluation of $3 + 4 \cdot 5$.

$$3 + 4 \cdot 5$$
$$3 + 20$$

Therefore, all multiplication is performed before addition.

$$23$$

*Focus on
expressions
with no
parentheses*

In evaluating $(8 + 2) \cdot 3$ the parentheses indicate that the first step is to evaluate $8 + 2$. In the expression $8 + 2 \cdot 3$ there are no parentheses. Consequently, the first step is to multiply 2 and 3. The two expressions $(8 + 2) \cdot 3$ and $8 + 2 \cdot 3$ are evaluated below. Notice the different steps in the evaluations.

$$(8 + 2) \cdot 3 \qquad\qquad 8 + 2 \cdot 3$$
$$10 \cdot 3 \qquad\qquad\quad 8 + 6$$
$$30 \qquad\qquad\qquad 14$$

Notice also that $(8 + 2) \cdot 3$ and $8 + 2 \cdot 3$ are not equivalent because $(8 + 2) \cdot 3$ has an evaluation of 30 and $8 + 2 \cdot 3$ has an evaluation of 14.

An Equivalence Involving Addition and Multiplication

As a general guideline, two numerical expressions involving both addition and multiplication will not be equivalent even when they contain the same numbers.

For example, $7 \cdot 5 + 4$ is not equivalent to $7 \cdot (5 + 4)$

$7 \cdot 5 + 4$	$7 \cdot (5 + 4)$
$35 + 4$	$7 \cdot 9$
39	63

There is, however, one major exception to this general guideline. The expressions $8 \cdot 5 + 8 \cdot 2$ and $8 \cdot (5 + 2)$ are equivalent.

$8 \cdot 5 + 8 \cdot 2$	$8 \cdot (5 + 2)$
$40 + 16$	$8 \cdot 7$
56	56

Study the pattern of the numerical expressions shown above. This pattern will become important in later units.

Focus on an equivalence involving addition and multiplication

The numerical expressions $6 \cdot (8 + 2)$ and $6 \cdot 8 + 6 \cdot 2$ involve both addition and multiplication. The numbers 6, 8, and 2 are in both expressions.

$6 \cdot (8 + 2)$	$6 \cdot 8 + 6 \cdot 2$
$6 \cdot 10$	$48 + 12$
60	60

Notice that $6 \cdot (8 + 2)$ and $6 \cdot 8 + 6 \cdot 2$ are equivalent because they both have 60 as their evaluations. If we maintain the same pattern using the numbers 48, 912, and 63 another pair of equivalent expressions can be written.

$48 \cdot (912 + 63)$ is equivalent to $48 \cdot 912 + 48 \cdot 63$

Both have an evaluation of 46,800 and are equivalent expressions.

Unit 2 Exercise

This exercise reviews the preceding unit. The exercise is divided into three parts. Part A reviews the foci of Unit 2. Part B offers opportunities to practice the skills and concepts of Unit 2. Part C contains problems that review your previous work in this text. All answers for Parts A and B are at the back of the book. Each problem of Part C is accompanied by a notation «CU» that refers to the Chapter (C) and Unit (U) in which that type of problem is studied.

Part A: Reviewing the foci of Unit 2.

Evaluate.

1. 7 + 13

2. 19 − 7

3. 41 • 5

4. 72 ÷ 6

5. Does the placement of parentheses in an expression involving only the operation addition affect the final evaluation?

6. Write a numerical expression to indicate the product of 12 and 7.

7. Write a numerical expression to indicate the sum of 11 and 16.

8. Is 4 • (15 ÷ 3) a numerical expression?

9. The first step in evaluating (5 + 4) • 7 is to evaluate _____.

10. The first step in evaluating 12 + 8 • 5 is to evaluate _____.

11. Evaluate the expression 6 + [8 • 3].

12. Evaluate [6 + 8] • 3.

13. What should be evaluated first in 8 + [9 • (3 + 1)]

14. What is the result of the first step in evaluating 8 + [(3 + 4) • 2]?

15. Evaluate [(6 • 5) + 4] + 3.

16. Evaluate 10 + (6 + [3 • 5]).

Use the order indicated by the grouping symbols to evaluate.

17. $(40 \div 8) + 2$

18. $40 \div (8 + 2)$

19. $19 - (6 \cdot 2)$

20. $[19 - 6] \cdot 2$

21. The first step in evaluating the expression $7 + 6 \cdot 3$ is to evaluate

    _____.

22. Do $(3 + 4) \cdot 5$ and $3 + 4 \cdot 5$ have the same evaluation?

23. Do $3 \cdot 5 + 7$ and $[3 \cdot 5] + 7$ have the same evaluation?

24. Do $5 + (3 \cdot 7)$ and $5 + 3 \cdot 7$ have the same evaluation?

Evaluate using the rule that if a numerical expression contains no grouping symbols, all multiplication is completed before any addition.

25. $3 + 7 \cdot 2$

26. $8 \cdot 9 + 8 \cdot 3$

27. $5 \cdot (10 - 6) + 4$

28. $3 \cdot 7 + 3 \cdot 2$

In problems 29-38 determine if the expressions are equivalent.

29. $10 + 7$ and $7 + 10$

30. $10 - 7$ and $7 - 10$

31. $10 \cdot 7$ and $7 \cdot 10$

32. $10 \div 7$ and $7 \div 10$

33. $83 + (46 + 713)$ and $[83 + 46] + 713$

34. $83 - (46 - 13)$ and $[83 - 46] - 13$

35. $83 \cdot (46 \cdot 713)$ and $[83 \cdot 46] \cdot 713$

36. $(400 \div 8) \div 2$ and $400 \div (8 \div 2)$

37. $9 + 4 \cdot 3$ and $(9 + 4) \cdot 3$

38. $9 \cdot 6 + 4$ and $9 \cdot (6 + 4)$

Part B: Drill and Practice

Evaluate.

1. $8 + 4 \cdot 3$

2. $5 \cdot (1 + 4)$

3. $12 + 10 \cdot 5$

4. $(10 + 4) \cdot 2$

5. $15 + 13 \cdot 2$

6. $(9 + 4) \cdot 2$

7. $7 \cdot 3 + 9$

8. $4 + 3 \cdot 5$

9. $(2 + 6) \cdot [5 + 3]$

10. $[7 + 2] \cdot [8 + 2]$

11. $6 \cdot 3 + 2 \cdot 7$

12. $10 \cdot 3 + 8 \cdot 5$

13. $4 \cdot 9 + 2 \cdot 5$

14. $5 \cdot 3 + 2 \cdot 4$

15. $7 + [8 \cdot 2 + 1]$

16. $(4 + 3 \cdot 6) + 1$

17. $2 + [4 \cdot (1 + 5)]$

18. $[(9 + 2) \cdot 4]$

19. $5 + [8 \cdot 3 + 2]$

20. $6 + ([13 + 2] \cdot 4)$

21. $5 \cdot [4 \cdot 2 + 3]$

22. $2 \cdot (3 \cdot [4 + 1])$

Decide whether each of the following pairs of numerical expressions are equivalent.

23. $48 + (19 + 3)$ and $(48 + 19) + 3$

24. $76 + (43 + 512)$ and $[76 + 43] + 512$

25. $48 - (19 - 3)$ and $(48 - 19) - 3$

26. $96 - (45 - 17)$ and $[96 - 45] - 17$

27. $48 \cdot (19 \cdot 3)$ and $(48 \cdot 19) \cdot 3$

28. $57 \cdot (86 \cdot 511)$ and $[57 \cdot 86] \cdot 511$

29. $(400 \div 8) \div 2$ and $400 \div (8 \div 2)$

30. $48 \div (18 \div 3)$ and $(48 \div 18) \div 3$

31. $9 + 4 \cdot 3$ and $(9 + 4) \cdot 3$

32. $9 \cdot 6 + 4$ and $9 \cdot (6 + 4)$

33. $4 \cdot 9 + 4 \cdot 8$ and $4 \cdot (9 + 8)$

34. $7 \cdot 3 + 7 \cdot 8$ and $7 \cdot (3 + 8)$

35. $4 \cdot (7 + 3)$ and $4 \cdot 7 + 4 \cdot 3$

Part C: Review. Answers for the problems of Part C are not given. However, the notation «C,U» refers to the Chapter (C) and Unit (U) in which problems of the same type were presented.

1.   49,835 + 6,453 + 31,978   «1,2»

2.   4,836 x 8   «2,1»

3.   Simplify $\frac{18}{30}$ «3,1»

4.   Write $\frac{38}{9}$ as a mixed number.   «3,2»

5.   Find the numerator of $\frac{3}{4} = \frac{?}{32}$ «3,2»

6. $8\frac{7}{8} + 3\frac{7}{12}$ «3,3»

7. $8\frac{1}{3}$ x $\frac{4}{5}$ «3,5»

8.   Write $\frac{78}{1000}$ as a decimal fraction.   «4,1»

9.   Add 4,715, 23.91, and 0.415.   «4,2»

10. Find the sum of 652 and 408. «1,4»

11. Find the difference of 73.72 and 31.5   «4,2»

12. Find the product of 3,680 and 23. «2,2»

13. Find 0.07 of 3,850.   «4,3»

14. How many $\frac{1}{3}$ gallons are there in 7 gallons?. «3,7»

15. Evaluate 16 ÷ (40 ÷ 5)   «5,1»

16. Evaluate 48 ÷ [(18 ÷ 9) ÷ 2]   «5,1»

17. What is the difference between 8.75 and 4.9? «4,2»

18. Each can in a case of 24 cans of beans contains 4.6 ounces of beans. How many ounces of beans are in the full case?   «4,3»

19. Change $\frac{5}{9}$ to a decimal fraction rounded off at two decimal places.   «4,4»

20. Phelps estimates that he spends 0.22 of his salary on housing. If his monthly salary is $1,200, how much does he estimate he spends on housing each month?   «4,4»

Unit 3:   EQUIVALENT OPEN EXPRESSIONS

Definition	Open Expression	
An expression is an **open** expression	**if and only if**	it is a numerical expression in which one or more of the positions designated for a number has been left blank.

$8 + 7$ is a numerical expression. If the position now filled by 8 is left blank, it becomes ____ $+ 7$ and this is called an **open expression**.

Open Expressions

$\square + 3$ is *not* a numerical expression, because the box is not a number. $5 + 3$ *is* a numerical expression, but $\square + 5$ is not a numerical expression.

$\square + 5$ is an **open expression**. Similarly, $\square + 13$ is not a numerical expression, but is called an **open expression**.

$(3 + 4) + 2$ is a numerical expression. $(5 + 6) + \square$ is an open expression.

The box in $\square + 3$ represents a place for a number and when a number replaces the box in the open expression the result is a numerical expression.

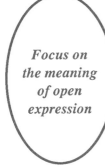

Focus on the meaning of open expression

7 • 5 + 14 is a numerical expression.

7 • □ + 14 is an open expression. The box represents a place for a number.

12 + (19 + 12) is a numerical expression in which the number 12 appears two times.

□ + (19 + □) is an open expression in which the box appears two times. The box represents a place for a number and whatever number goes in one box must also go in the other box in the same open expression.

Evaluating Open Expressions

Whenever the box in the open expression 5 • □ + 2 is replaced by a counting number, the result is a numerical expression. The numerical expression can then be evaluated.

For example, when the box in 5 • □ + 2 is replaced by 8, the result is the numerical expression 5 • 8 + 2 which has an evaluation of 42.

> The open expression 5 • □ + 2 has an evaluation of 42 when the box is replaced by 8.

When the box in 5 • □ + 2 is replaced by 3, the result is the numerical expression 5 • 3 + 2 which has an evaluation of 17.

> The open expression 5 • □ + 2 has an evaluation of 17 when the box is replaced by 3.

The evaluation of an open expression depends upon the specific number chosen to replace the box.

Focus on substituting a number in an open expression

Any number can replace the box of □ + 9. If the box is replaced by the counting number 4, the numerical expression 4 + 9 is obtained and this numerical expression has an evaluation of 13.

If the box of □ + 9 is replaced by 97, the numerical expression 97 + 9 with an evaluation of 106 is obtained.

Any number can replace the box of □ • (2 + □), but whatever number replaces one box must also replace the other. If the box is replaced by the counting number 4, the numerical expression 4 • (2 + 4) is obtained and this numerical expression has an evaluation of 24.

If the box of □ • (2 + □) is replaced by 7, the numerical expression 7 • (2 + 7) with an evaluation of 63 is obtained.

Variables

In mathematics we commonly use *letters* rather than boxes to write open expressions. For example, the open expression □ + 3 would be seen more commonly as x + 3 or m + 3.

The letters we use instead of boxes are called **variables.**

For the open expression 4 • □ + 3, if the box is replaced by a **variable**, we would have the open expression 4 • a + 3 or 4 • w + 3.

The open expression $5 + 8 \cdot \square$ can be written with a variable as $5 + 8 \cdot z$.

The open expression $6 \cdot \square + 11 \cdot \square$ has two boxes which must be replaced by the same number. Therefore, the same letter must be used as a variable in both places. $6 \cdot x + 11 \cdot x$ is an open expression with one variable, x, and whatever number is used as a replacement for x must be used in both positions.

Focus on writing open expressions with variables

The open expression $k + 3 \cdot m$ is an open expression with two variables, k and m. The use of two variables allows different numbers to replace the variables. Any number can replace k and any number can replace m.

In the open expression, $k + 3 \cdot m$ there are two variables. If k is replaced by 2 and m is replaced by 5, the numerical expression $2 + 3 \cdot 5$ is obtained. It has an evaluation of 17 as shown below.

$$k + 3 \cdot m$$
$$2 + 3 \cdot 5$$
$$2 + 15$$
$$17$$

Omitting the Multiplication Dots

The multiplication dots are normally omitted except in the situation where it is between two numbers.

$5 \cdot x$ is written as 5x

$z \cdot y$ is written as zy

$4 \cdot (x + 3)$ is written as $4(x + 3)$

$6 \cdot 7$ needs the multiplication dot.

The multiplication dot should be used if the variable precedes the number. $x \cdot 6$

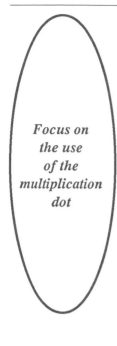

*Focus on
the use
of the
multiplication
dot*

The open expression 9x means that 9 is to be multiplied by the number that replaces x.

In 9x, if x is replaced by 7, the resulting numerical expression is $9 \cdot 7$.

The open expression 8(x + 3) means that 8 is to be multiplied by the sum of the number that replaces x and 3.

In 8(x + 3) if x is replaced by 6, the resulting numerical expression is 8(6 + 3) which means $8 \cdot (6 + 3)$.

The open expression 6x + 5y contains two variables. It means that 6 is to be multiplied by the number that replaces x and 5 is to be multiplied by the number that replaces y. Then the products are to be added.

In 6x + 5y if x is replaced by 7 and y is replaced by 4 the resulting numerical expression is $6 \cdot 7 + 5 \cdot 4$, which has an evaluation of 62.

Evaluating Open Expressions

To evaluate the open expression 3x + 4 when x is replaced by 5, the following steps are used.

$$3x + 4 \qquad \text{when } x = 5$$
$$(3 \cdot 5) + 4$$
$$15 + 4$$
$$19$$

To evaluate the open expression 5y + 7z when y is replaced by 2 and z is replaced by 4, the following steps are used.

$$5y + 7z \qquad \text{when } y = 2 \text{ and } z = 4$$
$$5 \cdot 2 + 7 \cdot 4$$
$$10 + 28$$
$$38$$

Focus on evaluating open expressions

To evaluate the open expression 3(x + 5) when x is replaced by 4, the following steps are used.

$$3(x + 5)$$

When x = 4, 3(x + 5) becomes 3(4 + 5)

$$3(4 + 5)$$
$$3 \cdot 9$$
$$27$$

To evaluate x(x + 3y) when x is replaced by 7 and y is replaced by 6, the following steps are used.

When x = 7 and y = 6
  x(x + 3y) becomes 7(7 + 3 · 6)

$$x(x + 3y)$$
$$7(7 + 3 \cdot 6)$$
$$7(7 + 18)$$
$$7 \cdot 25$$
$$175$$

Comparing the Evaluations of Two Open Expressions

It is a valuable skill to compare the evaluations of two open expressions with the same variable.

For example, if x is replaced by 9 the evaluations of (x + 8) + 3 and 11 + x can be compared.

if x = 9
(x + 8) + 3	11 + x
(9 + 8) + 3	11 + 9
17 + 3	20
20	

Notice that the evaluations were the same, 20. This gives rise to an interesting question. Will the evaluations be the same regardless of what number is chosen to replace x?

Focus on comparing evaluations

Below is shown the evaluations of $3x + 4$ and $7x$ when x is replaced by 2.

$$
\begin{array}{cc}
3x + 4 & 7x \\
3 \cdot 2 + 4 & 7 \cdot 2 \\
6 + 4 & 14 \\
10 &
\end{array}
$$

If x = 2

Notice that the evaluations of $3x + 4$ and $7x$ are different when x is replaced by 2.

Definition		Equivalent Open Expressions
Two open expressions are **equivalent**	**if and only if**	they have the same evaluations for any replacements of their variables.

Equivalent Open Expressions

Earlier in this unit it was shown that $(x + 8) + 3$ and $11 + x$ had the same evaluations when $x = 9$. Later we will see that these two open expressions would have the same evaluations for any replacements of x. The open expressions are **equivalent.**

A comparison of the evaluations of $3x + 4$ and $7x$ when $x = 2$ showed that the open expressions had different evaluations. It can definitely be stated that $3x + 4$ and $7x$ are **not equivalent.**

Notice that $(x + 8) + 3$ and $x + 11$ were not guaranteed to be equivalent just because one replacement of the variable resulted in equal evaluations. However, $3x + 4$ and $7x$ were shown **not** to be equivalent by just one replacement of the variable that showed they had different evaluations.

Focus on testing open expressions for equivalency

To test the open expressions 5(4x) and 20x for equivalency, any number may replace x.

If $x = 3$ then 5(4x) becomes

$$5(4 \cdot 3)$$
$$5 \cdot 12$$
$$60$$

If $x = 3$ then 5(4x) has an evaluation of 60 and also 20x has an evaluation of 60.

If $x = 3$ then 20x becomes

$$20 \cdot 3$$
$$60$$

5(4x) and 20x may be equivalent, but we would need to know that all other replacements of x would also give equal evaluations to guarantee the equivalency.

Focus on testing open expressions that are not equivalent

To test the open expressions 5(x + 4) and 5x + 4 for equivalency, any number may replace x.

If $x = 3$ then 5(x + 4) becomes

$$5(3 + 4)$$
$$5 \cdot 7$$
$$35$$

If $x = 3$ then 5(x + 4) has an evaluation of 35 and 5x + 4 has an evaluation of 19.

If $x = 3$ then 5x + 4 becomes

$$5 \cdot 3 + 4$$
$$15 + 4$$
$$19$$

5(x + 4) and 5x + 4 are definitely not equivalent because they have different evaluations.

Unit 3 Exercise

This exercise reviews the preceding unit. The exercise is divided into three parts. Part A reviews the foci of Unit 3. Part B offers opportunities to practice the skills and concepts of Unit 3. Part C contains problems that review your previous work in this text. All answers for Parts A and B are at the back of the book. Each problem of Part C is accompanied by a notation «CU» that refers to the Chapter (C) and Unit (U) in which that type of problem is studied.

Part A: Reviewing the foci of Unit 3.

1. $\square + 7$ is a(n) _____ expression.

2. $[6 + 3] \cdot (5 + 1)$ is a(n) _____ expression.

3.   The box in $13 + \square$ represents a place for a _____ .

4.   What numerical expression is obtained by replacing the box by 6 in $10 + \square$?

5.   Evaluate $\square + 10$ when the box is replaced by 5.

6.   Evaluate $2 \cdot \square + 5$ when the box is replaced by 8.

7.   Evaluate $(\square + 3) \cdot 5$ when the box is replaced by 4.

8.   Evaluate $(\square + 2) \cdot (1 + \square)$ when the box is replaced by 2.

9.   Write $16 + 5 \cdot \square$ using the letter z.

10. Write $(\square + 5) \cdot \square$ using the letter w.

11. In the open expression $3x + 17$, the letter x may be replaced by _____ counting number.

Evaluate.

12. $3x + 4$ when x is replaced by 8.

13. $2x + 5x$ when x = 10.

14. $(5 + x) + 3$ when x = 1.

15. $x(x + 5)$ when x = 5.

16. $3x + y$ when x = 10 and y = 9.

In problems 17-20 determine if the two expressions have the same evaluation.

17. $(x + 4) + 5$ and $x + 9$ when x = 7

18. $4x + 5$ and $9x$ when x = 7

19. 2(x + 3) and 2x + 3 when x = 5

20. xy and yx when x = 4 and y = 15

Part B: Drill and Practice

1. [4 + 7] • (□ + 2) is a(n) _____

    _____.

2. (4 + □) • (3 + 7) is a(n) _____
    expression.

3. (8 + 2) • (5 + 3) is a(n) _____
    expression.

4. [8 + 3•□] + 4 is a(n) _____

    _____.

5. 3 + 5 • [(4 + 2) • 7] is a(n) _____

    _____.

6. The box in 5 + □ represents a place for a
    _____ number.

7. Can 21 replace the box of 6•□ ?

8. Can 8 replace the box of 5 + □ ?

9. Replace the box of □ + 8 by 3 and evaluate
    the numerical expression.

10. Replace the box of 3 • □ by 5 and evaluate
    the numerical expression.

11. Replace the box of 8 • (□ + 1) by 3 and
    evaluate.

12. Replace the box of (□ + 3) • 2 by 6 and
    evaluate.

13. Replace the box of (1 + 2) • (□ + 3) by 4
    and evaluate.

14. Replace both boxes of □ • (5 + □) by 3
    and evaluate the numerical expression.

15. Replace the boxes of 8 • □ + □ by 9 and
    evaluate.

16. Replace the boxes of □ • □ + 3 by 2 and
    evaluate.

Evaluate problems 17-44 for the given replacement(s) of the variable(s).

17. $3x + 4$ when x is replaced by 14

18. $2x + 5$ when x is replaced by 4

19. $2x + 4$ when x is replaced by 15

20. $5t + 3$ when t is replaced by 7

21. $3(x + 5)$ when $x = 10$

22. $2x + 5x$ when $x = 7$

23. $5x + y$ when $x = 6$ and $y = 7$

24. $z + 9b$ when $b = 5$ and $z = 6$

25. $x + 5y$ when $x = 7$ and $y = 3$

26. $2x + 5y$ when $x = 7$ and $y = 3$

27. $(7 + 3a) + b$ when $a = 5$ and $b = 2$

28. $3x + 3y$ when $x = 4$ and $y = 5$

29. $10m + n$ when $m = 3$ and $n = 1$

30. $2x + (3y + 5)$ when $x = 3$ and $y = 5$

31. $x(5 + 2y)$ when $x = 3$ and $y = 2$

32. $(5r + 3s) + 6t$ when $r = 3$, $s = 4$, and $t = 1$

33. $2x + (y + 3z)$ when $x = 4$, $y = 2$, and $z = 7$

34. $2(x + 3y) + 4z$ when $x = 4$, $y = 2$, and $z = 7$

35. $x + [2x + (y + 3)]$ when $x = 3$ and $y = 7$

36. $c(3a + 2b)$ when $a = 2$, $b = 4$, and $c = 5$

37. $ab + 7$ when $a = 3$ and $b = 4$

38. $rs + r$ when $r = 4$ and $s = 5$

39. $2x + 4$ when $x = 7$

40. $5 \cdot (x + 4)$ when $x = 3$

41. $x + 3y$ when $x = 4$ and $y = 5$

42. $5 + xy$ when $x = 4$ and $y = 7$

43. a(5 + 2b) when a = 3 and b = 2

44. 5x + [2a + (5 + b)] when a = 3,
    b = 10, and x = 2

Evaluate each pair of expressions for problems
45-49 and state whether the evaluations are
equivalent.

45. (2 + y) + 9 and 11 + y when y = 3

46. (6 + a) + 7 and 13 + a when a = 3

47. 8x + 3x and 11x when x = 3

48. (2x + 3x) + 6 and 5x + 6 when x = 3

49. x + y and y + x when x = 2 and y = 5

Part C: Review. Answers for the problems of Part C are not given. However, the notation «C,U» refers to the Chapter (C) and Unit (U) in which problems of the same type were presented.

1. What is the hundred thousands digit of 4,825,369?   «1,1»

2. 86,407 − 27,614   «1,3»

3. $43 \overline{) 45348}$ «2,4»

4. Find the lcm for 7 and 15.   «3,2»

5. Select the smaller fraction from $\frac{5}{8}, \frac{3}{5}$.   «3,2»

6. Change $4\frac{3}{8}$ to an improper fraction.   «3,2»

7. $6\frac{1}{2} - 4\frac{5}{6}$ «3,4»

8. $4\frac{4}{5} \div 2\frac{3}{10}$ «3,6»

9. Write 0.4257 as a fraction.   «4,1»

10. Subtract 8.93 from 16.4   «4,2»

11. Evaluate [60 ÷ 3] ÷ 4   «5,1»

12. Evaluate 16 − [8 − (28 − 23)]   «5,1»

13. Evaluate (5 + 2 • 6) + 7   «5,2»

14. Evaluate (3 • 6 + 4) + 3   «5,2»

15. Find the total of 8,512 and 375.   «1,4»

16. Find the sum of 2.763 and 17.62   «4,2»

17. Divide 9.78 by 0.8 and round off the answer to 2 decimal places.   «4,4»

18. Betty has 785 milliliters of medicine to be placed in 5 milliliter bottles. How many bottles are needed for the medicine?   «4,4»

19. Mike is $\frac{2}{3}$ of the way through his 12 week course. How many weeks has Mike completed?   «3,7»

20. Change $\frac{7}{8}$ to a decimal fraction rounded off at two decimal places.   «4,4»

Unit 4:   ADDITION POSTULATES

Equivalent Open Expressions

In Unit 3, the definition for equivalent open expressions was given. By that definition, $x + 5$ would be equivalent to $5 + x$ if the two expressions had the same evaluations regardless of the number used to replace x.

In Unit 2, the equivalence of numerical expressions was studied. Recall that an addition expression such as

$$812 + (435 + 9,346)$$

was declared equivalent to any other addition expression involving the same three numbers, 812, 435, and 9,346.

$812 + (435 + 9,346)$ is equivalent to $(812 + 435) + 9,346$ and $812 + (9,346 + 435)$

In this unit, the knowledge of Units 2 and 3 will be used to study the basic properties of the counting numbers.

Definition	Postulate	
A mathematical statement is a **postulate**	**if and only if**	it makes a claim about numbers and/or their operations $(+, \bullet)$ and is assumed true.

Commutative Law of Addition

The Commutative Law of Addition is our first postulate for the algebra of the real numbers. It is concerned with expressions of the form x + y, in other words the addition of two numbers.

The Commutative Law of Addition

Any expression of the form x + y where x and y are variables is equivalent to the expression y + x.

$$x + y = y + x$$

We can actually compute the evaluations for numerical expressions such as

$$512 + 4,389 \qquad \text{and} \qquad 4,389 + 512$$

to show that they have the same evaluations and therefore are equivalent, but we cannot actually replace x with every possible number to show that

$$x + 31 \qquad \text{and} \qquad 31 + x$$

are equivalent.

The Commutative Law of Addition allows us to claim that x + 31 is equivalent to 31 + x.

Focus on the Commutative Law of Addition

The Commutative Law of Addition allows us to add two counting numbers by changing the order, with the assurance that the sum *will not* be changed.

For example, 3 + 7 will have the same sum as 7 + 3.

x + 5 is equivalent to 5 + x, because the Commutative Law of Addition states that the order of the addends in an addition problem has no effect on the sum.

2x + 7 is equivalent to 7 + 2x, because only the order of the addends for addition has been changed. 2x is the first addend of 2x + 7, and 7 is the first addend of 7 + 2x. The Commutative Law of Addition states that 2x + 7 is equivalent to 7 + 2x .

Associative Law of Addition

The Associative Law of Addition is our second postulate for algebra. It is concerned with expressions of the form $(x + y) + z$, in other words the addition of three numbers.

> **The Associative Law of Addition**
>
> **Any expression of the form $(x + y) + z$ where x, y, and z represent numbers is equivalent to the expression $x + (y + z)$.**
>
> $$(x + y) + z = x + (y + z)$$

We can actually compute the evaluations for numerical expressions such as

$$(36 + 12) + 89 \quad \text{and} \quad 36 + (12 + 89)$$

to show that they have the same evaluations and therefore are equivalent, but we cannot actually replace x with every possible number to show that

$$(x + 3) + 11 \quad \text{and} \quad x + (3 + 11)$$

are equivalent.

The Associative Law of Addition allows us to claim that $(x + 3) + 11$ is equivalent to $x + (3 + 11)$.

Focus on the Associative Law of Addition

The open expression $(3x + 2) + 5$ is equivalent to $3x + (2 + 5)$ by the Associative Law of Addition.

The Associative Law of Addition allows us to move parentheses in an addition expression without changing the evaluation.

$$(3x + 5) + 12 \text{ is equivalent to } 3x + (5 + 12)$$

$$9 + (x + 3y) \text{ is equivalent to } (9 + x) + 3y$$

Simplifying Addition Expressions

The Commutative Law of Addition allows the *order* of two
numbers to be *reversed.* $2 + 5 = 5 + 2$

The Associative Law of Addition allows the *grouping* of three
numbers to be changed. $(3 + 4) + 5 = 3 + (4 + 5)$

To simplify $(9 + x) + 3$ means to find an equivalent expression
in which there is only one addition indicated.

In general, *the simplification of open expressions means to find
an equivalent expression that requires fewer operations.*

*Focus on
simplifying
addition
expressions*

In simplifying open expressions with more than two terms, the Commutative
and Associative Laws of Addition allow us to change the order and grouping
of the addends.

To simplify $(7 + z) + 16$, we
change the order and the grouping
of the addends so that 7 can be added
to 16.

$$(7 + z) + 16$$
$$(z + 7) + 16$$
$$z + (7 + 16)$$
$$z + 23$$

The simplification of $(7 + z) + 16$ is $z + 23$. The two open expressions
are equivalent, but $z + 23$ requires only one addition and $(7 + z) + 16$
requires two additions.

More Simplifying of Addition Expressions

The open expression $3k + 8$ is not equivalent to $11k$. If
k is replaced by almost any number, $3k + 8$ and $11k$ will
have different evaluations.

Do not attempt to combine addends when one has a variable
and the other does not. Two numbers may be combined in addition,
but a number and a variable cannot be combined.

To simplify 5 + (3k + 8) the addends 5 and 8 need to be
combined. The Commutative and Associative Laws
allow the re-ordering and re-grouping shown below.

$$5 + (3k + 8)$$
$$(3k + 8) + 5$$
$$3k + (8 + 5)$$
$$3k + 13$$

5 + (3k + 8) simplifies to 3k + 13.

*Focus on
simplifying
addition
expressions*

To simplify any addition expression,
change the order and grouping of the
addends so that some combining can
occur.

$$7x + [5 + (4y + 6)]$$
$$7x + 4y + (5 + 6)$$
$$7x + 4y + 11$$

In the simplification shown above, the addition expression was re-arranged
so that 5 and 6 could be added. Notice that 7x + 4y could not be simplified.
Their variables are different. Just as 7 horses cannot be added to 4 cows,
7x cannot be added to 4y.

The Addition Law Of Zero

Besides the Commutative and Associative Laws of Addition,
there is one other important law of addition.

The whole number 0 has a very important addition property,
namely, that 0 added to any number will result in a sum of
that same number.

For example, 0 + 7 = 7 is a true statement. Similarly,
8 + 0 = 8 and 17 + 0 = 17 are true statements.

For all number replacements of x,

$$x + 0 \text{ is equivalent to } x$$

Focus
on the
Addition Law
of Zero

The Addition Law of Zero allows any variable to be written as 0 plus that variable.

Each of the equivalences shown at the right is justified by the Addition Law of Zero.

y is equivalent to $y + 0$

4x is equivalent to $4x + 0$

$(2x + 5)$ is equivalent to $(2x + 5) + 0$

xyz is equivalent to $xyz + 0$

$9 + z$ is equivalent to $(9 + z) + 0$

$(3 + 6k)$ is equivalent to $(3 + 6k) + 0$

Unit 4 Exercise

This exercise reviews the preceding unit. The exercise is divided into three parts. Part A reviews the foci of Unit 4. Part B offers opportunities to practice the skills and concepts of Unit 4. Part C contains problems that review your previous work in this text. All answers for Parts A and B are at the back of the book. Each problem of Part C is accompanied by a notation «CU» that refers to the Chapter (C) and Unit (U) in which that type of problem is studied.

Part A: Reviewing the foci of Unit 4.

1. Are $14 + x$ and $x + 14$ equivalent expressions?

2. Is $x + y$ equivalent to $y + x$ a true statement for all replacements for x and y?

3. According to the Commutative Law of Addition, $x + 93$ is equivalent to _____.

4. The Commutative Law of Addition states that $a + 3$ is equivalent to _____.

5. $11 + y$ is equivalent to _____ by the Commutative Law of Addition.

6. $17x + 3$ is equivalent to _____ by the Commutative Law of Addition.

7. The Commutative Law of Addition states that $3x + 9a$ is equivalent to _____.

8. Are $(x + 5) + 7$ and $x + (5 + 7)$ equivalent expressions?

9. $(8 + 17) + 3$ is equivalent to $8 + (17 + 3)$ because of the _____ Law of Addition.

10. The Associative Law of Addition states that $(x + 7) + p$ is equivalent to _____.

11. The Associative Law of Addition states that $(5x + 3) + 2b$ is equivalent to _____.

12. Is x equivalent to $x + 0$ for all counting-number replacements for x?

13. By the Addition Law of Zero, x is equivalent to _____.

Part B: Drill and Practice

1. $5x + (7 + 4)$ is equivalent to _____ by the Associative Law of Addition.

2. $(7 + 3) + 9b$ is equivalent to _____ by the Associative Law of Addition.

3. $x + 7$ is equivalent to _____ by the Commutative Law of Addition.

4. $(7x + 2x) + 9$ is equivalent to _____ by the Associative Law of Addition.

5. $(6 + x) + 5$ is equivalent to $(x + 6) + 5$ by the _____ Law of Addition.

6. $(8 + x) + 3$ is equivalent to $(x + 8) + 3$ because of the _____ Law of Addition.

7. $9 + a$ is equivalent to $a + 9$ because of the _____ Law of Addition.

8. $(a + b) + c$ is equivalent to $(b + a) + c$ because of the _____ Law of Addition.

9. $b + 10$ is equivalent to $10 + b$ because of the _____ Law of Addition.

10. $(7 + a) + 6$ is equivalent to $7 + (a + 6)$ because of the _____ Law of Addition.

11. $(x + y) + z$ is equivalent to $x + (y + z)$ is a statement of the _____ Law of Addition.

12. The Commutative Law of Addition states $x + y$ is equivalent to _____.

13. The Associative Law of Addition states $(x + y) + z$ is equivalent to _____.

14. By the Addition Law of Zero, $8x + 0$ is equivalent to _____.

15. By the Addition Law of Zero,
    $(2x + 3)$ is equivalent to _____.

16. By the Addition Law of Zero,
    $5x$ is equivalent to _____.

17. By the Addition Law of Zero,
    $0 + (3x + 4)$
    is equivalent to _____.

Simplify.

18. $7 + (b + 15)$

19. $(x + 12) + 6$

20. $(4 + x) + 7$

21. $9 + (g + 8)$

22. $(5 + r) + 17$

23. $(d + 8) + 5$

24. $(9 + a) + 3$

25. $9 + (r + 3)$

26. $(a + 4) + 9$

27. $(6 + a) + 11$

28. $8 + (c + 4)$

29. $(23 + b) + 15$

30. $8 + (3h + 16)$

31. $(4 + x) + 8$

32. $3 + (5 + y)$

33. $4 + (3x + 7)$

34. $(5 + 2x) + 7$

35. $5 + (2x + 7)$

36. $(x + 13) + 8$

37. $(8 + 2x) + 5$

38. $(y + 8) + 12$

39. $5 + (3x + 8)$

40. $(5 + 3x) + 8$

Part C: Review. Answers for the problems of Part C are not given. However, the notation «C,U» refers to the Chapter (C) and Unit (U) in which problems of the same type were presented.

1. 58,346 + 8,353 + 22,978   «1,2»

2. 4,714 x 73   «2,2»

3. Simplify $\frac{18}{36}$ «3,1»

4. Write $\frac{29}{7}$ as a mixed number.   «3,2»

5. Find the numerator of $\frac{3}{4} = \frac{?}{36}$ «3,2»

6. $5\frac{5}{6} + 3\frac{7}{12}$ «3,3»

7. $5\frac{1}{3} \times \frac{3}{4}$ «3,5»

8. Write $\frac{579}{10000}$ as a decimal fraction.   «4,1»

9. Add 3.817, 45.19, and 0.843   «4,2»

10. Evaluate 6 + [9 • 2 + 7]   «5,2»

11. Evaluate 6 • (x + 4) when x = 3   «5,3»

12. Evaluate x + 3y when x = 5 and y = 3   «5,3»

13. Find the sum of 8,403 and 986. «1,4»

14. Find the difference of 56.09 and 65.8   «4,2»

15. Find the product of 4,008 and 59. «2,2»

16. Find 0.15 of 4,850. «4,3»

17. How many $\frac{1}{5}$ miles are there in 12 miles?   «3,7»

18. Evaluate 18 ÷ (12 ÷ 4)   «5,1»

19. Evaluate 36 ÷ [(40 ÷ 5) ÷ 2]   «5,1»

20. What is the difference between 6.93 and 3.059?   «4,2»

21. Each Boy Scout in Troop 412 sold 95 candy bars during their fund drive. If Troop 412 has 23 Boy Scouts, how many candy bars were sold?   «4,3»

22. Change $\frac{4}{7}$ to a decimal fraction rounded off at two decimal places.   «4,4»

23. Carla finds that she owes 0.28 of her income in taxes. If her yearly salary is $22,000, how much does she owe in taxes for the year?   «4,4»

Unit 5:   MULTIPLICATION POSTULATES

Equivalent Multiplication Expressions

In Unit 2, the equivalence of numerical expressions involving
multiplication was studied. Recall that multiplication expressions
such as

$$812 \cdot (435 \cdot 9{,}346)$$

were declared equivalent to any other multiplication expression
involving the same three factors, 812, 435, and 9,346.

$812 \cdot (435 \cdot 9{,}346)$ is equivalent to $812 \cdot (9{,}346 \cdot 435)$ and
$(812 \cdot 435) \cdot 9{,}346$

In this unit, the postulates for algebra are expanded
as we add statements concerning multiplication.

Commutative Law of Multiplication

The Commutative Law of Multiplication is our first postulate
of multiplication for algebra. It is concerned with expressions
of the form $x \cdot y$, in other words the multiplication of two numbers.

The Commutative Law of Multiplication

**Any expression of the form xy where x and y
are variables is equivalent to the expression yx.**

$$xy \; = \; yx$$

We can actually compute the evaluations for numerical
expressions such as

$512 \cdot 4{,}389$ and $4{,}389 \cdot 512$

to show that they have the same evaluations and therefore are
equivalent, but we cannot actually replace x with every possible
number to show that

$x \cdot 31$ and $31x$

are equivalent.

The Commutative Law of Multiplication allows us to claim
that $x \cdot 31$ is equivalent to $31x$.

Focus on the Commutative Law of Multiplication

The Commutative Law of Multiplication allows us to multiply two counting numbers by changing the order of the numbers without changing the result.

The Commutative Law of Multiplication states that the order of the factors can be changed with the assurance that the product will not be changed. For example, $3 \cdot 7$ will have the same product as $7 \cdot 3$.

$x \cdot 5$ is equivalent to $5x$, because the Commutative Law of Multiplication states that the order of the factors in a multiplication problem has no effect on the product.

$2x \cdot 7$ is equivalent to $7 \cdot 2x$, because only the order of the factors for multiplication has been changed. $2x$ is the first factor of $2x \cdot 7$, and 7 is the first factor of $7 \cdot 2x$. The Commutative Law of Multiplication states that $2x \cdot 7$ is equivalent to $7 \cdot 2x$.

Associative Law of Multiplication

The Associative Law of Multiplication is concerned with expressions of the form $(xy)z$, in other words the multiplication of three numbers.

The Associative Law of Multiplication

Any expression of the form $(xy)z$ where x, y, and z represent numbers is equivalent to the expression $x(yz)$.

$$(xy)z = x(yz)$$

We can actually compute the evaluations for numerical expressions such as

$\qquad (36 \cdot 12) \cdot 89 \qquad$ and $\qquad 36 \cdot (12 \cdot 89)$

to show that they have the same evaluations and therefore are equivalent, but we cannot actually replace x with every possible number to show that

$\qquad (x \cdot 3) \cdot 11 \qquad$ and $\qquad x \cdot (3 \cdot 11)$

are equivalent.

The Associative Law of Multiplication allows us to claim that $(x \cdot 3) \cdot 11$ is equivalent to $x \cdot (3 \cdot 11)$.

Focus on the Associative Law of Multiplication

The open expression $(3x \cdot 2) \cdot 5$ is equivalent to $3x \cdot (2 \cdot 5)$ by the Associative Law of Multiplication.

The Associative Law of Multiplication allows us to move parentheses in an multiplication expression without changing the final evaluation.

$(3x \cdot 5) \cdot 12$ is equivalent to $3x \cdot (5 \cdot 12)$

$9 \cdot (x \cdot 3y)$ is equivalent to $(9x) \cdot 3y$

Simplifying Multiplication Expressions

The Commutative Law of Multiplication allows the *order* of two factors to be *reversed.* $xy = yx$

The Associative Law of Multiplication allows the *grouping* of three factors to be changed. $x(yz) = (xy)z$

To simplify $(9 \cdot x) \cdot 3$ means to find an equivalent expression in which there is only one multiplication indicated.

In general, *the simplification of open expressions means to find an equivalent expression that requires fewer operations.*

Focus on simplifying multiplication expressions

To simplify the expression $8 \cdot (4x)$, we use the Associative Law of Multiplication as shown below.

$$8 \cdot (4x)$$
$$(8 \cdot 4)x$$
$$32x$$

The Commutative Law is used to change the order of the numbers to be multiplied. The Associative Law is used to change the grouping of the numbers to be multiplied. To simplify the expression $(3r) \cdot 5$, the following steps are used.

$$(3r) \cdot 5$$
$$5 \cdot (3r)$$
$$(5 \cdot 3)r$$
$$15r$$

More Simplifying of Multiplication Expressions

To simplify $5 \cdot (3k \cdot 8)$ the factors 5, 3, and 8 need to be combined. The Commutative and Associative Laws allow the re-ordering and re-grouping shown below.

$$5 \cdot (3k \cdot 8)$$
$$5 \cdot (8 \cdot 3k)$$
$$(5 \cdot 8) \cdot 3k$$
$$40 \cdot 3k$$
$$(40 \cdot 3)k$$
$$120k$$

$5 \cdot (3k \cdot 8)$ simplifies to 120k.

Focus on simplifying multiplication expressions

To simplify any multiplication expression, change the order and grouping of the factors so that some multiplication can occur.

$$6x \cdot (4y \cdot 2)$$
$$[6 \cdot 4 \cdot 2] \cdot (x \cdot y)$$
$$[48] \cdot xy$$
$$48xy$$

In the simplification shown above, the multiplication expression was re-arranged so that 6, 4, and 2 could be multiplied. The variable factors are multiplied just by arranging them as xy.

The Multiplication Law of One

Besides the Commutative and Associative Laws of Multiplication, there is one other important law of multiplication.

The counting number 1 has a very important multiplication property, namely, that 1 multiplied by any number will result in a product of that same number.

For example, $1 \cdot 7 = 7$ is a true statement. Similarly, $1 \cdot 8 = 8$ and $1 \cdot 17 = 17$ are true statements.

For all number replacements of x,

$1 \cdot x$ or 1x is equivalent to x.

Focus on the Multiplication Law of One

The Multiplication Law of One allows any variable to be written as 1 times that variable.

Each of the equivalences shown at the right is justified by the Multiplication Law of One.

y is equivalent to 1y
4x + x is equivalent to 4x + 1x
(2x + 5) is equivalent to 1(2x + 5)
xyz is equivalent to 1xyz
9 + z is equivalent to 9 + 1z
6 + (3 + k) is equivalent to 6 + 1(3 + k)

Note that any expression or any term of an expression can be multiplied by 1 and the evaluation of the expression is not changed.

Unit 5 Exercise

This exercise reviews the preceding unit. The exercise is divided into three parts. Part A reviews the foci of Unit 5. Part B offers opportunities to practice the skills and concepts of Unit 5. Part C contains problems that review your previous work in this text. All answers for Parts A and B are at the back of the book. Each problem of Part C is accompanied by a notation «CU» that refers to the Chapter (C) and Unit (U) in which that type of problem is studied.

Part A: Reviewing the foci of Unit 5.

1. Are 14x and x • 14 equivalent expressions?

2. Is xy equivalent to yx for all replacements for x and y?

3. According to the Commutative Law of Multiplication, x • 93 is equivalent to _____.

4. The Commutative Law of Multiplication states that a • 3 is equivalent to _____.

5. 11y is equivalent to _____ by the Commutative Law of Multiplication.

6. 17x • 3 is equivalent to _____ by the Commutative Law of Multiplication.

7. The Commutative Law of Multiplication states that 3x • 9a is equivalent to _____.

8. Are (x • 5) • 7 and x • (5 • 7) equivalent expressions?

9. $(8 \cdot 17) \cdot 3$ is equivalent to $8 \cdot (17 \cdot 3)$ because of the _____ Law of Multiplication.

10. The Associative Law of Multiplication states that $(x \cdot 7) \cdot p$ is equivalent to _____.

11. The Associative Law of Addition states that $(5x \cdot 3) \cdot 2b$ is equivalent to _____.

12. Is $1x$ equivalent to x for all counting-number replacements for x?

13. By the Multiplication Law of One 1, x is equivalent to _____.

Part B: Drill and Practice

1. $5x \cdot (7 \cdot 4)$ is equivalent to _____ by the Associative Law of Multiplication.

2. $(7 \cdot 3) \cdot 9b$ is equivalent to _____ by the Associative Law of Multiplication.

3. $x \cdot 7$ is equivalent to _____ by the Commutative Law of Multiplication.

4. $(7x \cdot 2x) \cdot 9$ is equivalent to _____ by the Associative Law of Multiplication.

5. $(6 \cdot x) \cdot 5$ is equivalent to $(x \cdot 6) \cdot 5$ by the _____ Law of Multiplication.

6. $(8 \cdot x) \cdot 3$ is equivalent to $(x \cdot 8) \cdot 3$ because of the _____ Law of Multiplication.

7. $9 \cdot a$ is equivalent to $a \cdot 9$ because of the _____ Law of Multiplication.

8. $(a \cdot b) \cdot c$ is equivalent to $(b \cdot a) \cdot c$ because of the _____ Law of Multiplication.

9. $b \cdot 10$ is equivalent to $10 \cdot b$ because of the _____ Law of Multiplication.

10. $(7 \cdot a) \cdot 6$ is equivalent to $7 \cdot (a \cdot 6)$ because of the _____ Law of Multiplication.

11. $(x \cdot y) \cdot z$ is equivalent to $x \cdot (y \cdot z)$ is a statement of the _____ Law of Multiplication.

12. The Commutative Law of Multiplication states $x \cdot y$ is equivalent to _____.

13. The Associative Law of Multiplication states $(x \cdot y) \cdot z$ is equivalent to _____.

14. By the Multiplication Law of One, $8x + x$ is equivalent to _____.

15. By the Multiplication Law of One, $1(2x + 3)$ is equivalent to _____.

16. By the Multiplication Law of One, $x + 5x$ is equivalent to _____.

17. By the Multiplication Law of One, $8 + (3x + 4)$ is equivalent to _____.

Simplify.

18. $7 \cdot (b \cdot 15)$

19. $(x \cdot 12) \cdot 6$

20. $(4 \cdot x) \cdot 7$

21. $9 \cdot (g \cdot 8)$

22. $(5 \cdot r) \cdot 17$

23. $(d \cdot 8) \cdot 5$

24. $8 \cdot (c \cdot 4)$

25. $(23 \cdot b) \cdot 15$

26. $8 \cdot (3h \cdot 16)$

27. $(4 \cdot x) \cdot 8$

28. $3 \cdot (5 \cdot y)$

29. $4 \cdot (3x \cdot 7)$

30. $9 \cdot (5x)$

31. $(4r) \cdot 6$

32. $6 \cdot (8x)$

33. $(5x) \cdot 9$

Part C: Review. Answers for the problems of Part C are not given. However, the notation «C,U» refers to the Chapter (C) and Unit (U) in which problems of the same type were presented.

1. What is the ten thousands digit of 4,825,369? «1,1»

2. 86,407 − 36,084   «1,3»

3. 27) 45348   «2,4»

4. Find the lcm for 8 and 10.   «3,2»

5. Select the smaller fraction from $\frac{3}{8}, \frac{2}{5}$ «3,2»

6. Change $3\frac{5}{11}$ to an improper fraction.   «3,2»

7. $6\frac{5}{9} - 4\frac{5}{6}$ «3,4»

8. $4\frac{4}{7} \div 1\frac{3}{5}$ «3,6»

9. Write 0.087 as a fraction.   «4,1»

10. Subtract 6.04 from 15.4   «4,2»

11. Evaluate 27 − [9 − (25 − 19)]   «5,1»

12. Evaluate (9 + 3 • 6) + 4   «5,2»

13. Evaluate 2x + 4 when x = 7   «5,3»

14. Evaluate 5 + xy when x = 4 and y = 7   «5,3»

15. Simplify (4 + x) + 8   «5,4»

16. Simplify 3 + (5 + y)   «5,4»

17. Find the sum of 4,870 and 537.   «1,4»

18. Find the total of 4.803 and 19.62   «4,2»

19. Divide 6.08 by 0.9 and round off the answer to 2 decimal places.   «4,4»

20. Mike is on a diet and this week ate only 25,500 calories? What was his average daily calory intake?   «4,4»

21. Shirley estimates that she spends $\frac{7}{10}$ of her waking hours at work. In a week when she was awake 100 hours, how many hours does she estimate she works? «3,7»

22. Change $\frac{2}{3}$ to a decimal fraction rounded off at two decimal places.   «4,4»

Unit 6:   THE DISTRIBUTIVE LAW OF MULTIPLICATION OVER ADDITION

Equivalent Numerical Expressions

Recall from Unit 2 that expressions such as

$$8 \cdot 2 + 8 \cdot 5 \quad \text{and} \quad 8(2 + 5)$$

were evaluated.

$$
\begin{array}{ccc}
8 \cdot 2 + 8 \cdot 5 & \text{and} & 8(2 + 5) \\
16 + 40 & & 8 \cdot 7 \\
56 & & 56
\end{array}
$$

It is no coincidence that $8 \cdot 2 + 8 \cdot 5$ and $8(2 + 5)$ have the same evaluations. Later in this unit there will be another postulate for algebra based on the patterns of these two expressions.

Focus on evaluating a numerical expression

To evaluate $7(6 + 4)$ the first step is the addition of 6 and 4.

$$
\begin{array}{c}
7(6 + 4) \\
7 \cdot 10 \\
70
\end{array}
$$

To evaluate $7 \cdot 6 + 7 \cdot 4$ the first step is the multiplications of 7 and 6 and 7 and 4.

$$
\begin{array}{c}
7 \cdot 6 + 7 \cdot 4 \\
42 + 28 \\
70
\end{array}
$$

Notice that $7(6 + 4)$ and $7 \cdot 6 + 7 \cdot 4$ have the same evaluation, 70. They both involve the same numbers, but in $7(6 + 4)$ the addition is done first and in $7 \cdot 6 + 7 \cdot 4$ the multiplications are done first.

The Distributive Law of Multiplication Over Addition

The pattern of the numbers in

$$7(6 + 4) \quad \text{and} \quad 7 \cdot 6 + 7 \cdot 4$$

can be described by the open expressions

$$x(y + z) \quad \text{and} \quad xy + xz$$

The one and only postulate which deals with expressions involving both addition and multiplication is the Distributive Law of Multiplication over Addition.

> **The Distributive Law of Multiplication over Addition**
>
> $x(y + z)$ **is equivalent to** $xy + xz$
> **for all number replacements**
> **of x, y, and z.**

Focus on equivalent open expressions

The Distributive Law of Multiplication over Addition allows many open expressions in algebra to be simplified. Six examples are shown at the right.

$5(x + 7)$ is equivalent to $5x + 5 \cdot 7$
$x(8 + 3)$ is equivalent to $x \cdot 8 + x \cdot 3$
$4(3x + 5)$ is equivalent to $4 \cdot 3x + 4 \cdot 5$
$6x + 2x$ is equivalent to $x(6 + 2)$
$8x + 12$ is equivalent to $4 \cdot 2x + 4 \cdot 3$
$3x + 5x$ is equivalent to $x(3 + 5)$

Removing Parentheses

The Distributive Law of Multiplication over Addition allows us to remove the parentheses from open expressions such as $7(x + 4)$.

To remove the parentheses from the expression $7(x + 4)$, the following steps are used.

$$7(x + 4)$$
$$7x + 7 \cdot 4$$
$$7x + 28$$

$7(x + 4)$ and $7x + 28$ are equivalent expressions.

Focus on removing parentheses

To remove the parentheses from the expression $3(2x + 5)$, the Distributive Law of Multiplication over Addition is used.

$$3(2x + 5)$$

$$3 \cdot 2x + 3 \cdot 5$$

$$6x + 15$$

Each term of $(2x + 5)$ is multiplied by 3.

Definition   Like Terms

In an addition expression two terms are **like terms**   **If and only if**   the terms have exactly the same variable factors.

For example, $6x + 2x$ involves the **like terms** $6x$ and $2x$. The terms are **like** because they both contain the same variable factor, x.

Adding Like Terms

The open expression $8x + 3x$ can be simplified using the Distributive Law of Multiplication over Addition. This is because 8 is being multiplied by x and 3 is being multiplied by x. The two multiplications have x as a common multiplier.

The simplification of $8x + 3x$ is shown by the following steps.

$$8x + 3x$$
$$(8 + 3)x$$
$$11x$$

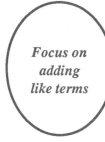

Focus on adding like terms

To simplify addition expressions look for like terms, group them, and then use the Distributive Law to simplify those groupings.

In the example at the right,
  1. The like terms were grouped
  2. The like terms were simplified.

$$5x + 3y + 7x + 8y$$

1. $(5x + 7x) + (3y + 8y)$

2. $(5 + 7)x + (3 + 8)y$

$$12x + 11y$$

Using the Multiplication Law of One

To simplify $7x + x$, the Distributive Law of Multiplication over Addition and the Multiplication Law of One are used. x is the same as 1x by The Multiplication Law of One. Consequently, $7x + x$ is equivalent to $7x + 1x$.

$$7x + x$$
$$7x + 1x$$
$$(7 + 1)x$$
$$8x$$

$7x + x$ is simplified to 8x.

Focus on simplifying using 1x instead of x

To simplify the expression shown at the right, the like terms are grouped, x is written as 1x, and the terms are added.

$$9x + 3y + x + 4y$$

$$(9x + 1x) + (3y + 4y)$$

$$10x + 7y$$

Simplifying Addition Expressions

To simplify $3y + 5 + 4y + 6$ the like terms are grouped.
Notice that 5 and 6 are considered like terms.

$$3y + 5 + 4y + 6$$
$$(5 + 6) + (3y + 4y)$$
$$11 + 7y$$

Notice that $11 + 7y$ *cannot* be further simplified because 11
and 7y are not like terms.

Focus on simplifying addition expressions

The addition expression shown
at the right has four different
types of terms. Like terms
are grouped together and those
terms that are unlike any others
must be left alone.

$$6x + 7 + 3y + 4z + 2x + 9$$

$$(6x + 2x) + 3y + 4z + (7 + 9)$$

$$8x + 3y + 4z + 16$$

Simplifying Expressions With Parentheses

To simplify the expression $6 + 3(2x + 4) + 5x$, the
parentheses should be removed first as shown in the
following steps.

$$6 + 3(2x + 4) + 5x$$
$$6 + 3 \cdot 2x + 3 \cdot 4 + 5x$$
$$6 + 6x + 12 + 5x$$
$$(6x + 5x) + (6 + 12)$$
$$11x + 18$$

Focus on simplifying expressions with parentheses

To simplify an open expression involving parentheses, use the Distributive Law of Multiplication over Addition to remove the parentheses.

$$3x + 5(x + 4) + 7$$

$$3x + 5x + 20 + 7$$

$$8x + 27$$

Focus on expressions with two parentheses

The open expression
$$3(2x + 5) + (5x + 1)$$
contains two sets of parentheses.

The first parentheses has 3 as its multiplier. The second parentheses has 1 as its multiplier.

The steps in simplifying the expression are shown at the right.

$$3(2x + 5) + (5x + 1)$$

$$3(2x + 5) + 1(5x + 1)$$

$$3 \cdot 2x + 3 \cdot 5 + 1 \cdot 5x + 1 \cdot 1$$

$$6x + 15 + 5x + 1$$

$$(6x + 5x) + (15 + 1)$$

$$11x + 16$$

The open expression
$$(4x + 7) + 3(2x + 5)$$
contains two sets of parentheses.

The first parentheses has 1 as its multiplier. The second parentheses has 3 as its multiplier.

The steps in simplifying the expression are shown at the right.

$$(4x + 7) + 3(2x + 5)$$

$$1(4x + 7) + 3(2x + 5)$$

$$1 \cdot 4x + 1 \cdot 7 + 3 \cdot 2x + 3 \cdot 5$$

$$4x + 7 + 6x + 15$$

$$(4x + 6x) + (7 + 15)$$

$$10x + 22$$

Unit 6 Exercise

This exercise reviews the preceding unit. The exercise is divided into three parts. Part A reviews the foci of Unit 6. Part B offers opportunities to practice the skills and concepts of Unit 6. Part C contains problems that review your previous work in this text. All answers for Parts A and B are at the back of the book. Each problem of Part C is accompanied by a notation «CU» that refers to the Chapter (C) and Unit (U) in which that type of problem is studied.

Part A: Reviewing the foci of Unit 6.

1. Is the evaluation of $3 \cdot 4 + 3 \cdot 6$ the same as the evaluation of $3(4 + 6)$?

2. Is the open expression $a(b + c)$ equivalent to $ab + ac$?

3. The Distributive Law of Multiplication over Addition guarantees that $7(8 + 5)$ has the same evaluation as _____.

4. The Distributive Law of Multiplication over Addition guarantees that $6 \cdot 8 + 6 \cdot 3$ has the same evaluation as _____.

5. Simplify $9z + 6z$ by using the Distributive Law of Multiplication over Addition.

6. Simplify $5a + a$ using the Multiplication Law of One and the Distributive Law of Multiplication over Addition.

7. Simplify $2 + 6x + 3x$ by adding like terms.

8. Simplify $7r + 4 + 5r$ by adding like terms.

9. Simplify $4 + 9x + 8$ by adding like terms.

10. Simplify $8x + 4 + 2x$ by adding like terms.

11. Remove the parentheses from $3(y + 5)$ by multiplying both y and 5 by 3.

12. Remove the parentheses from $7(x + 3)$ by multiplying each term of $(x + 3)$ by 7.

13. Remove the parentheses from $5(4x + 3)$.

14. Remove the parentheses from $(x + 7)$.

15. Simplify $2x + 5(3x + 2) + 3$ by first removing the parentheses.

16. Simplify $9 + 2(9x + 1) + 3x$ by first removing the parentheses.

Part B: Drill and Practice

Remove the parentheses.

1. $7(2 + y)$

2. $10(m + 2)$

3. $7(y + 9)$

4. $3(1 + c)$

5. $6(6 + x)$

6. $8(3 + a)$

7. $(5 + y)$

8. $6(m + 5)$

9. $9(2 + e)$

10. $6(7 + r)$

11. $3(z + 6)$

12. $2(3 + 7x)$

13. $4(3x + 2)$

14. $(4x + 7)$

15. $6(9x + 4)$

Simplify.

16. $12a + 7a$.

17. $18x + 13x$

18. $4x + x$

19. $r + 12r$

20. $19x + 4x$

21. $x + x$

22. $3m + 9m$

23. $8x + 4 + 3x + 9$

24. $9 + 3x + 2 + 5x$.

25. $12x + 3(7 + 2x) + 9$

26. $5(4x + 1) + 4(6x + 8)$

27. $(2a + 7) + 6a$

28. $5x + (3x + 7)$

29. $(17a + 6a) + 2a$

30. $3m + (8m + 6m)$

31. $12x + (8x + 3x)$

32. $3a + (2a + 7)$

33. $(5y + 6y) + 4y$

34. $(18t + 3t) + 6t$

35. $(5 + 7x) + 9$

36. $8 + 3x + 5$

37. $8 + 12 + 9x$

38. $5x + 7 + 12$

39. $9t + 7 + 4t$

40. $9x + 3 + 7x$

41. $5x + 7x + 15$

42. $17 + 3x + 5x$

43. $9 + 3x + 2 + 5x$

44. $2x + 7 + 3x + 2$

45. $4 + 8x + x + 3$

46. $5x + 7 + 9 + x$

47 $3x + 2 + 4x + 5 + 6x$

48. $7 + (2x + 3) + 5x$

49. $3 + 7(2x + 4) + 3x$

50. $5(3x + 1) + 2(4x + 3)$

51. $8(x + 7) + 5(2x + 3)$

52. $3(3x + 4) + 2(x + 7)$

53. $6(3x + 2) + (x + 3)$

54. $(2x + 5) + 3(x + 4)$

55. $(2 + x) + 5$

56. $(2x + 5) + 8x$

57. $5(x + 6)$

58. $2x + 3(5 + x)$

59. $2 + (4x + 3) + 7x$

Part C: Review. Answers for the problems of Part C are not given. However, the notation «C,U» refers to the Chapter (C) and Unit (U) in which problems of the same type were presented.

1. 4,976 + 48,353 + 2,978   «1,2»

2. 5,614 x 47   «2,2»

3. Simplify $\frac{15}{50}$ «3,1»

4. Write $\frac{23}{5}$ as a mixed number.   «3,2»

5. Find the numerator of $\frac{5}{9} = \frac{?}{36}$ «3,2»

6. $5\frac{2}{3} + 3\frac{7}{12}$ «3,3»

7. $6\frac{1}{4} \times \frac{1}{5}$ «3,5»

8. Write $\frac{509}{10000}$ as a decimal fraction.   «4,1»

9. Add 4.877, 46.89, and 0.045   «4,2»

10. Evaluate 6 + [4 + 9 • 2]   «5,2»

11. Evaluate 6 • (x + 2) when x = 3   «5,3»

12. Evaluate 2x + 3y when x = 4 and y = 3
    «5,3»

13. Find the difference of 85.09 and 63.8   «4,2»

14. Find the product of 4.008 and 5.9   «2,2»

15. Find 0.28 of 350.   «4,3»

16. Evaluate 48 ÷ [(52 ÷ 13) ÷ 2]   «5,1»

17. Simplify (23 + b) + 15   «5,4»

18. Simplify 8 + (3h + 16)   «5,4»

19. Simplify (4 • x) • 7   «5,5»

20. Simplify 9 • (g • 8)   «5,5»

21. How many $\frac{1}{4}$ inches are there in one foot (12 inches)?.   «3,7»

22. What is the difference between 14.6 and 3.059? «4,2»

23. Find the average of 6.83 and 9.51   «4,4»

24. Change $\frac{1}{9}$ to a decimal fraction rounded off at two decimal places.   «4,4»

Unit 7:   SOLVING EQUATIONS

> **Definition   Mathematical Statement**
>
> A sentence is a   **If and**   the sentence is true
> **mathematical**   **only if**   or false, but not
> **statement**   both.

Mathematical Statements

By the definition, any sentence that makes a claim, true or false
is a **mathematical statement.**

Sentences such as:

> Today is Tuesday.
> George Washington was the first President of the U.S.
> The number 5 is greater than 3.

are, by the definition, **mathematical statements**.

Sentences such as:

> Go to the store.
> Is it cold outside?
> Wow!

are, by the definition, not mathematical statements because
they are neither true nor false.

> **Definition   Equality**
>
> A mathematical   **If and**   the action verb
> statement is an   **only if**   is "equals."
> **equality**

Equalities

The sentence "5 equals 3 plus 2" is a mathematical statement which uses the action verb equals.

"5 equals 3 plus 2" is an **equality**.

Commonly, we write equalities in symbols.

"5 equals 3 plus 2"
would be written as:
5 = 3 + 2

Another equality written in symbols is: $4 + 6 = 2 \cdot 5$
This equality consists of two numerical expressions,
$4 + 6$ and $2 \cdot 5$, joined by the action verb "equals."

$4 + 6 = 2 \cdot 5$
left side   right side

$4 + 6$ is called the left side of the equality, and
$2 \cdot 5$ is called the right side of the equality.

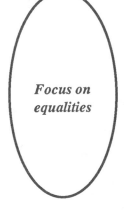

*Focus on
equalities*

An equality can be either true or false. The equality $4 + 6 = 2 \cdot 5$ is true because the numerical expression on the left side has the same evaluation as the numerical expression on the right side.

The equality $8 \cdot 3 + 7 = 8(3 + 7)$
is a false equality. As the work
at the right shows, $8 \cdot 3 + 7$ has
an evaluation of 31 and $8(3 + 7)$
has an evaluation of 80.

$8 \cdot 3 + 7 = 8(3 + 7)$

$24 + 7$ $8(10)$

31 80

The equality $8 \cdot 3 + 7 = 8(3 + 7)$ claims that the evaluation of $8 \cdot 3 + 7$ is equal to the evaluation of $8(3 + 7)$. The claim is false.

> **Definition   Equation**
>
> A sentence is   **If and**   it is an equality in
> an **equation**   **only if**   which one or more of
>   the numbers have been
>   replaced by variables.
>
> $8 + 5 = 13$ is an equality. If 8 is replaced by the
> variable x, the result is an equation: $x + 5 = 13$

Equations

$4 + 6 = 10$ is an equality, a true equality. If 4 is replaced
by x the result is an equation: $x + 6 = 10$.

$x + 6 = 10$ is neither true nor false. When x is replaced by
a number the equation becomes an equality. Then it will be
true or false depending upon which number is chosen.

If $x = 7$ then $x + 6 = 10$ becomes the
false equality $7 + 6 = 10$.

If $x = 4$ then $x + 6 = 10$ becomes the
true equality $4 + 6 = 10$.

*Focus on
equations*

$3 + w = 12$ is an equation.

If w is replaced by 9, the equation becomes the equality $3 + 9 = 12$.
The equality $3 + 9 = 12$ is true.

If w is replaced by 6 in the equation $3 + w = 12$, the statement $3 + 6 = 12$
is obtained, which is false.

For the equation $2x + 4 = 14$ if x is replaced by 5, the true statement
$2 \cdot 5 + 4 = 14$ is obtained.

Solving Equations

To solve an equation is to find a number or numbers that can replace the variable and result in a true equality.

To solve $x + 7 = 10$ a number replacement for the variable x is needed so that the number plus 7 will give 10. 3 is the needed replacement because $3 + 7 = 10$ is a true equality.

To solve $6x = 24$ a number replacement for the variable x is needed so that 6 times the number will give 24. 4 is the needed replacement because $6 \cdot 4 = 24$ is a true equality.

Focus on finding a replacement for x to solve simple equations

To find a replacement for x that will make the open sentence $x + 7 = 12$ true, we ask the question "What number added to 7 will give 12?" The correct answer is 5.

$$x + 7 \qquad = \qquad 12$$

What plus 7   gives   12?

To find the replacement for x to make $3x = 18$ a true statement, we ask the question "3 multiplied by what number will give a product of 18?" The correct number is 6.

$$3x \qquad = \qquad 18$$

3 times what   gives   18?

Definition		**Root or Solution**
A number is a **root** or **solution** of an equation	**if and only if**	the number used as a replacement for the variable results in a true equality.

6 is a **root** of $3x = 18$ because $3 \cdot 6 = 18$ is true.

5 is a **solution** of $x + 7 = 12$ because $5 + 7 = 12$ is true.

Using Simplification to Solve an Equation

The question associated with $(x + 7) + 3 = 19$ is:

> What number can be added to 7 and then
> the sum added to 3 to give 19?

That question is unnecessarily difficult. By
simplifying the left side of the equation a far easier
question can be answered.

To solve $(x + 7) + 3 = 19$, first simplify the expression
$(x + 7) + 3$. The following steps are used.

$$(x + 7) + 3 = 19$$
$$x + 10 = 19$$

The question now is:

> What number plus 10 gives 19?

The correct solution is 9.

*Focus on
simplifying
to solve
equations*

To solve $27 = 7x + 2x$, we first
must simplify the right member
of the equation, $7x + 2x$.

$$27 = 7x + 2x$$
$$27 = 9x$$

This simplification results in a far easier equation, $27 = 9x$, which has 3
as its solution.

To solve $2(3x) = 4 \cdot 7 + 2$, both
sides of this equation must first be
simplified.

$$2(3x) = 4 \cdot 7 + 2$$
$$6x = 28 + 2$$
$$6x = 30$$

The simplification results in a far easier equation, $6x = 30$, which has 5
as its solution.

Unit 7 Exercise

This exercise reviews the preceding unit. The exercise is divided into three parts. Part A reviews the foci of Unit 7. Part B offers opportunities to practice the skills and concepts of Unit 7. Part C contains problems that review your previous work in this text. All answers for Parts A and B are at the back of the book. Each problem of Part C is accompanied by a notation «CU» that refers to the Chapter (C) and Unit (U) in which that type of problem is studied.

Part A: Reviewing the foci of Unit 7.

1. $5x + 2 = 19$ is a(n) _____.

2. $x + 3 = 9$ is a(n) _____ .

3. If c is replaced by 6 in the equation $12 + c = 18$, the equality $12 + 6 = 18$ is _____. (true or false)

4. If x is replaced by 8 in the equation $2x = 12$, the equality is _____. (true or false)

5. If x is replaced by 2 in the equation $4x + 3 = 7 + 2x$, the equality is _____. (true or false)

6. If x is replaced by 3 in the equation $7x + 6 = 27$, is the equality obtained true?

7. Solve the equation $9 + x = 13$, by correctly answering the question "9 plus what number will give a sum of 13?"

8. Solve the equation $2x = 10$ by correctly answering the question "2 multiplied by what number will give 10?"

9. Solve the equation $36 = 9x$ by asking the question "36 is equal to 9 times what number?"

10. For the equation $15 + x = 18$, find a replacement for x to make a true equality.

11. For the equation $5x = 20$, find a replacement for x that will make the equality true.

12. Find the number replacement for y in the equation $y + 7 = 9$ that will make the equality true.

13. Solve $2 + r + 7 = 12$ by first simplifying $2 + r + 7$.

14. Solve $3x + 2x = 30$ by first simplifying $3x + 2x$.

15. Solve $(5x) \cdot 2 = 8 \cdot 7 + 4$ by first simplifying both sides of the equation.

16. Solve $36 = 8x + 4x$ by first simplifying the right side of the equation.

Part B: Drill and Practice

Solve.

1. $x + 3 = 19$

2. $3x = 33$

3. $47 + x = 60$

4. $24 = 4x$

5. $13 + x = 17$

6. $9 = x + 4$

7. $47 = x + 9$

8. $6x = 42$

9. $10x = 150$

10. $63 = 7x$

11. $8x = 64$

12. $93 = x + 84$

13. $945 + x = 1000$

14. $425 + x = 600$

15. $343 + x = 450$

16. $x + 15 = 23$

17. $5x = 40$

18. $5x = 80$

19. $4 + x = 17$

20. $3x = 9$

21. $7x = 42$

22. $8x = 24$

23. $32 = 4x$

24. $13 + x = 18$

25. $12x = 180$

26. $8 + z + 5 = 21$

27. $x + 2x = 12$

28. $7x + x = 24$

29. $4(3x) = 24$

30. $5(3x) = 45$

31. $2(4x) = 40$

32. $3x + 2(5x) = 13$

33. $x + x = 8$

34. $x + 5 = 14$

35. $2x = 26$

36. $x + 4x = 500$

37. $200 + x = 283$

38. $5x + x = 24$

39. $2(3x) = 4 + 5 \cdot 10$

Part C: Review. Answers for the problems of Part C are not given. However, the notation «C,U» refers to the Chapter (C) and Unit (U) in which problems of the same type were presented.

1. What is the tens digit of 4,825,369?   «1,1»

2. 67,007 − 41,084   «1,3»

3. 24 $\overline{)67348}$ «2,4»

4. Find the lcm for 12 and 20.   «3,2»

5. Select the smaller fraction from: $\frac{1}{4}$, $\frac{3}{13}$ «3,2»

6. Change $2\frac{12}{13}$ to an improper fraction.   «3,2»

7. $9\frac{1}{3} - 4\frac{1}{2}$ «3,4»

8. $4\frac{5}{8} \div \frac{2}{3}$ «3,6»

9. Write 0.053 as a fraction.   «4,1»

10. Subtract 8.36 from 13.4   «4,2»

11. Evaluate [30 ÷ 3] ÷ 5   «5,1»

12. Evaluate (7 + 4 • 6) + 5   «5,2»

13. Evaluate 9 + xy when x = 3 and y = 7.   «5,3»

14. Simplify (7 + x) + 8   «5,4»

15. Simplify (x • 12) • 6   «5,5»

16. Simplify 8(x + 7) + 5(2x + 3)   «5,6»

17. Find the sum of 6.703 and 13.62.   «4,2»

18. Divide 16.08 by 0.7 and round off the answer to 2 decimal places.   «4,4»

19. John works on a commission. When sales are good his income is high. When sales are poor, his income is very low. Last month, with 30 days, John earned $3,600 dollars. What was his daily average income?   «4,4»

20. Carl's business spends $\frac{1}{3}$ of its income on advertising. In a month when income is $20,400, what amount is spent on advertising?   «3,7»

21. Change $\frac{5}{7}$ to a decimal fraction rounded off at two decimal places.   «4,4»

Unit 8:   WORD PROBLEMS STATED USING ALGEBRA

Writing Open Expressions Showing Addition Relationships

The addition expression 8 + 5 represents a number that is 5 more than 8.

Similarly, the expression 4 + 17 represents the number that is 4 increased by 17, or the total of 4 and 17, or the sum of 4 and 17.

The words "more than," "increased," "total," and "sum" are clue words that always indicate that two quantities are to be added. This understanding of clue words for addition is helpful in writing open expressions for quantities related by the operation addition.

Writing Open Expressions for Addition Situations

A skill at writing open expressions is valuable in solving word problems such as:

> Manny had his weekly pay increased by $10.
> Write an open expression for Manny's new
> salary in terms of his old salary.

If the letter **s** is used to represent Manny's old **salary** then his new salary could be shown by the open expression s + 10.

> If s represents Manny's old weekly salary,
> then s + 10 represents Manny's new weekly salary.

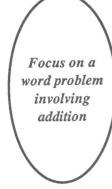

Focus on a word problem involving addition

Word Problem: Pete has $21 more than Bill. Represent
the amount of money Pete has in terms of
the money Bill has.

The exact amount of money that Pete or Bill has is not given, but the use
of the words "more than" indicate that there is an addition relationship
between the two amounts. Consequently, if the letter **b** is used to represent
the amount of money **Bill** has, then the open expression b + 21 can be used
to represent the amount of money Pete has.

      If b represents the dollars Bill has,

      then b + 21 represents the dollars Pete has.

Interpretations of Open Expressions

The use of letters and open expressions can indicate a
numerical relationship between two quantities. The letter n
and the open expression n + 3 could be used to show all the
following addition relationships between two quantities.

a number plus 3
3 more than a number
a number increased by 3
3 added to a number
the sum of a number and 3
the total of a number and 3

 (n + 3)

Open Expressions Involving Subtraction

n − 3 is an open expression that will have an evaluation of three
less than the number used to replace n.

The open expression n − 10 could be used to indicate the difference
between some number and 10, a number decreased by 10, or a
number diminished by 10.

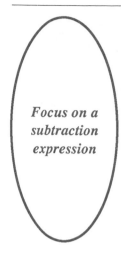

Focus on a subtraction expression

Word Problem: If one angle of a triangle is 31 degrees less than another, represent the first angle in terms of the second angle.

If **s** represents the size of the **second** angle then the first angle may be shown by the open expression s − 31 because the first angle is 31 degrees less.

If the size of the second angle replaces s in the open expression s − 31, then the evaluation of s − 31 will be the size of the smaller first angle.

If s equals the size of the second angle, then
s − 31 equals the size of the first angle.

Open Expressions Involving Subtraction

Open expressions involving subtraction will arise from a number of different word relationships. The open expression n − 8 could show any of the following relationships.

8 subtracted from a number
8 less than a number
a number decreased by 8
a number less 8   } (n − 8)
a number diminished by 8
a number minus 8
the difference between a number and 8

Multiplication Problems

Multiplication expressions can be used to show the relationships between two quantities in problems such as the following.

John receives 3 times as much allowance as Jim.
The City of Dundee collected twice as much trash
        this year as last year.
The number of absences at Couch Senior High has
        increased sevenfold this semester.
The distance a car travels is equal to the product of its
        speed per hour and the number of hours traveled.

Writing Open Expressions for Multiplication Situations

A skill at writing open expressions is valuable in solving
word problems such as:

> The interest on a one year loan at City Bank is
> equal to the product of the amount of the loan
> and the rate of interest. Represent one year's
> interest on a $100 loan in terms of the rate of
> interest.

For a $100 loan, if **i** represents the rate of **interest**, then
100i would represent the year's interest.

> If i represents the rate of interest, then
> 100i represents the amount of interest.

*Focus on a
word problem
involving
multiplication*

Word Problem: Henry takes twice as many courses as Mike this
semester. Represent the number of courses Henry
has in terms of the number of Mike's courses.

The letter **m** is used to represent the number of courses **Mike** is taking.
Then the open expression 2m will represent the relationship between
the courses taken by Henry and Mike.

> If m represents the number of courses taken by Mike,
> then 2m represents the number of courses taken by Henry.

Unit 8 Exercise

This exercise reviews the preceding unit. The exercise is divided into three parts. Part A reviews the foci of Unit 8. Part B offers opportunities to practice the skills and concepts of Unit 8. Part C contains problems that review your previous work in this text. All answers for Parts A and B are at the back of the book. Each problem of Part C is accompanied by a notation «CU» that refers to the Chapter (C) and Unit (U) in which that type of problem is studied.

Part A: Reviewing the foci of Unit 8.

1. The Ito Company makes 90,000 more transistors per day than the Smyth Company. If the letter s is used to represent the Smyth Company production per day, what open expression would represent the production of the Ito Company?

2. Clay Banks increased the number of its cash registers to 87 more than the number Vanderbilt had. Write an open expression for the number of cash registers at Clay when v represents the number at Vanderbilt.

3. Rosie the Riveter drives 803 rivets each day. If the letter r represents the number of rivets driven by Ralph each day, what open expression would indicate the total number of rivets driven each day by both Rosie and Ralph?

4. Only two companies in the country, Universal and Exceptional, produce flatcars. Universal manufactures 37 cars per week. If the letter e is used to show the weekly production of Exceptional, what open expression would show the sum of all flatcars produced each week?

5. If Mr. Jones has his rent decreased by $17 and the letter r is used to represent his original rent, what open expression represents his present rent?

6. The width of a rectangle is 20 inches less than its length. If the letter e represents the length, what is the open expression which represents the width?

7. The area of the jib of a sailboat is 215 square feet. If the letter m is used to show the area of the larger mainsail, what open expression would show the difference in their areas?

8. The Citric Orange Plant diminished the overhead expenses by $305 per month. If the overhead expenses of last month are represented by the letter e, what open expression represents the present monthly expenses?

9. Each worker at an assembly plant produces 2 toasters per hour. If w represents the number of workers in the plant, what open expression would represent the plant production of toasters per hour?

10. One side of a triangle is 6 times as long as another side. If s represents the length of the shorter side, what open expression would represent the length of the longer side?

11. The perimeter of a square is the product of 4 and the length of one side. If s represents the length of one side of a square, what open expression would represent the perimeter?

12. Taxes in Doublee City have increased threefold since 1950. If the letter t represents the taxes in 1950, what open expression would represent the taxes today?

Part B: Drill and Practice

1. If n represents a number and another number is 5 larger than n, write an open expression for the larger number.

2. One number is 9 less than another number. Write an open expression for the smaller number if n represents the other number.

3. One number is diminished by another to give 13. If e represents the larger number, write an open expression for the other number.

4. Write an open expression for a number if it is 5 times a number represented by n.

5. A first number is the same as 10 decreased by a second number. If s represents the second number, write an open expression for the first number.

6. One number is double a second number. If s represents the second number, write an open expression for the first number.

7. A first number is the same as the sum of 11 and a second number. If s represents the second number, write an open expression for the first number.

8. 4 added to a number gives a second number. If f represents the first number, write an open expression for the second number.

9. One number was 45 less than a second number. Write an open expression for the first number when s represents the second number.

10. Mary had 5 more eggs than Sue. Write an open expression for the number of eggs Mary had if s represents the number of eggs that Sue had.

11. A taxi driver made 6 times as much a week in the winter as he did in the summer. Write an open expression for the money he made in the winter if s represents the amount he made a week in the summer.

12. Jefferson Junior High has 40 more boys than Bay Junior High. If b represents the number of boys at Bay, write an open expression for the number of boys at Jefferson.

13. Mr. Stone had 40 less accounts than Mr. Bryant. If b represents the number of Bryant's accounts, write an open expression for the number of Stone's accounts.

14. If deposits in the Red Bank were decreased by $11,000 and d represents the amount before the decrease, write an open expression to represent the present deposits.

15. A sample in a chemistry experiment increased in weight by 42 grams under heat. If the original weight is represented by w, write an open expression to indicate its weight during the experiment.

16. The flow of blood in an artery is 5 times the rate of flow in a capillary. If c represents the rate of flow in the capillary, write an open expression representing the rate of flow in an artery.

17. The cost of manufacturing a watch is 4 times the cost of advertising the watch. If the advertising cost is represented by a, write an open expression representing the manufacturing cost.

18. Alpha Manufacturing Company produced 150 more units than the Norton Company. If the production of the Norton Company is represented by the letter n, what open expression should represent the production of the Alpha Company?

19. The Lo-Cal Company makes 300 less boxes of candy per day than the Smith Company. Write an open expression for the daily production of Lo-Cal when the letter s represents the daily production of the Smith Company.

20. 4 times as much iron ore as coke is to be placed with the coke in a smelter furnace. If the letter c represents the number of tons of coke put into the furnace, find an open expression to represent the number of tons of iron ore that should be used.

21. Of two numbers the smaller number increased by 8 becomes the larger number. Write an open expression that represents the larger number when s represents the smaller number.

22. If e represents a larger number and a smaller number is 50 less than the larger number, write an open expression for the smaller number.

Part C: Review. Answers for the problems of Part C are not given. However, the notation «C,U» refers to the Chapter (C) and Unit (U) in which problems of the same type were presented.

1. 78,435 + 6,353 + 2,436   «1,2»

2. 5,143 x 86   «2,2»

3. Simplify $\frac{14}{21}$ «3,1»

4. Write $\frac{23}{9}$ as a mixed number.   «3,2»

5. Find the numerator of: $\frac{7}{12} = \frac{?}{36}$ «3,2»

6. $5\frac{2}{3} + 3\frac{7}{10}$ «3,3»

7. $5\frac{2}{3} \times \frac{3}{8}$ «3,5»

8. Write $\frac{407}{100}$ as a decimal fraction.   «4,1»

9. Add 5.877, 46.38, and 0.402   «4,2»

10. Evaluate 7 + [4 + 3 • 2]   «5,2»

11. Evaluate 2x + y when x = 4 and y = 5   «5,3»

12. Find the difference of 73.09 and 63.4   «4,2»

13. Find the product of 4.278 and 6.3   «2,2»

14. Find 0.18 of 478   «4,3»

15. Evaluate 60 ÷ [(42 ÷ 7) ÷ 2]   «5,1»

16. Simplify (3 + 4x) + 15   «5,4»

17. Simplify (5 • x) • 7   «5,5»

18. Simplify 6(3x + 2) + (x + 3)   «5,6»

19. Solve 63 = 7x   «5,7»

20. How many $\frac{1}{2}$ feet are there in 5 yards?   «3,7»

21. What is the difference between 91.6 and 31.059?   «4,2»

22. Find the average of 7.84 and 6.96   «4,4»

23. Change $\frac{8}{9}$ to a decimal fraction rounded off at two decimal places.   «4,4»

CHAPTER 5 TEST

«5,U» shows the unit in which the problem is found in the chapter.

Perform the first step in evaluating each of the following.

1. $(6 + 4) + 3$ «5,1»

2. $(4 + 7) \cdot 9$ «5,2»

3. $6 \cdot (2 \cdot 4)$ «5,1»

4. $4 + 8 \cdot 3$ «5,2»

Evaluate the following expressions.

5. $6 + 4x$ when $x = 5$ «5,3»

6. $8(2 + x)$ when $x = 7$ «5,3»

7. $3x + 5y$ when $x = 4$ and $y = 3$ «5,3»

8. $x(5 + x)$ when $x = 3$ «5,3»

Simplify the following expressions.

9. $(4 + x) + 11$ «5,4»

10. $5x \cdot 7$ «5,5»

11. $6y + 2(5y + 3)$ «5,6»

12. $7(3 + 2x)$ «5,6»

13. $8x + 1 + 6x + 4$ «5,6»

14. $4 + 3x + 5(2x + 6)$ «5,6»

Solve the following equations.

15. $x + 8 = 17$ «5,7»

16. $5x = 35$ «5,7»

17. $x + 3 = 10$ «5,7»

18. $3x + 4x = 56$ «5,7»

19. $x + x = 2 + 2 \cdot 2$ «5,7»

20. $15 + x = 21$ «5,7»

21. If n represents a number and another number is 65 less than the number, write an open expression for the smaller number.   «5,8»

22. Mr. Largon had his salary increased by $110 a month. Write an open expression for his increased salary when s represents his old salary.   «5,8»

6
Algebra of the Integers

Unit 1:   ADDING INTEGERS

In the first five chapters of this book we have been concerned with
the numbers of arithmetic.

We began with the counting numbers, $\{1, 2, 3, \ldots\}$, and later included
zero, and then decimals and fractions, $\{\frac{x}{y} \mid x$ and y whole numbers, $y \neq 0\}$.

Along the way, we encountered two types of problems which could
not be done with these numbers:

1. $7 - 9$ is a subtraction problem which had no answer
       because 9 is greater than 7.
2. $x + 8 = 5$ is an equation which had no solution because
       we had no number to add to 8 that would give 5.

In this chapter we introduce a new set of numbers, the negative integers,
which will yield answers for these problems.

> **Definition**   **Opposites**
>
> Two numbers   **if and**   their sum
> are **opposites**   **only if**   is zero.

Opposites of the Whole Numbers

The opposite of a whole number is indicated by
placing a minus sign preceding the whole number symbol.

For example, the opposite of 3 is written as -3 and, by the definition
of opposites,

$$3 + (-3) = 0$$

The opposite of 14 is written as -14. Since 14 and -14 are
opposites, their sum is zero.

$$-14 + 14 = 0$$

> **Definition**   **Integers**
>
> A number is   **if and**   it is in the set
> an **integer**   **only if** $\{\ldots, -3, -2, -1, 0, 1, 2, 3, \ldots\}$.
>
> The set of positive integers is $\{1, 2, 3, \ldots\}$.
> The set of negative integers is $\{-1, -2, -3, \ldots\}$.
> Zero is neither positive nor negative.
>
> Notice that zero is an integer.

The set {. . . , -3, -2, -1, 0, 1, 2, 3, . . .} combines the counting numbers, zero, and the negative integers. This set is called the *set of integers.*

   4 is in the set of integers.
   -8 is in the set of integers.
   0 is in the set of integers.

Every counting number, zero, and the opposite of every counting number is in the set of integers, {. . . , -3, -2, -1, 0, 1, 2, 3, . . .}.

958 is a counting number, and since every counting number is in the set of integers, 958 is in the set of integers, {. . ., -3, -2, -1, 0, 1, 2, 3, . . .}.

Addition of Integers With the Same Sign

The sum of two counting numbers is always a counting number. Therefore, the sum of two positive integers is always a positive integer.

For example,

   7 + 2 = 9 and 5 + 6 = 11.

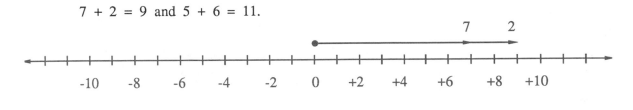

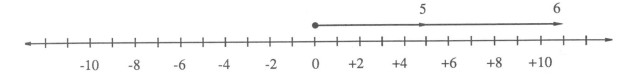

The sum of two negative integers is a negative integer.

-4 + (-6) = -10 and -3 + (-5) = -8.

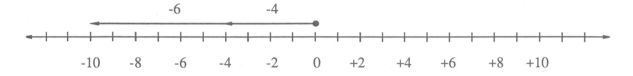

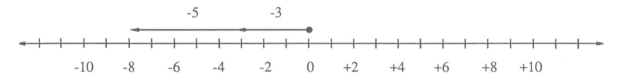

When two negative integers are added, the sum is negative.

-12 + (-8) = -20 and -17 + (-6) = -23.

Focus on adding integers with the same sign

The number line is normally shown with the positive numbers extending to the right and the negative numbers extending to the left.

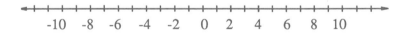

Positive addends are associated with moves to the right on the number line. The addition of 3 + 4 on the number line is the same as moving 3 to the right followed by moving 4 more to the right.   3 + 4 = 7

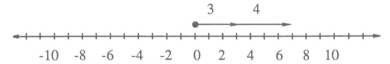

Negative addends are associated with moves to the left on the number line. The addition of -6 + (-4) on the number line is the same as moving 6 to the left followed by moving 4 more to the left.   -6 + (-4) = -10

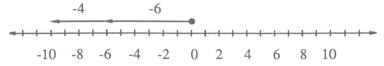

Reviewing the Associative Law of Addition

Recall that one postulate for algebra is the Associative Law
of Addition which states that

$$(x + y) + z \text{ is equivalent to } x + (y + z)$$

If the Associative Law of Addition is applied to the
addition of $(8 + 7) + [-7]$ it is equivalent to $8 + (7 + [-7])$

Since $7 + [-7] = 0$ (the sum of opposites is zero), the
Associative Law of Addition shows that $15 + [-7] = 8$.

$$
\begin{aligned}
15 + [-7] &= (8 + 7) + [-7] \\
&= 8 + (7 + [-7]) \\
&= \quad 8 + 0 \\
&= \quad 8
\end{aligned}
$$

*Focus
on the
Associative
Law of
Addition*

The Associative Law of Addition
may by used to show that the sum
of -9 and 2 is -7.

$$
\begin{aligned}
-9 + 2 &= (-7 + [-2]) + 2 \\
&= -7 + (-2 + 2) \\
&= \quad -7 + 0 \\
&= \quad -7
\end{aligned}
$$

On the number line the sum of -9 and 2 is found by moving 9 to the left
and then moving 2 to the right. The result is -7.

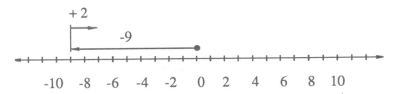

The Associative Law of Addition
may by used to show that the sum
of 10 and -6 is 4.

$$
\begin{aligned}
10 + (-6) &= (4 + 6) + (-6) \\
&= 4 + (6 + [-6]) \\
&= \quad 4 + 0 \\
&= \quad 4
\end{aligned}
$$

On the number line the sum of 10 and -6 is found by moving 10 to the right
and then moving 6 to the left. The result is 4.

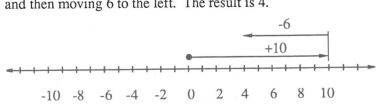

Adding Integers With Opposite Signs

To find the sum of 9 + (-5), write 9 as 4 + 5.
The problem is done as follows.

$$9 + (-5)$$
$$(4 + 5) + (-5)$$
$$4 + (5 + [-5])$$
$$4 + 0$$
$$4$$

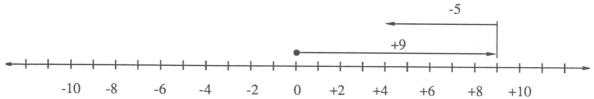

Focus on addition using opposites

To add 4 + -1, first write
4 as the sum of 3 + 1.

$$4 + (-1)$$
$$(3 + 1) + (-1)$$
$$3 + (1 + [-1])$$
$$3 + 0$$
$$3$$

To add -8 + 3, first write -8
as -5 + -3.

$$-8 + 3$$
$$(-5 + -3) + 3$$
$$-5 + (-3 + 3)$$
$$-5 + 0$$
$$-5$$

Summary of Addition of Integers

It is important to have an understanding of the following
facts about adding integers.

1. If two integers are positive, their sum is positive.
$$7 + 12 = 19$$

2. If two integers are negative, their sum is negative.
$$-6 + (-8) = -14.$$

3. If a positive integer is added to a negative integer, the
    sum may be positive, negative, or zero.
$$8 + (-8) = 0$$
$$-21 + 6 = -15$$
$$17 + (-8) = 9$$

Unit 1 Exercise

This exercise reviews the preceding unit. The exercise is divided into three parts. Part A reviews the foci of Unit 1. Part B offers opportunities to practice the skills and concepts of Unit 1.
Part C contains problems that review your previous work in this text. All answers for Parts A
and B are at the back of the book. Each problem of Part C is accompanied by a notation «CU»
that refers to the Chapter (C) and Unit (U) in which that type of problem is studied.

Part A: Reviewing the foci of Unit 1.

1. The opposite of -27 is _____.

2. -9 is the opposite of _____.

3. Is -29 in the set {-1, -2, -3, . . .}?

4. -37 is in the set of _____ integers.

5. Is every counting number in the set of integers
    {. . . , -3, -2, -1, 0, 1, 2, 3, . . .}?

6. Is every negative integer in the set of integers {..., -3, -2, -1, 0, 1, 2, 3, ...}?

7. $27 + (-27) =$ _____

8. _____ $+ 21 = 0$

9. What number can replace x as a solution for $x + 8 = 0$?

10. What number can replace x as a solution for $x + (-15) = 0$?

11. Is 19 in the set of integers {..., -3, -2, -1, 0, 1, 2, 3, ...}?

12. Is -12 in the set of integers {..., -3, -2, -1, 0, 1, 2, 3, ...}?

13. Is 0 in the set of integers {..., -3, -2, -1, 0, 1, 2, 3, ...}?

14. Is -93 in the set of integers?

15. Is 958 in the set of integers?

16. $61 +$ _____ $= 0$

17. _____ $+ 39 = 0$

18. _____ $+ (-29) = 0$

19. _____ $+ (-14) = 0$

20. $56 +$ _____ $= 0$

21. _____ $+ 0 = 0$

22. _____ $+ 17 = 0$

23. _____ $+ (-47) = 0$

24. $84 +$ _____ $= 0$

25. 0 is the opposite of _____.

26. {..., -3, -2, -1, 0, 1, 2, 3, ...} is called the set of _____.

27. -5 is the opposite of _____.

28. 13 is the opposite of _____.

29. Whenever two opposites are added, their sum is _____.

30. In the set of integers, the opposites of the positive integers are called _____ _____.

31. 34 is a _____ integer.

32. -156 is a _____ integer.

33. -6 + (-3) + (-9)

34. -8 + (-12) + (-9)

35. 9 + (-3)

36. 3 + (-7)

37. -6 + 4

38. -2 + 9

39. -14 + (-1)

40. -4 + 7

41. 0 + (-6)

42. -15 + 7

Part B: Drill and Practice

Add.

1.   21 + 3

2.   -7 + (-8)

3.   -15 + (-17)

4.   14 + 3

5.   5 + 0

6.   0 + (-7)

7.   8 + (-3)

8.   10 + (-2)

9.   -8 + 3

10. -17 + 5

11. -24 + (-7)

12. 2 + 0

13. -7 + (-9)

14. 41 + 8

15. -2 + 8

16. -3 + 10

17. 15 + (-7)

18. 14 + (-3)

19. 8 + (-5)

20. 13 + (-5)

21. -8 + 3

22. -13 + 2

23. 15 + (-7)

24. 12 + (-17)

25. -9 + 14

26. -4 + 7

27. -14 + 11

28. 4 + (-11)

29. 9 + (-3)

30. 17 + 16

31. 7 + (-3)

32. -5 + (-4)

33. 4 + (-19)

34. -9 + (-3)

35. -12 + 5

36. -8 + 3

37. -15 + (-6)

38. 13 + 7

39. -5 + (-8)

40. -2 + 6

41. 7 + (-7)

42. 3 + (-11)

Part C: Review. Answers for the problems of Part C are not given. However, the notation «C,U» refers to the Chapter (C) and Unit (U) in which problems of the same type were presented.

1. What is the thousands digit of 6,285,731? «1,1»

2. 58,707 − 35,084   «1,3»

3. 21) 43348   «2,4»

4. Find the lcm for 15 and 20.   «3,2»

5. Select the smaller fraction from $\frac{1}{5}$, $\frac{4}{19}$ «3,2»

6. Change $2\frac{3}{13}$ to an improper fraction.   «3,2»

7. $7\frac{3}{8} - 4\frac{1}{2}$ «3,4»

8. $4\frac{1}{5} \div \frac{3}{10}$ «3,6»

9. Write 0.723 as a fraction.   «4,1»

10. Subtract 4.96 from 13.4   «4,2»

11. Evaluate $[20 \div 2] \div 5$ «5,1»

12. Evaluate $(9 + 4 \cdot 5) + 6$ «5,2»

13. Evaluate $7 + xy$ when $x = 2$ and $y = 9$ «5,3»

14. Simplify $(13 + x) + 2$ «5,4»

15. Simplify $3(b \cdot 15)$ «5,5»

16. Simplify $(2x + 5) + 3(x + 4)$ «5,6»

17. Solve $8x = 64$ «5,7»

18. A first number is the same as 9 decreased by a second number. If x represents the second number, write an open expression for the first number.   «5,8»

19. Find the sum of 4.923 and 8.62   «4,2»

20. Divide 9.07 by 0.4 and round off the answer to 2 decimal places.   «4,4»

21. Mike weighed himself each Monday last month. His weights were: 133 pounds, 130 pounds, 129 pounds, and 136 pounds. What was his average weight?   «4,4»

22. Tom's business requires that he spend $\frac{1}{4}$ of his work time each week in the office. In a week in which he works 42 hours, how much time must he be in the office?   «3,7»

23. Change $\frac{5}{11}$ to a decimal fraction rounded off at two decimal places.   «4,4»

Unit 2:   SUBTRACTION OF INTEGERS

The common notion of "take away" is not always sufficient
to answer subtraction problems such as 3 − 9. The more
general notion of changing every subtraction problem to
an addition problem, by adding the opposite of the second
number (subtrahend) to the first number, will help simplify all
subtraction problems.

The subtraction problem 3 − 9 is the same as the addition
problem 3 + (-9).

5 − 12 has the same answer as 5 + (-12).

$$5 + (-12) = -7.$$

Therefore, 5 − 12 = -7.

Definition:	Subtraction	
The subtraction of y from x is z (x − y = z)	If and only if	x + (-y) = z. (The opposite of y is added to x.)

The Meaning of Subtraction for Integers

A minus sign between two integers should be read as an
addition problem in which the opposite of the second
integer (subtrahend) is added to the first integer.

5 − 3 means 5 + (-3)

4 − 8 means 4 + (-8)

15 − 23 means 15 + (-23)

A minus sign between two integers should be read as an addition problem in which the opposite of the second integer (subtrahend) is added to the first integer.

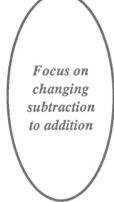

Focus on changing subtraction to addition

$$9 - (-4) \text{ means } 9 + 4$$

$$5 - (-2) \text{ means } 5 + 2$$

$$-13 - (-12) \text{ means } -13 + 12$$

$$4 - 7 \text{ means } 4 + (-7)$$

$$-8 - 9 \text{ means } -8 + (-9)$$

Evaluating Subtraction Expressions

To evaluate any subtraction expression change the subtrahend to its opposite and add as signed numbers.

To evaluate $(-7) - 8 + 7$, the following steps are used.

$$(-7) - 8 + 7$$
$$-7 + [-8] + 7 \qquad \text{Change 8 to -8 and add}$$
$$(-7 + [-8]) + 7$$
$$-15 + 7$$
$$-8$$

Focus on evaluating subtraction

To evaluate $-4 - 7 - (-12)$, the steps shown at the right are used.

Note: $- 7$ means $+ (-7)$
$ - (-12)$ means $+ 12$

$$-4 - 7 - (-12)$$
$$-4 + (-7) + 12$$
$$(-4 + [-7]) + 12$$
$$-11 + 12$$
$$1$$

The Sign of an Integer Precedes It

Because subtraction is a form of addition, it simplifies all evaluations if they are read as additions rather than subtractions. This is possible by reading each integer as including the sign that precedes it.

For example,

$-7 - 12 = \boxed{-7}\boxed{-12}$ is read as the addition of -7 and -12.

$5 - 8 = \boxed{5}\boxed{-8}$ is read as the addition of +5 and -8.

$-3 + 7 = \boxed{-3}\boxed{+7}$ is read as the addition of -3 and +7.

$6 - (-9) = \boxed{6}\boxed{-(-9)}$ is read as the addition of +6 and +9.

Notice that in every case each addend was read as the integer including the sign preceding it. Notice also that the last example shows that a double negative 9 is a positive 9.

Focus on evaluating expressions

The expression $5 - 8 + 3 - 9 - (-7)$ is evaluated as the addition of:

$\boxed{5}\boxed{-8}\boxed{+3}\boxed{-9}\boxed{-(-7)}$

		$5 - 8 \;+3\; -9\; -(-7)$
1. Add 5 and -8	1.	$-3 \;+3\; -9\; -(-7)$
2. Add -3 and 3	2.	$0 \;-9\; -(-7)$
3. Add 0 and -9	3.	$-9 \;-(-7)$
4. Add -9 and 7	4.	-2

Unit 2 Exercise

This exercise reviews the preceding unit. The exercise is divided into three parts. Part A reviews the foci of Unit 2. Part B offers opportunities to practice the skills and concepts of Unit 2. Part C contains problems that review your previous work in this text. All answers for Parts A and B are at the back of the book. Each problem of Part C is accompanied by a notation «CU» that refers to the Chapter (C) and Unit (U) in which that type of problem is studied.

Part A: Reviewing the foci of Unit 2.

1. The problem 15 – 8 means 15 + (-8). Write 12 – 5 as an addition problem.

2. The problem 6 – 8 means 6 + (-8). Write 3 – 5 as an addition problem.

3. The problem -9 – 7 means -9 + (-7). Write -8 – 3 as an addition problem.

4. The problem 13 – (-3) means 13 + 3. Write 12 – (-4) as an addition problem.

5. Evaluate 9 – 12 by first writing it as an addition problem.

6. Evaluate 10 – (-5) by first writing it as an addition problem.

7. Evaluate -12 – (-4) by first writing it as an addition problem.

8. Evaluate -4 – (-10) by first writing it as an addition problem.

Evaluate.

9. 7 – 3

10. 5 – (-3)

11. 9 – 0

12. 17 – (-4)

13. 7 – 6

14. 14 – (-17)

15. -9 – 3

Part B: Drill and Practice

Evaluate.

1. $7 - (-7)$

2. $-17 - 13$

3. $15 - 15$

4. $0 - 2$

5. $27 - 9$

6. $8 - 3$

7. $-21 - 7$

8. $-9 - (-4)$

9. $9 - 13$

10. $-8 - (-10)$

11. $-6 - 7 + 12$

12. $-4 - 3 + 8$

13. $2 - 5 - 1$

14. $-7 - 10 - (-3)$

15. $14 - 3 - 22$

16. $5 - 8 - 3$

17. $12 - (-3) - 5$

18. $-5 - 4 - 8$

19. $4 - 3 - 8$

20. $-7 - 6 - 5$

21. $-9 - 3$

22. $14 + -5$

23. $4 - (-7)$

24. $2 - 5 - 3$

25. $-3 - (-2) + 4$

Part C: Review. Answers for the problems of Part C are not given. However, the notation «C,U» refers to the Chapter (C) and Unit (U) in which problems of the same type were presented.

1. 41,835 + 6,209 + 2,376   «1,2»

2. Simplify $\frac{15}{24}$ «3,1»

3. Find the numerator of $\frac{5}{12} = \frac{?}{36}$ «3,2»

4. $5\frac{5}{8}$ x $\frac{4}{9}$ «3,5»

5. Add 4.977, 35.38, and 0.204   «4,2»

6. Evaluate x + 5y when x = 4 and y = 0   «5,3»

7. Find the difference of 19.69 and 12.4   «4,2»

8. Find the product of 9.378 and 4.7 «2,2»

9. Find 0.12 of 905.   «4,3»

10. Evaluate 20 ÷ [(60 ÷ 4) ÷ 3]   «5,1»

11. Simplify (8 + 3x) + 11   «5,4»

12. Simplify (9x) • 2   «5,5»

13. Simplify 3(3x + 4) + 2(x + 7)   «5,6»

14. Solve 345 + x = 864   «5,7»

15. One number is four times a second number. If s represents the second number, write an open expression for the first number.   «5,8»

16. Evaluate -5 + (-8)   «6,1»

17. Evaluate -2 + 6   «6,1»

18. How many $\frac{7}{8}$ inches are there in 28 inches?.   «3,7»

19. Fred's test scores in mathematics were 93, 75, 86, and 84. Find his average on these four tests rounded off to the nearest whole number.   «4,4»

20. Change $\frac{7}{12}$ to a decimal fraction rounded off at two decimal places.   «4,4»

Unit 3:   MULTIPLYING INTEGERS

Multiplying Positive Integers

The multiplication of positive integers is the same as multiplying counting numbers. 3 and 5 are counting numbers. 3 and 5 are also integers.

$3 \cdot 5 = 15$ and $7 \cdot 4 = 28$

Since counting numbers are positive integers, the rule for multiplying positive integers is

A positive integer multiplied by a positive integer always gives a positive integer.

Focus on multiplying positive integers

Positive integers are the same as counting numbers.

Consequently, the product of two positive integers is the same as if we had multiplied two counting numbers.

$8 \cdot 7 = 56$

$3 \cdot 7 = 21$

$9 \cdot 6 = 54$

$6 \cdot 8 = 48$

Multiplication as Shortcut Addition

Multiplication is a shortcut for addition.

$3 \cdot 8$ is equivalent to $8 + 8 + 8$

$3 \cdot -5$ is equivalent to $-5 + (-5) + (-5)$

Since $-5 + (-5) + (-5)$ has an evaluation of -15 then $3 \cdot -5 = -15$.

Focus on multiplication of integers

The multiplication of integers, whether the integers are positive or negative, is always a shortcut for a special addition problem.

$2 \cdot 3$ means $3 + 3$. Since $3 + 3 = 6$ then $2 \cdot 3 = 6$

$2 \cdot \text{-}3$ means $\text{-}3 + (\text{-}3)$. Since $\text{-}3 + (\text{-}3) = \text{-}6$ then $2 \cdot \text{-}3 = \text{-}6$

$2 \cdot \text{-}5$ means $\text{-}5 + (\text{-}5)$. Since $\text{-}5 + (\text{-}5) = \text{-}10$ then $2 \cdot \text{-}5 = \text{-}10$

Multiplying A Positive Times a Negative

$3 \cdot \text{-}5 = \text{-}15$

$2 \cdot \text{-}7 = \text{-}14$

Notice that in both of the examples, a positive integer multiplied by a negative integer resulted in a negative integer. A positive integer times a negative integer is a case of adding negative integers, and their sum would be negative.

For example, $3 \cdot \text{-}7 = \text{-}7 + (\text{-}7) + (\text{-}7) = \text{-}21$.

Consequently,

A positive integer times a negative integer is always a negative integer.

Focus on multiplying a positive times a negative

The product of any positive integer times a negative integer will always be a negative integer.

Consequently, each of the problems shown at the right has a negative product.

$6 \cdot \text{-}7 = \text{-}42$

$5 \cdot \text{-}9 = \text{-}45$

$8 \cdot \text{-}3 = \text{-}24$

$9 \cdot \text{-}8 = \text{-}72$

Applying the Commutative Law of Multiplication

Recall that the Commutative Law of Multiplication states that
the product of two numbers is the same regardless of the order
in which they are multiplied.

Therefore, $3 \cdot -7$ is equal to $-7 \cdot 3$. This means that the product
of a positive integer and a negative integer will be the same as the
product of the negative integer and the positive integer.

Since $3 \cdot -7 = -21$, then $-7 \cdot 3$ must also be -21.

Consequently,

> **A negative integer times a positive integer
> is always a negative integer.**

*Focus on
multiplying a
negative times
a positive*

Whenever a negative and a positive
integer are multiplied, the product is
negative.

$-3 \cdot 12 = -36$ and $12 \cdot -3 = -36$

$8 \cdot -10 = -80$ and $-10 \cdot 8 = -80$

The order of the integers has no effect
on their product.

$-6 \cdot 11 = -66$ and $11 \cdot -6 = -66$

Multiplying Zero and Any Integer

The multiplication of any integer and zero gives zero as the
product.

$$0 \cdot 8 = 0$$

$$-19 \cdot 0 = 0$$

$$0 \cdot -53 = 0$$

$$14 \cdot 0 = 0$$

**Any integer multiplied by zero gives
zero as the product.**

The multiplication of zero by any other integer always results in a product of zero.

$$15 \cdot 0 = 0$$

$$-18 \cdot 0 = 0$$

Focus on multiplying zero and an integer

This result is not surprising, but later we will see that it is useful in arriving at the conclusion that a negative times a negative is positive.

$$0 \cdot -31 = 0$$

$$6(5 + [-5]) = 6 \cdot 0 = 0$$

$$-3(-8 + 8) = -3 \cdot 0 = 0$$

Two Equivalent Expressions

According to the Distributive Law of Multiplication over Addition the numerical expressions $8(5 + [-7])$ and $8 \cdot 5 + 8 \cdot -7$ have the same evaluations.

$$8(5 + [-7]) \qquad 8 \cdot 5 + 8 \cdot -7$$
$$8 \cdot -2 \qquad\qquad 40 + (-56)$$
$$-16 \qquad\qquad\quad -16$$

Notice that the evaluations are the same, -16.

Focus on expressions with the same evaluations

The Distributive Law of Multiplication over Addition states that $3(5 + [-5])$ and $3 \cdot 5 + 3 \cdot -5$ should have the same evaluation.

The evaluations are the same, 0.

$$3(5 + [-5])$$
$$3 \cdot 0$$
$$0$$

$$3 \cdot 5 + 3 \cdot -5$$
$$15 + [-15]$$
$$0$$

Multiplying a Negative Times a Negative

Using the Distributive Law of Multiplication over Addition, it is now possible to show that a negative times a negative must result in a positive product.

-2(3 + [-3]) and -2 • 3 + -2 • -3 should have the same evaluations.

$$
\begin{array}{ccc}
-2(3 + [-3]) & -2 \cdot 3 + -2 \cdot -3 & \\
-2 \cdot 0 & -6 + ??? & [-6 + 6 = 0] \\
0 & 0 &
\end{array}
$$

The product of -2 and -3 must replace the question marks, ??? , and be added to -6 to give 0. The only way to accomplish this is to replace the question marks by +6. Therefore, the product of -2 and -3 is +6.   (-2 • -3 = 6)

The product of two negative integers is always a positive integer.

Focus on multiplying a negative times a negative

The product of two negatives is a positive.

This rule, which sometimes surprises students, is a direct result of the postulates studied earlier.

-3 • -4 = 12

-9 • -5 = 45

-8 • -6 = 48

-6 • -5 = 30

Summary of Multiplication Rules

Three rules are necessary when multiplying integers

The product of two integers which have the same signs is a positive integer.

$$8 \cdot 5 = 40 \text{ and } 16 \cdot 3 = 48$$
$$-6 \cdot -9 = 54 \text{ and } -2 \cdot -21 = 42$$

The product of a positive integer and a negative integer is a negative integer.

$$7 \cdot -9 = -63 \text{ and } -11 \cdot 5 = -55$$
$$-4 \cdot 6 = -24 \text{ and } 13 \cdot 3 = -39$$

The product of zero and any other integer is zero.

$$0 \cdot 47 = 0, \ -83 \cdot 0 = 0, \text{ and } 0 \cdot 0 = 0$$

Focus on multiplying signed numbers

In multiplying signed numbers, the following sign rules apply.

positive x positive = positive
negative x negative = positive

positive x negative = negative
negative x positive = negative

positive x zero = zero
negative x zero = zero
zero x zero = zero

Unit 3 Exercise

This exercise reviews the preceding unit. The exercise is divided into three parts. Part A reviews the foci of Unit 3. Part B offers opportunities to practice the skills and concepts of Unit 3. Part C contains problems that review your previous work in this text. All answers for Parts A and B are at the back of the book. Each problem of Part C is accompanied by a notation «CU» that refers to the Chapter (C) and Unit (U) in which that type of problem is studied.

Part A: Reviewing the foci of Unit 3.

1. Find the sum of three -8's. -8 + (-8) + (-8)

2. Find the sum of four -7's.
    -7 + (-7) + (-7) + (-7)

3. According to the Commutative Law of Multiplication 5 • -7 = _____.

4. According to the Commutative Law of Multiplication 3 • -9 = _____.

5. Separately find the evaluations of 8(4 + [-7]) and 8 • 4 + 8 • -7.

6. Separately find the evaluations of 4(9 + [-5]) and 4 • 9 + 4 • -5.

7. Separately find the evaluations of 6(7 + [-7]) and 6 • 7 + 6 • -7.

8. Separately find the evaluations of 3(9 + [-9]) and 3 • 9 + 3 • -9.

9. Separately find the evaluations of -6(8 + [-8]) and -6 • 8 + -6 • -8.

10. Separately find the evaluations of -3(4 + [-4]) and -3 • 4 + -3 • -4.

11. The product of two positive integers is a _____ integer.

12. The product of a positive integer and a negative integer is a _____ integer.

13. The product of two negative integers is a _____ integer.

14. The product of zero and any integer is _____.

Multiply.

15. 9 • 8

16. 8 • -4

17. -3 • 5

18. 0 • -9

19. -8 • -5

20. -12 • -10

21. -3 • -9

22. -7 • -3

23. 6 • -4

24. 5 • 12

25. -8 • 3

Part B: Drill and Practice

Multiply.

1. 5 • -9

2. 10 • -6

3. 0 • 5

4. -6 • -12

5. 5 • -17

6. -6 • 3

7. -7 • 8

8. 0 • 0

9. 8 • -4

10. -1 • -8

11. -12 • 7

12. 6 • -5

13. -8 • -3

14. 4 • 8

15. -7 • 1

16. -3 • -2

17. 0 • -17

18. -1 • 71

19. -4 • -6

20. 5 • 7

21. 3 • -9

22. 8 • 3

23. -12 • -7

24. -9 • -6

25. 7 • 4

26. -14 • 2

27. -23 • 2

28. 4 • -1

29. 0 • -5

30. 4 • -5

Part C: Review. Answers for the problems of Part C are not given. However, the notation «C,U» refers to the Chapter (C) and Unit (U) in which problems of the same type were presented.

1. What is the hundreds digit of 6,285,731? «1,1»

2. $35 \overline{)18348}$ «2,4»

3. Select the smaller fraction from $\frac{3}{7}, \frac{4}{9}$ «3,2»

4. $7\frac{1}{3} - 3\frac{1}{2}$ «3,4»

5. Write 0.917 as a fraction.   «4,1»

6. Evaluate $[40 \div 8] \div 5$ «5,1»

7. Evaluate $7 + xy$ when $x = 3$ and $y = 4$ «5,3»

8. Simplify $6(b \cdot 7)$ «5,5»

9. Solve $6x = 54$ «5,7»

10. A first number is the same as 7 increased by a second number. If s represents the second number, write an open expression for the first number.   «5,8»

11. Harry has read $\frac{3}{4}$ of his 564 page book. How many pages has Harry read?   «3,7»

12. Solve $47 = x + 15$ «5,7»

13. Evaluate $7 + (-7)$ «6,1»

14. Evaluate $3 + (-11)$ «6,1»

15. Evaluate $-7 - 6 - 5$ «6,2»

16. Evaluate $-9 - 3$ «6,2»

17. Evaluate $-9 \cdot -6$ «6,3»

18. Harold owns 57.8 acres of farmland and his brother owns 95.6 acres of farmland. Together, how much do they own?   «4,2»

19. A store advertises that any item may be purchased for 0.78 of its marked price. What is the sales price of an item marked for $57.95? «4,3»

20. What is the average, to two decimal places, of 47.6, 54.9, and 40.8?   «4,4»

Unit 4:   DIVIDING INTEGERS

Dividing Integers With the Same Signs

The division of two integers is similar to their multiplication.

The division of integers with the same sign will always give a quotient which is positive.

$$40 \div 8 = 5 \qquad \text{because } 5 \cdot 8 = 40$$

$$-36 \div -9 = 4 \qquad \text{because } 4 \cdot -9 = -36$$

$$48 \div 3 = 16 \qquad \text{because } 16 \cdot 3 = 48$$

$$-60 \div -5 = 12 \qquad \text{because } 12 \cdot -5 = -60$$

The quotient of two integers with the same sign will always be positive.

Focus on dividing integers with the same sign

The division of two positive integers will result in a positive quotient.

$$24 \div 8 = 3$$
$$35 \div 7 = 5$$

The division of two negative integers will result in a positive quotient.

$$-18 \div -6 = 3$$
$$-72 \div -9 = 8$$

Dividing Integers With Opposite Signs

The division of two integers is similar to their multiplication.

The division of integers with different signs will always give a quotient which is negative.

$$40 \div -8 = -5 \qquad \text{because } -5 \cdot -8 = 40$$

$$-36 \div 9 = -4 \qquad \text{because } -4 \cdot 9 = -36$$

$$-48 \div 3 = -16 \qquad \text{because } -16 \cdot 3 = -48$$

$$60 \div -5 = -12 \qquad \text{because } -12 \cdot -5 = 60$$

The quotient of two integers with different signs will always be negative.

Focus on dividing integers with different signs

The division of a positive integer by a negative integer will result in a negative quotient.

$$24 \div -8 = -3$$
$$35 \div -7 = -5$$

The division of a negative integer by a positive integer will result in a negative quotient.

$$-18 \div 6 = -3$$
$$-72 \div 9 = -8$$

Divisions That Result in Fractions

Many division problems do not result in integer quotients. For example, the division of 10 by 20 is the fraction $\frac{1}{2}$.

$$10 \div 20 = \frac{10}{20} = \frac{1}{2}$$

When a division does not have a quotient that is an integer, leave it as a fraction, proper or improper, with the sign of the quotient shown in the numerator.

$$-15 \div 20 = \frac{-15}{20} = \frac{-3}{4}$$

$$8 \div -10 = \frac{8}{-10} = \frac{-4}{5}$$

$$-12 \div -8 = \frac{-12}{-8} = \frac{3}{2}$$

In dividing integers, the sign of the quotient is written with the numerator. The denominator of a fraction should always be positive.

Focus on dividing integers with non-integer quotients

The division of integers follows the same sign rules as multiplication of integers.

$$36 \div 20 = \frac{36}{20} = \frac{9}{5}$$

$$20 \div -24 = \frac{20}{-24} = \frac{-5}{6}$$

When the division has a fraction as its quotient, the sign of the quotient goes with the numerator of the fraction.

$$-16 \div 12 = \frac{-16}{12} = \frac{-4}{3}$$

$$-6 \div -10 = \frac{-6}{-10} = \frac{3}{5}$$

Notice that the denominator of every quotient is positive.

Zero in a Division Problem

When zero is divided by any integer except itself the quotient is always zero.

$0 \div 3 = 0$ because $0 \cdot 3 = 0$ is true.

$0 \div -73 = 0$ because $0 \cdot -73 = 0$ is true.

Division by zero is undefined. It is not possible.

$10 \div 0$ is undefined   because $0 \cdot \underline{\quad} = 10$ is impossible.

$-23 \div 0$ is undefined   because $0 \cdot \underline{\quad} = -23$ is impossible.

$0 \div 0$ is undefined   because $0 \cdot \underline{\quad} = 0$ is true for any number.

Focus on division with zero

There are two important facts to remember about division with zero.

1. Zero divided by any number except zero gives 0.

2. It is not possible to divide any number by zero.

$0 \div 13 = 0$
$0 \div -2 = 0$

$31 \div 0$ is undefined
$-7 \div 0$ is undefined
$0 \div 0$ is undefined

Evaluating Addition/Subtraction Expressions

To evaluate an expression involving addition and/or subtraction:
1. Think of the expression as all addition
2. Group and order the addends in any way

For example, the addition/subtraction expression
$5 - 17 + 8 + 13 - 6 - 3$ can be evaluated in many ways including the following three possibilities:

$$(5 - 17) + (8 + 13) + (-6 - 3)$$
$$(5 + [-17 + 8]) + ([13 - 6] - 3)$$
$$(5 + 8 + 13) + (-17 - 6 - 3)$$

Notice that each of the possibilities preserves the sign of each addend integer. All of the above expressions have 5, -17, 8, 13, -6, and -3 as addends.

The addition/subtraction of integers may be accomplished in any order and grouping as long as the sign of each integer is maintained.

Focus on the evaluation of addition expressions

To evaluate
$$5 - 17 + 8 + 13 - 6 - 3$$
any order and grouping can be used if it preserves the sign of each addend. In the evaluation shown at the right, the addends were ordered and grouped according to their signs. The evaluation is 0.

$5 - 17 + 8 + 13 - 6 - 3$

$(5 + 8 + 13) + (-17 - 6 - 3)$

$26 + (-26)$

0

Evaluating Multiplication/Division Expressions

To evaluate an expression involving multiplication and/or division
 1. If there are grouping symbols follow them.
 2. In the absence of grouping symbols always work the expression from left to right.

For example, the multiplication/division expression $10 \cdot 4 \div 8 \cdot 3 \div 20$ is evaluated as:

$$\underline{10 \cdot 4} \div 8 \cdot 3 \div 20$$
$$\underline{40 \div 8} \cdot 3 \div 20$$
$$\underline{5 \cdot 3} \div 20$$
$$\underline{15 \div 20}$$
$$\frac{15}{20} = \frac{3}{4}$$

Notice that the evaluation proceeded from left to right taking each operation as it appeared first in the line.

Focus on the evaluation of multiplication or division expressions

To evaluate $12 \div 16 \cdot 4 \div 3 \cdot 10$ the absence of any grouping symbols means that the evaluation moves from left to right.

In the evaluation shown at the right, each line is completed by doing the first operation reading from left to right.

$$\underline{12 \div 16} \cdot 4 \div 3 \cdot 10$$
$$\underline{\frac{3}{4} \cdot 4} \div 3 \cdot 10$$
$$\underline{3 \div 3} \cdot 10$$
$$\underline{1 \cdot 10}$$
$$10$$

Evaluating Expressions Involving Grouping Symbols

To evaluate a numerical expression, any expression inside grouping symbols (parentheses or square brackets) is to be evaluated first.

To evaluate -5(2 − 3) the following steps are used.

$$-5(2 - 3)$$
$$-5 \cdot -1$$
$$5$$

To evaluate -2(5 + [3 − 7]) the following steps are used.

$$-2(5 + [3 - 7])$$
$$-2(5 + [-4])$$
$$-2(1)$$
$$-2$$

Focus on evaluating expressions with grouping symbols

To evaluate 8 + (3 • -4), the steps shown at the right are used.

$$8 + (3 \cdot -4)$$
$$8 + -12$$
$$-4$$

To evaluate [-19 + 17] ÷ 4, the steps shown at the right are used.

$$[-19 + 17] \div 4$$
$$-2 \div 4$$
$$\frac{-2}{4} = \frac{-1}{2}$$

Evaluating Expressions With No Grouping Symbols

When there are no grouping symbols in a numerical expression, all multiplication/division is performed from left to right. Then the expression becomes an addition/subtraction expression and is evaluated accordingly.

The evaluation of $5 \cdot -3 + 12 \div 6$ is done as shown.

$$5 \cdot -3 + 12 \div 6$$
$$-15 + 12 \div 6$$
$$-15 + 2$$
$$-13$$

Focus on evaluating expressions with no grouping symbols

To evaluate $-21 \div -3 - 2 \cdot -5$, the process requires that

1. All multiplication/division be completed from left to right.

   $-21 \div -3 = 7$ and $-2 \cdot -5 = +10$

2. All addition/subtraction is completed

$$-21 \div -3 - 2 \cdot -5$$

$$7 - 2 \cdot -5$$

$$7 + 10$$

$$17$$

Unit 4 Exercise

This exercise reviews the preceding unit. The exercise is divided into three parts. Part A reviews the foci of Unit 4. Part B offers opportunities to practice the skills and concepts of Unit 4. Part C contains problems that review your previous work in this text. All answers for Parts A and B are at the back of the book. Each problem of Part C is accompanied by a notation «CU» that refers to the Chapter (C) and Unit (U) in which that type of problem is studied.

Part A: Reviewing the foci of Unit 4.

For each of the problems in 1-8, state the first operation to be performed.

1. $8 \cdot (13 \div 6) + 9$

2. $4 + 9 \cdot 3 - 7 \div 10$

3. $15 \div -3 \div 10 \cdot 4$

4. $6 + 9 - 7 - 5 + 4$

5. $8 + 6(10 - 3) \div 7$

6. $4(-6 + 13) - 3(-2 - 7)$

7. $6[13 - (3 + 8)] - 14$

8. $15 \cdot 3 \div 7 - 3 \div 6 \cdot 2$

9.   Evaluate $4(6 + -4)$

10. Evaluate $9 + (2 \cdot -7)$

11. Evaluate $7 + 8 \cdot -3$

12. Evaluate $[3 - (-5)] - (-4)$

13. In evaluating an expression with grouping symbols always first evaluate any expression in the_____ symbols.

14. In evaluating an expression with no grouping symbols always first evaluate any _____.

15. In dividing two integers with the same sign the quotient will be _____.

16. In dividing two integers with different signs the quotient will be _____.

17. In dividing zero by any integer except zero the quotient will be _____.

18. In dividing any integer by zero the quotient will

Part B: Drill and Practice

Evaluate.

1. $34 \div -17$

2. $-42 \div -6$

3. $24 \div 8$

4. $-28 \div 7$

5. $0 \div 6$

6. $14 \div -21$

7. $40 \div 25$

8. $5 \div 0$

9. $-36 \div -8$

10. $-44 \div 33$

11. $4 \cdot -7 \div -2$

12. $3 \cdot (-2 + 9)$

13. $2 - 5 \cdot -12$

14. $8 \div 4 \cdot -5 \cdot -2 \div 4$

15. $5(-2 - 3)$

16. $-2 \cdot 4 \div -3 \cdot 6$

17. $8 - 5 \cdot -2$

18. $(-2 \cdot -4) + 3 \cdot -7$

19. $-3 \cdot -3 \cdot -3$

20. $2 \cdot 2 \cdot 2$

21. $-2(8 - 11)$

22. $-6(-2 - 4)$

23. $-7 \cdot 2 - 5.$

24. $-3 + 6 \cdot 4$

25. $-3 \cdot -4 - 5$

26. $-7 + 6 \cdot -2$

27. $-9 \cdot -5 - 3$

28. $2 \cdot -7 + 4$

29. $-7 - 4 \cdot -2$

30. $-4 \cdot 5 - (-3)$

31. $-3 \cdot 2 - 10 \div -5$

32. $3 - (-3) \cdot -2 \div 6$

33. $-1 \cdot -8 \cdot -9$

34. $-5(-3 - 4)$

35. $[3 - (-4)] - (-6)$

36. $2 - 6 \cdot 3$

Part C: Review. Answers for the problems of Part C are not given. However, the notation «C,U» refers to the Chapter (C) and Unit (U) in which problems of the same type were presented.

1.   7,569 x 8   «2,1»

2.   Write $\frac{27}{6}$ as a mixed number.   «3,2»

3. $5\frac{1}{2} + 3\frac{7}{10}$ «3,3»

4.   Write $\frac{707}{10000}$ as a decimal numeral.
     «4,1»

5.   Evaluate 4 + [7 + 3 • 5]   «5,2»

6.   Evaluate 2x + 5y when x = 2 and
     y = 4.   «5,3»

7.   Find the difference of 96.09 and 38.4
     «4,2»

8.   Find the product of 6.978 and 5.3   «2,2»

9.   Find 0.31 of 618.   «4,3»

10. Evaluate 12 ÷ [(54 ÷ 6) ÷ 3]   «5,1»

11. Simplify (4 + 7x) + 12   «5,4»

12. Simplify (6x) • 2   «5,5»

13. Simplify 5(3x + 1) + (x + 4)   «5,6»

14. Solve 21 = 3x   «5,7»

15. Evaluate -2 + 6   «6,1»

16. Evaluate 4 − (-7)   «6,2»

17. Evaluate -14 • 2   «6,3»

18. How many $\frac{1}{2}$ feet are there in 7 feet?   «3,7»

19. Jan's family spends 0.35 of its income on rent. If the family income per year is $21,600, how much does the family pay in rent each year? «4,3»

20. Each student at Morel College pays a $7 activity fee. If there are 2,600 students at Morel College, how much money is paid as activity fees?   «4,3»

21. One number is double a second number. If s represents the second number, write an open expression for the first number.   «5,8»

Unit 5:   POSTULATES FOR ADDITION AND MULTIPLICATION

Evaluating Open Expressions

$2x + 5$ is an open expression. The letter x represents a place for an integer. If x is replaced by 3, the open expression $2x + 5$ becomes the numerical expression $2 \cdot 3 + 5$ and has an evaluation of 11.

Notice that there are no parentheses in the expression $2 \cdot 3 + 5$. Whenever a numerical expression has no parentheses, multiplication is performed before addition.

Focus on evaluating an open expression with no parentheses

If 5 is used as a replacement in the open expression $-2 + 3x$, the numerical expression $-2 + 3 \cdot 5$ is obtained.

$$-2 + 3x \quad \text{when } x = 5$$
$$-2 + 3 \cdot 5$$
$$-2 + 15$$
$$13$$

$-2 + 3 \cdot 5$ has an evaluation of 13 as shown. $-2 + 3x$ has an evaluation of 13 when $x = 5$.

The evaluation of $3 - 3x$ when $x = -4$ is shown at the right.

$$3 - 3x \quad \text{when } x = -4$$
$$3 - 3 \cdot -4$$
$$3 - (-12)$$
$$3 + 12$$
$$15$$

Notice the manner in which the minus sign is used. $3 - 3x$ has an evaluation of 15 when $x = -4$.

Evaluating Open Expressions With Parentheses

The parentheses in $2(y + 4)$ indicate that when y is replaced by an integer, the addition should be performed first.

The evaluation of $2(y + 4)$ when $y = 3$ is shown below.

$$2(y + 4) \quad \text{when } y = 3$$
$$2(3 + 4)$$
$$2 \cdot 7$$
$$14$$

Focus on evaluating an open expression with parentheses

If 5 is used as a replacement in the open expression $-2(x - 6)$, the numerical expression $-2(5 - 6)$ is obtained.

$$-2(x - 6) \quad \text{when } x = 5$$
$$-2(5 - 6)$$
$$-2 \cdot -1$$
$$2$$

$-2(5 - 6)$ has an evaluation of 2 as shown. $-2(x - 6)$ has an evaluation of 2 when $x = 5$.

The evaluation of $5 - (7 - x)$ when $x = -4$ is shown at the right.

$$5 - (7 - x) \quad \text{when } x = -4$$
$$5 - (7 - [-4])$$
$$5 - (11)$$
$$5 - 11$$
$$-6$$

Notice the manner in which the minus sign is used. $5 - (7 - x)$ has an evaluation of -6 when $x = -4$.

The Commutative Law of Addition

The equation $x + y = y + x$ becomes a true equality when x is replaced by -7 and y is replaced by -8.

$$-7 + (-8) = -8 + (-7)$$

$x + y = y + x$ will become a true equality for all number replacements for x and y. This is the Commutative Law of Addition.

The Commutative Law of Addition states $x + y$ is equivalent to $y + x$ for all number replacements of x and y. The order of two addends does not effect the sum.

*Focus
on the
Commutative
Law of
Addition*

The Commutative Law of Addition states that the sum of two numbers is not affected by the order in which they are added.

By the Commutative Law of Addition the open expression a + 9 is equivalent to 9 + a. In other words, for any number replacement of a, a + 9 and 9 + a will have the same evaluation.

a + 9 and 9 + a may be substituted for each other.

The Associative Law of Addition

The Associative Law of Addition states that $(x + y) + z$
is equivalent to $x + (y + z)$ for all number replacements of
x, y, and z.

The Associative Law of Addition states that the sum of three numbers is not affected by the grouping.

$(x + y) + z = x + (y + z)$ will become a true equality
for all number replacements of x, y, and z.

*Focus
on the
Associative
Law of
Addition*

By the Associative Law of Addition $(x + 5) - 2$ is equivalent to $x + (5 - 2)$. Each expression contains x, 5, and -2.

This means that $x + (5 - 2)$ can
be substituted for $(x + 5) - 2$.
Such a substitution leads to the
simplification of $(x + 5) - 2$
shown at the right.

$$(x + 5) - 2$$

$$x + (5 - 2)$$

$$x + 3$$

$(x + 5) - 2$ has been simplified to $x + 3$. Any number can replace x and the two expressions will have the same evaluations.

Simplifying Addition Expressions

To simplify $(4 + x) - 3$, both the Commutative and Associative Laws of Addition are used.

By the Commutative Law of Addition:

$(4 + x) - 3$ is equivalent to $(x + 4) - 3$

$(4 + x) - 3$
$(x + 4) - 3$

By the Associative Law of Addition:

$(x + 4) - 3$ is equivalent to $x + (4 - 3)$

$x + (4 - 3)$
$x + 1$

Therefore, $(4 + x) - 3$ can be simplified to $x + 1$.

Focus on simplifying addition expressions

To simplify $(3x - 9) + 4$, first note that the expression contains 3x, -9, and 4.

$(3x - 9) + 4$

$(3x + [-9]) + 4$

Using the Associative Law of Addition to regroup the terms leads to the simplification of $(3x - 9) + 4$.

$3x + (-9 + 4)$

$3x + (-5)$

$(3x - 9) + 4$ simplifies to $3x - 5$.

$3x - 5$

To simplify $(3 - 5x) - 7$, first note that the expression contains 3, -5x, and -7.

$(3 - 5x) - 7$

$(3 + -5x) + [-7]$
$(-5x + 3) + [-7]$

Using the Commutative and Associative Laws of Addition to reorder and regroup the terms leads to the simplification of $(3 - 5x) - 7$.

$-5x + (3 + [-7])$

$-5x + (-4)$

$(3 - 5x) - 7$ simplifies to $-5x - 4$.

$-5x - 4$

The Addition Law of Zero

$x + 0 = x$ will become a true equality for all number replacements for x. This is the Addition Law of Zero.

The Addition Law of Zero states $x + 0$ is equivalent to x for all number replacements of x.

$5x + 0$ is equivalent to $5x$

$0 - 3x$ is equivalent to $-3x$

$-2x + (-6 + 6)$ is equivalent to $-2x$

Focus on the Addition Law of Zero

The Addition Law of Zero states that the sum of zero and any number is equal to that same number. For this reason, zero is sometimes called the Identity Element for Addition.

By the Addition Law of Zero the open expression $7x + 0$ is equivalent to 7x. In other words, for any number replacement of x, $7x + 0$ and 7x will have the same evaluation.

$7x + 0$ and 7x may be substituted for each other.

A New Addition Postulate

In the set of integers, {. . . ,-3, -2, -1, 0, 1, 2, 3, . . .} every number has an opposite.

The opposite of 7 is -7.

The opposite of -32 is 32.

Because the use of signed numbers gives every number an opposite a new postulate for algebra can now be stated.

The Law of Additive Inverses: Every number has an opposite and the sum of any number and its opposite is zero.

The Law of Additive Inverses states that every number has an opposite.

Furthermore, the Law of Additive Inverses states that the sum of any number and its opposite is zero.

5 has an opposite, -5. $5 + (-5) = 0$

-41 has an opposite, 41 $-41 + 41 = 0$

x has an opposite, -x. $x + (-x) = 0$

-4x has an opposite, 4x. $-4x + 4x = 0$

The Commutative Law of Multiplication

The equation $xy = yx$ becomes a true equality when x is replaced by -7 and y is replaced by -8.

$$-7 \cdot -8 = -8 \cdot -7$$

$xy = yx$ will become a true equality for all number replacements for x and y. This is the Commutative Law of Multiplication.

The Commutative Law of Multiplication states xy is equivalent to yx for all number replacements of x and y. The order of two factors does not effect the product.

The Commutative Law of Multiplication states that the product of two numbers is not affected by the order in which they are multiplied.

By the Commutative Law of Multiplication the open expression $5x \cdot 9$ is equivalent to $9 \cdot 5x$. In other words, for any number replacement of x, $5x \cdot 9$ and $9 \cdot 5x$ will have the same evaluation.

$5x \cdot 9$ and $9 \cdot 5x$ may be substituted for each other.

The Associative Law of Multiplication

The Associative Law of Multiplication states that (xy)z is
equivalent to x(yz) for all number replacements of x, y, and z.

The Associative Law of Multiplication states that the product of
three numbers is not affected by the grouping.

(xy)z = x(yz) will become a true equality for all number
replacements of x, y, and z.

*Focus
on the
Associative
Law of
Multiplication*

To simplify the expression (-4x) • 3, both the Commutative and Associative
Laws of Multiplication are used.

	(-4x) • 3
First write 3(-4x) by using the Commutative Law of Multiplication.	3(-4x)
3(-4x) is equivalent to (3 • -4)x by the Associative Law of Multiplication.	(3 • -4)x
	(-12)x
(3 • -4)x is simplified to -12x. Therefore, (-4x) • 3 = -12x	-12x

To simplify (-8x) • -3, the
following steps are used.

	(-8x) • -3
First, change the order of the factors, -8x and -3.	-3(-8x)
Second, change the grouping so that -3 and -8 are together.	(-3 • -8)x
	(24)x
Multiply -3 and -8. Therefore, (-8x) • -3 = 24x	24x

The Multiplication Laws of 1 and -1

The integers 1 and -1 have special properties in multiplication.

$1x = x$ is a true equality for any number replacement of x.
This is the statement of the Multiplication Law of One.

> **The Multiplication Law of 1: 1 times
> any number results in that number.**

$$1x = x$$

When any number is multiplied by -1, the product is the opposite
of the number.
   -1 • -10 is 10 and the opposite of -10 is also 10.
   -1 • 13 is -13 and the opposite of 13 is also -13.

> **The Multiplication Law of -1: -1 times
> any number results in the opposite of
> that number.**

$$-1x = -x$$

*Focus on
multiplication
properties of 1
and -1*

The use of 1 and -1 as factors arises in a number of open expression. For
example, by the Multiplication Law of One $4x + x$ is equivalent to $4x + 1x$,
and consequently, can be simplified to 5x.

-1 is often important as a factor in an open expression. $8x - x$ is equivalent
to $8x - 1x$ and consequently, can be simplified to 7x.

$6 + (3x - 7)$ is equivalent to $6 + 1(3x - 7)$.

$5 - (4x - 3)$ is equivalent to $5 - 1(4x - 3)$.

Unit 5 Exercise

This exercise reviews the preceding unit. The exercise is divided into three parts. Part A reviews the foci of Unit 5. Part B offers opportunities to practice the skills and concepts of Unit 5. Part C contains problems that review your previous work in this text. All answers for Parts A and B are at the back of the book. Each problem of Part C is accompanied by a notation «CU» that refers to the Chapter (C) and Unit (U) in which that type of problem is studied.

Part A: Reviewing the foci of Unit 5.

1. Evaluate $-2a - 3$ when $a = 7$

2. Evaluate $-5(x - 2)$ when $x = 6$

3. Evaluate $2a + 3b$ when $a = -2$ and $b = 5$

4. According to the Commutative Law of Addition $x - 8$ is equivalent to _____.

5. According to the Commutative Law of Addition $5 - 3x$ is equivalent to _____.

6. According to the Associative Law of Addition $(x - 7) + 4$ is equivalent to _____.

7. According to the Associative Law of Addition $-5 + (x - 3)$ is equivalent to _____.

8. Simplify $(z - 4) + 8$ using the Commutative and/or Associative Laws of Addition.

9. Simplify $(-3 + x) + 5$ using the Commutative and/or Associative Laws of Addition.

10. According to the Addition Law of Zero, $5x + 0$ is equivalent to _____.

11. According to the Addition Law of Zero, $3x + (-5 + 5)$ is equivalent to _____.

12. According to the Inverse Law of Addition, there is a number that can be added to 17 to give 0. What is this number?

13. According to the Inverse Law of Addition, there is an expression that can be added to $5x$ to give 0. What is this expression?

14. By the Commutative Law of Multiplication $y \cdot -7$ is equivalent to _____.

15. By the Commutative Law of Multiplication $(4z) \cdot 8$ is equivalent to _____.

16. By the Associative Law of Multiplication $5(-3x)$ is equivalent to _____.

17. By the Associative Law of Multiplication $7(2x)$ is equivalent to _____.

18. Simplify $(5x) \cdot -4$ using the Commutative and/or Associative Laws of Multiplication.

19. Simplify $(-6x) \cdot -2$ using the Commutative and/or Associative Laws of Multiplication.

20. Is $1x$ equivalent to x for any number replacement for x?

21. Is $x + 7x$ equivalent to $1x + 7x$?

22. Is $-1x$ equivalent to $-x$ for any number replacement for x?

23. Is $-1(3 - x)$ equivalent to $-(3 - x)$?

24. Is $6 - 5x$ equivalent to $1(6 - 5x)$?

Part B: Drill and Practice

Evaluate.

1. $3x + 2$ when $x = -3$

2. $4x - 3$ when $x = 4$

3. $a + 7$ when $a = -3$

4. $-2z - 3$ when $z = -5$

5. $-5 - 4r$ when $r = -3$

6. $6 + 2c$ when $c = -1$

7. $6x + 2$ when $x = -4$

8. $-3x + 5y + 2z$ when $x = -2, y = 3,$ and $z = -4$

9. $a - 3b$ when $a = -7$ and $b = -4$

10. $4x + 3c$ when $x = -1$ and $c = 4$

11. $2ab$ when $a = -2$ and $b = 5$

12. $-17x + 3y$ when $x = 0$ and $y = -2$

Simplify.

13. $(8x \quad 3) + 7$

14. $(3x - 4) + 8$

15. $(3x - 8) + 10$

16. $(9 + 4x) - 9$

17. $7 + (5 - 3x)$

18. $(4 + x) + 9$

19. $12 + (8x - 7)$

20. $(7x + 8) - 3$

21. $(3 + 5x) - 9$

22. $(2x - 7) + 9$

23. $(4 - 3x) - 9$

24. $(3x - 4) + 4$

25. $-3(6x)$

26. $8(3x)$

27. $-2(-3b)$

28. $5(-3r)$

29. $-9(-7x)$

30. $(3 + x) + -5$

31. $9(-2x)$

Part C: Review. Answers for the problems of Part C are not given. However, the notation «C,U» refers to the Chapter (C) and Unit (U) in which problems of the same type were presented.

1. 47,907 − 35,084   «1,3»

2. Find the lcm for 18 and 10.   «3,2»

3. Change $2\frac{5}{6}$ to an improper fraction.   «3,2»

4. $4\frac{1}{4} \div \frac{3}{10}$ «3,6»

5. Subtract 6.26 from 11.4   «4,2»

6. Evaluate (7 + 2 • 5) + 9   «5,2»

7. Simplify (19 + x) + 3   «5,4»

8. Simplify (5x + 2) + 3(x + 2)   «5,6»

9. Evaluate 7 + (-7)   «6,1»

10. Evaluate -7 − 6 − 5   «6,2»

11. Evaluate -9 − 3   «6,2»

12. Evaluate -12 • -7   «6,3»

13. Evaluate 9 − (-3) • -2 ÷ 2   «6,4»

14. A first number is the same as 6 decreased by a second number. If s represents the second number, write an open expression for the first number.   «5,8»

15. Find the sum of 2.973 and 5.62   «4,2»

16. Divide 14.97 by 0.7 and round off the answer to 2 decimal places.   «4,4»

17. Sam earned $285, $314, and $412 in three successive weeks. What was his average weekly earnings during that 3-week period? «4,4»

18. Change $\frac{5}{14}$ to a decimal fraction rounded off at two decimal places.   «4,4»

Unit 6:   THE DISTRIBUTIVE LAW OF MULTIPLICATION OVER ADDITION

The open sentence $xy + xz = x(y + z)$ is a true equality for all number replacements of x, y, and z. This is the Distributive Law of Multiplication over Addition.

To simplify the open expression $4x - 9x$, the Distributive Law of Multiplication over Addition is used as follows.

$$4x - 9x$$
$$(4 - 9)x$$
$$-5x$$

$4x$ and $-9x$ are called **like terms** in the expression $4x - 9x$.

Focus on the addition of like terms

To simplify $7x - 4x$, the steps shown at the right are used.
$7x$ and $-4x$ are like terms.

$$7x - 4x$$
$$(7 - 4)x$$
$$3x$$

To simplify $7x + 5x - 8x$ there are three like terms.

$$7x + 5x - 8x$$
$$12x - 8x$$
$$4x$$

Removing Parentheses from Open Expressions

The parentheses of the open expression $3(3x + 5)$ may be removed using the Distributive Law of Multiplication over Addition as follows.

$$3(3x + 5)$$
$$3 \cdot 3x + 3 \cdot 5$$
$$9x + 15$$

Focus on removing parentheses

To remove the parentheses from $-4(y - 5)$, each term of $(y - 5)$ is multiplied by -4.

1. y is multiplied by -4.
2. -5 is multiplied by -4.

$$-4(y - 5)$$

$$-4y + 20$$

To remove the parentheses from $7(2x - 3)$, each term of $(2x - 3)$ is multiplied by 7.

1. 2x is multiplied by 7.
2. -3 is multiplied by 7.

$$7(2x - 3)$$

$$14x - 21$$

Removing Parentheses with -1 as the Multiplier

The parentheses of $-(x + 5)$ are removed by using the Distributive Law of Multiplication over Addition and -1 as the multiplier.

The following steps are used to remove parentheses from $-(x + 5)$.

$$-(x + 5)$$

$-(x + 5)$ is equivalent to $-1(x + 5)$ $-1(x + 5)$

$$-1x + [-1 \cdot 5]$$

$$-x + [-5]$$

$$-x - 5$$

Focus on removing parentheses with -1 as the multiplier

$-(x - 2)$ is equivalent to $-1(x - 2)$. The parentheses may be removed by

1. multiplying x by -1
2. multiplying -2 by -1

$$-(x - 2)$$

$$-1(x - 2)$$

$$-x + 2$$

$-(5x + 7)$ is equivalent to $-1(5x + 7)$. The parentheses may be removed by

1. multiplying 5x by -1
2. multiplying 7 by -1

$$-(5x + 7)$$

$$-1(5x + 7)$$

$$-5x - 7$$

Simplifying Open Expressions with Parentheses

To simplify $6x - 3(2x - 5)$, the parentheses must be removed first.

$$6x - 3(2x - 5)$$

$$6x - 6x + 15$$

Since $6x - 6x = 0$, the final simplification is 15.

$$15$$

To simplify $8 - (3x + 7)$, the parentheses must be removed first.

$$8 - (3x + 7)$$

$$8 - 1(3x + 7)$$

Then the simplification is completed by adding the like terms, 8 and -7.

$$8 - 3x - 7$$

$$-3x + 1$$

Focus on simplifying open expressions with parentheses

To simplify $9x - 4(3x - 5)$, first multiply both terms of $(3x - 5)$ by -4. Then add like terms.

$$9x - 4(3x - 5)$$
$$9x - 12x + 20)$$
$$-3x + 20$$

To simplify $9 - (5x - 2)$, first multiply both terms of $(5x - 2)$ by -1. Then add like terms.

$$9 - (5x - 2)$$
$$9 - 1(5x - 2)$$
$$9 - 5x + 2$$
$$-5x + 11$$

Simplifying Open Expressions With Two Parentheses

To simplify $(5x - 3) - (2x + 2)$, the following steps are used.

$$(5x - 3) - (2x + 2)$$
$$1(5x - 3) - 1(2x + 2)$$
$$5x - 3 - 2x - 2$$
$$3x - 5$$

Notice that $(5x - 3)$ is equivalent to $1(5x - 3)$, and $-(2x + 2)$ is equivalent to $-1(2x + 2)$.

Focus on simplifying open expressions with two sets of parentheses

To simplify $3(2x - 1) - 2(x + 3)$,
1. $(2x - 1)$ is multiplied by 3.
2. $(x + 3)$ is multiplied by -2.
3. Like terms are added.

$$3(2x - 1) - 2(x + 3)$$

$$6x - 3 - 2x - 6$$

$$4x - 9$$

To simplify $5(4x + 7) + (2x - 3)$,
1. $(4x + 7)$ is multiplied by 5.
2. $(2x - 3)$ is multiplied by 1.
3. Like terms are added.

$$5(4x + 7) + (2x - 3)$$

$$20x + 35 + 2x - 3$$

$$22x + 32$$

Unit 6 Exercise

This exercise reviews the preceding unit. The exercise is divided into three parts. Part A reviews the foci of Unit 6. Part B offers opportunities to practice the skills and concepts of Unit 6. Part C contains problems that review your previous work in this text. All answers for Parts A and B are at the back of the book. Each problem of Part C is accompanied by a notation «CU» that refers to the Chapter (C) and Unit (U) in which that type of problem is studied.

Part A: Reviewing the foci of Unit 6.

1. For any number replacement of x, y, and z, is $xy + xz$ equivalent to $x(y + z)$?

2. In the open expression $12x - 8x$ the 12x and -8x are called like terms. Simplify $12x - 8x$ by adding the like terms.

3. Simplify $8x + 5x$ by adding the like terms.

4. Remove the parentheses from $-3(x - 8)$ by multiplying both x and -8 by -3.

5. Remove the parentheses from $-(4x - 7)$ by multiplying both 4x and -7 by -1.

6. To remove the parentheses of $8 + (3x - 4)$, both 3x and -4 should be multiplied by ____.

7. To remove the parentheses of $5 - (2x + 3)$, both 2x and 3 should be multiplied by ____.

9. Simplify $8 + 3(5x - 2)$ by first removing the parentheses and then adding like terms.

8. Simplify $7x - 6(x + 4)$ by first removing the parentheses and then adding like terms.

10. Simplify $-4(3x - 2) + 3(x - 5)$ by first removing the parentheses.

Part B: Drill and Practice

Simplify.

1. $7x - 9x$

2. $5x + x$

3. $-3x - 6x$

4. $4x - x$

5. $x - x$

6. $8x + 3x - 7x$

7. $2x - 9x - 3x$

8. $-(2x - 3)$

9. $4(8x - 3)$

10. $3(2x - 7)$

11. $-9(5 - 2x)$

12. $-(x + 7)$

13. $-(4x - x)$

14. $-(3 - 5x)$

15. $-(6x + 4)$

16. $-4(x - 3)$

17. $16 - (5x + 9)$

18. $9 - (x - 5)$

19. $(2x + 3) - (x + 5)$

20. $4(2x - 1) - 3(x - 4)$

21. $-6 - (7x + 3)$

22. $5 + (2x - 3)$

23. $7x - (3x + 4)$

24. $12 + (7 - 3x)$

25. $(5 - 4x) - (x + 3)$

26. $-(3x + 7) - (x + 3)$

27. $(3x - 4) - (2x + 5)$

28. $2x - (5 - 3x)$

29. $(4x - 3) - 2x$

30. $3x + (x - 9)$

Part C: Review. Answers for the problems of Part C are not given. However, the notation «C,U» refers to the Chapter (C) and Unit (U) in which problems of the same type were presented.

1. 57,080 + 34,099 + 4,766   «1,2»

2. Simplify $\frac{12}{32}$ «3,1»

3. Find the numerator of $\frac{5}{7} = \frac{?}{35}$ «3,2»

4. $6\frac{3}{4}$ x $\frac{4}{9}$ «3,5»

5. Add 5.267, 27.38, and 0. 042   «4,2»

6. Evaluate 3x + 5y when x = 4 and y = 2   «5,3»

7. Find the difference of 19.47 and 17.4 «4,2»

8. Find the product of 6.778 and 9.7   «2,2»

9. Find 0.42 of 612.   «4,3»

10. Simplify (4 + 5x) + 11   «5,4»

11. Simplify 2(3x + 5) + 2(x + 6)   «5,6»

12. Evaluate -7 + (-8)   «6,1»

13. Evaluate -2 + 7   «6,1»

14. Evaluate 6 − (-7)   «6,2»

15. Evaluate -7 • 4   «6,3»

16. Evaluate 6 − (-4) • -4 ÷ 8   «6,4»

17. Simplify (4x − 1) + 4   «6,5»

18. How many $\frac{3}{4}$ inches are there in 9 inches? «3,7»

19. Carl's test scores in mathematics were 88, 92, 86, and 78. Find his average on these four tests rounded off to the nearest whole number.   «4,4»

20. Change $\frac{7}{9}$ to a decimal fraction rounded off at two decimal places.   «4,4»

21. One number is three times a second number. If s represents the second number, write an open expression for the first number.   «5,8»

Unit 7:   SOLVING SIMPLE EQUATIONS

In Chapter 5, we solved equations for their roots or solutions,
but in that chapter it was not possible to solve x + 3 = 1 because
there are no negative numbers in the set of counting numbers, {1, 2, 3, . . . }.

Now we have the set of integers, { . . . , -3, -2, -1, 0, 1, 2, 3, . . . } and the
equation x + 3 = 1 has -2 as its root or solution because:

$$-2 + 3 = 1 \text{ is a true equality}$$

In this unit, we begin the solving of equations using signed
numbers. We shall see that the use of signed numbers makes
the solving of equations much easier.

Solving Simple Equations

x + 4 = 7 is an equation which is solved by asking:

What number can be added to 4 to give 7?

The answer to the question is 3. Therefore 3 is the **root** or **solution**
of x + 4 = 7.

7x = 42 is an equation which is solved by asking:

What number can be multiplied by 7 to give 42?

The answer to the question is 6. Therefore 6 is the **root** or **solution**
of 7x = 42.

*Focus on
finding
the root of
an equation*

To find the root for
$x - 5 = -3$, ask the question:

    What integer added to -5
      gives -3?

Since $2 + -5 = -3$ is true the root is 2.

$$x - 5 = -3$$

What plus -5 gives -3?

To find the truth set for
$-3x = 12$, ask the question:

    What integer multiplied by
      -3 gives 12?

Since $-3 \cdot -4 = 12$ is true the root is -4.

$$-3x = 12$$

What times -3 gives 12?

Definition   **Basic Equation**

An equation is a **basic equation**	**if and only if**	it is of the form $ax = b$ where x is a variable and a and b are numbers.

For example, $3x = 27$, $5x = -30$, and $-7x = 14$ are basic equations.

Solving Basic Equations

Every basic equation asks a simple multiplication question.

$3x = 27$ asks, "3 times what number gives 27?" Answer is 9.

$5x = -30$ asks, "5 times what number gives -30?" Answer is -6.

$-7x = 14$ asks, "-7 times what number gives 14?" Answer is -2.

A basic equation always asks a simple multiplication question and this makes basic equations easy to solve.

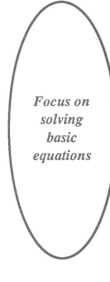

8x = 48 is a basic equation. It is solved by asking its multiplication question. The root of 8x = 48 is 6.

$$8x = 48$$

8 times what number gives 48?

-5x = -20 is a basic equation. It is solved by asking its multiplication question. The root of -5x = -20 is 4.

$$-5x = -20$$

-5 times what number gives -20?

28 = -4x can be written as -4x = 28 (the two sides of an equation can be interchanged). Then it is solved as the other basic equations. Its root is -7.

$$28 = -4x$$

$$-4x = 28$$

-4 times what number gives 28?

The Simplest Type of Basic Equation

Any equation of the form ax = b is a basic equation but the simplest basic equation occurs when the multiplier of x is 1.

For example, three of the simplest basic equations are:

$$x = 7, \quad x = -3, \text{ and } x = 19$$

and the roots are 7, -3, and 19 respectively.

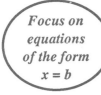

Any equation of the form

$$x = b$$

is easily solved.

x = 13 has 13 as its root.

x = -47 has -47 as its root.

x = 512 has 512 as its root.

Basic Equations of the Form -x = b

Basic equations of the form $-x = b$ always ask the question,

The opposite of what number is b?

For example, the equations $-x = 5$ and $-x = -8$ ask:

$-x = 5$ The opposite of what number is 5? The answer is -5.

$-(-5) = 5$ is true.

$-x = -8$ The opposite of what number is -8? The answer is 8.

$-(8) = -8$ is true.

*Focus on
equations
of the form
-x = b*

Any equation of the form

$$-x = b$$

has a root which is the
opposite of b.

$-x = 13$ has -13 as its root.

$-x = -47$ has 47 as its root.

$-x = 512$ has -512 as its root.

Definition	Equivalent Equations	
Two equations are **equivalent**	**if and only if**	they have exactly the same roots.

Generating Equivalent Equations

$x + 7 = 10$ and $x + 5 = 8$ are **equivalent** equations because both have 3 as their root.

In this text there are two ways taught for generating equivalent equations. The first method is:

> **Any number or open expression can be added to both sides of an equation to generate a new equivalent equation.**

For example, the equation $x + 5 = 8$ can be used to generate another equivalent equation by adding 2 to both sides of $x + 5 = 8$.

left side		right side
x + 5	=	8
+ 2		+ 2
x + 7	=	10

$x + 5 = 8$ and its generated equation $x + 7 = 10$ are equivalent (they have the same solution).

Focus on generating equivalent equations

Any number can be added to both sides of an equation to generate a new equivalent equation. Below are shown three examples of generating equivalent equations from $3x + 8 = 20$

$$3x + 8 = 20 \qquad 3x + 8 = 20 \qquad 3x + 8 = 20$$
$$\underline{+\ 3 \quad +\ 3} \qquad \underline{-\ 6 \quad -\ 6} \qquad \underline{-\ 8 \quad -\ 8}$$
$$3x + 11 = 23 \qquad 3x + 2 = 14 \qquad 3x + 0 = 12$$

All of the equations above are equivalent because they have 4 as their root. $3x + 0 = 12$ is the easiest equation to use because it is equivalent to the basic equation $3x = 12$.

Solving Equations of the Form   ax + b = c

To solve $5x + 6 = -4$, first generate an equivalent basic equation by adding an integer to both sides of $5x + 6 = -4$.

To decide what integer should be added to both sides of $5x + 6 = -4$, ask:

> **What term of $5x + 6 = -4$ must be eliminated to make a basic equation?**

The answer to that question is that 6 needs to be eliminated to generate a basic equation from $5x + 6 = -4$. Consequently, -6 is added to both sides of the equation to eliminate the 6.

$$5x + 6 = -4$$
$$\underline{\quad -6 \quad -6 \quad}$$
$$5x + 0 = -10 \quad \text{or } 5x = -10 \text{ or } x = -2$$

The basic equation $5x = -10$ has -2 as its root. Since the equations $5x + 6 = -4$ and $5x = -10$ are equivalent, -2 is also the root of $5x + 6 = -4$. $5 \cdot -2 + 6 = -4$ is true.

Focus on solving an equation of the form ax + b = c

The equation $2x + 5 = 13$ is used to generate an equivalent basic equation.

Since $2x = 8$ has 4 as its root then $2x + 5 = 13$ must also have 4 as its root. Notice that $2 \cdot 4 + 5 = 13$ is true.

$$2x + 5 = 13$$
$$\underline{\quad -5 \quad -5 \quad}$$
$$2x + 0 = 8 \quad \text{or } 2x = 8$$

Generating Basic Equations

To solve $7x - 8 = 27$, first generate an equivalent basic equation by adding an integer to both sides of $7x - 8 = 27$.

To decide what integer should be added to both sides of $7x - 8 = 27$, ask:

> **What term of $7x - 8 = 27$ must be eliminated to make a basic equation?**

The answer to that question is that -8 needs to be eliminated to generate a basic equation from $7x - 8 = 27$. Consequently, 8 is added to both sides to eliminate the -8.

$$
\begin{aligned}
7x - 8 &= 27 \\
+ 8 \quad\ &\ \ + 8 \\
\hline
7x + 0 &= 35 \quad \text{or } 7x = 35 \text{ or } x = 5
\end{aligned}
$$

The basic equation has 5 as its root. Since the equations are equivalent, 5 is also the root of $7x - 8 = 27$.

Focus on generating a basic equation

The steps in the solution of $5x - 2 = 13$ are shown at the right.

Since $5x = 15$ has 3 as its root then $5x - 2 = 13$ must also have 3 as its root.

$$
\begin{aligned}
5x - 2 &= 13 \\
+ 2 \quad &\ \ + 2 \\
\hline
5x + 0 &= 15 \\
5x &= 15 \\
x &= 3
\end{aligned}
$$

Solving Equations by Adding Opposites

The first step in solving $2x + 7 = -3$ is to add -7 to both sides of the equation.

The following steps are used to find the truth set of $2x + 7 = -3$.

$$
\begin{aligned}
2x + 7 &= -3 \\
- 7 \quad &\ \ - 7 \\
\hline
2x + 0 &= -10 \quad \text{or } 2x = -10 \text{ or } x = -5
\end{aligned}
$$

The basic equation $2x = -10$ has -5 as its root and $2x + 7 = -3$ also has -5 as its root.

Focus on adding opposites to find the truth set

The first step in solving $5 = 2x - 3$ is to notice that the equation is equivalent to $2x - 3 = 5$.

$2x = 8$ has 4 as its root and, therefore, $5 = 2x - 3$ also has 4 as its root.

$5 = 2x - 3$ is equivalent to $2x - 3 = 5$

$$
\begin{aligned}
2x - 3 &= 5 \\
+\,3 \quad &\;\; +\,3 \\
\hline
2x + 0 &= 8 \\[6pt]
2x &= 8 \\[6pt]
x &= 4
\end{aligned}
$$

Selecting the Term to be Eliminated

To generate a basic equation from $8 - 3x = 2$, notice that a basic equation always has an x term on the left side of the equation and a number, by itself, on the right side of the equation. The term of $8 - 3x = 2$ that needs to be eliminated is 8.

$$
\begin{aligned}
8 - 3x &= 2 \\
-\,8 \qquad &\;\; -\,8 \\
\hline
-3x = -6 \quad &\text{or} \quad x = 2
\end{aligned}
$$

The solution of $-3x = -6$ is 2 and, therefore, the solution of $8 - 3x = 2$ is also 2.

Focus on solving an equation of the form $b + ax = c$

The solution of $5 - 7x = 33$ is shown at the right. Notice that 5 was eliminated from both sides of the original equation to generate a new, equivalent, basic equation.

$-7x = 28$ has -4 as its solution and $5 - 7x = 33$ also has -4 as its root.

$$
\begin{aligned}
5 - 7x &= 33 \\
-\,5 \qquad &\;\; -\,5 \\
\hline
-7x &= 28 \\[6pt]
x &= -4
\end{aligned}
$$

Checking a Root for an Equation

To check 11 as the root of $5x - 30 = 2x + 3$,

1. Replace x in the
    equation by 11.

    $$5x - 30 \quad = \quad 2x + 3$$
    $$5 \cdot 11 - 30 \qquad 2 \cdot 11 + 3$$

2. Separately evaluate
    $5 \cdot 11 - 30$ and $2 \cdot 11 + 3$.

    $$55 - 30 \qquad 22 + 3$$
    $$25 \qquad\qquad 25$$

3. If the evaluations are the
    same, the root 11 checks.

    The evaluations are the same.
    The root 11 checks.

*Focus on
checking
a possible
root*

To check 7 as the root of $4x + 3 = 8x - 23$,

1. Replace x in the
    equation by 7.

    $$4x + 3 \quad = \quad 8x - 23$$
    $$4 \cdot 7 + 3 \qquad 8 \cdot 7 - 23$$

2. Separately evaluate
    $4 \cdot 7 + 3$ and $8 \cdot 7 - 23$.

    $$28 + 3 \qquad 56 - 23$$
    $$31 \qquad\qquad 33$$

3. If the evaluations are the
    same, the root 7 checks.

    The evaluations are different.
    The root 7 does not check.

Unit 7 Exercise

This exercise reviews the preceding unit. The exercise is divided into three parts. Part A reviews the foci of Unit 7. Part B offers opportunities to practice the skills and concepts of Unit 7. Part C contains problems that review your previous work in this text. All answers for Parts A and B are at the back of the book. Each problem of Part C is accompanied by a notation «CU» that refers to the Chapter (C) and Unit (U) in which that type of problem is studied.

Part A: Reviewing the foci of Unit 7.

1. Is $-2x = 12$ a basic equation?

2. Is $5x + 3 = 13$ a basic equation?

3. For the equation $3x - 7 = 8$, what number is added to both sides to give the basic equation $3x = 15$?

4. For the equation $15 - x = 7$, what number is added to both sides to give the basic equation $-x = -8$.

5. Solve $5x - 2 = 18$ by first adding 2 to both sides of the equation.

6. Solve $8x + 3 = 35$ by first adding -3 to both sides of the equation.

7. Solve $5 - 7x = -16$ by adding -5 to both sides of the equation.

8. Solve $14 = 2x + 4$ by adding -4 to both sides of the equation.

Solve (find the root).

9. $x - 3 = -5$

10. $x - 7 = -4$

11. $9x = 18$

12. $-2x = 12$

13. $-7x = -21$

14. $6x = -36$

15. Check -5 as a root for $7x + 13 = 4x - 2$.

16. Check -3 as a root for $6x + 5 = 2x - 9$.

Part B: Drill and Practice

Solve (find the root).

1. $x + 8 = 11$

2. $x + 2 = 5$

3. $5 = x + 6$

4. $x + 4 = 2$

5. $x + 5 = 1$

6. $x - 1 = -4$

7. $3x = 6$

8. $-15 = -3x$

9. $8x = 0$

10. $-3x = -12$

11. $7 = -7x$

12. $5x = -30$

13. $-15 = -5x$

14. $-5x + 3 = 13$

15. $-3x + 7 = 22$

16. $-17 = 3x - 8$

17. $14 = 6 - 2x$

18. $4 - 5x = -11$

19. $9x - 3 = 15$

20. $8x - 3 = 21$

21. $9 = 2x - 5$

22. $x - 15 = 63$

23. $-8x = -32$

24. $-x = 47$

25. $7x = -63$

26. $19x = 0$

27. $4x - 5 = 19$

28. $-9 = 3 + 6x$

Solve and Check.

29. $7 = 3 - 2x$

30. $19 = 12 + 7x$

31. $x + 7 = -3$

32. $5x = -40$

33. $3x - 6 = 18$

Part C: Review. Answers for the problems of Part C are not given. However, the notation «C,U» refers to the Chapter (C) and Unit (U) in which problems of the same type were presented.

1. What is the tens digit of 6,285,731?   «1,1»

2. $41 \overline{)\ 19548}$ «2,4»

3. Select the smaller fraction from: $\frac{2}{3}, \frac{8}{13}$
    «3,2»

4. $7\frac{1}{4} - 3\frac{2}{3}$ «3,4»

5. Write 0.097 as a fraction.   «4,1»

6. Evaluate $[56 \div 4] \div 7$ «5,1»

7. Evaluate $2 + xy$ when $x = 9$ and $y = 2$
    «5,3»

8. Simplify $4(b \cdot 7)$ «5,5»

9. Solve $9x = 54$ «5,7»

10. Evaluate $13 + (-7)$ «6,1»

11. Evaluate $6 + (-7)$ «6,1»

12. Evaluate $-6 - 8 - 5$ «6,2»

13. Evaluate $-8 \cdot -5$ «6,3»

14. Evaluate $4 - (-5) \cdot -2 \div 2$ «6,4»

15. Simplify $-7(6x)$ «6,5»

16. Simplify $6(3x - 1) - 5(x - 2)$ «6,6»

17. Harry has completed $\frac{2}{3}$ of his course work at Ober College. If his course work requires 66 hours credit, how many hours has Harry completed? «3,7»

18. A first number is the same as 5 increased by a second number. If s represents the second number, write an open expression for the first number.   «5,8»

19. What is the average, to two decimal places, of 82.8, 79.9, and 91.6?   «4,4»

Unit 8:   GENERATING BASIC EQUATIONS

Solving Equations by Adding Opposites

In the equation $5x = 9 + 2x$, there is an x term on both sides. To generate a basic equation from $5x = 9 + 2x$, the 2x needs to be eliminated. This is accomplished by adding its opposite, -2x, to both sides of the equation.

$$\begin{array}{rl} 5x = & 9 + 2x \\ -2x & \quad -2x \\ \hline 3x = & 9 \quad \text{or} \quad x = 3 \end{array}$$

The solution of $3x = 9$ is 3. Therefore 3 is also the solution for $5x = 9 + 2x$.

The solution, 3, must be checked in the original equation.
$15 = 9 + 6$ is true.

$$5x = 9 + 2x$$
$$5 \cdot 3 = 9 + 2 \cdot 3$$
$$15 = 9 + 6$$

Focus on adding the opposite of an open expression

For the equation $2x = 6 - x$, the -x may be eliminated by adding its opposite, x, to both sides of the equation.

$$\begin{array}{rl} 2x & = 6 - x \\ + x & \quad + x \\ \hline 3x & = 6 \qquad \text{or} \quad x = 2 \end{array}$$

Therefore, the root of $2x = 6 - x$ is 2.

The solution, 2, is checked in the original equation.

$$2x = 6 - x$$
$$2 \cdot 2 = 6 - 2$$
$$4 = 6 - 2$$

$4 = 6 - 2$ is true.

Solving Equations with Four Terms

The equation $5x + 6 = 3x - 2$ has four terms. There is an x term on each side of the equation and there is a number term on each side of the equation. To generate a basic equation, two terms need to be eliminated.

First, the opposite of 3x is added to each side of the equation.

$$\begin{array}{r} 5x + 6 = 3x - 2 \\ \underline{- 3x \qquad - 3x} \\ 2x + 6 = \ \ 0 - 2 \ \text{ or } 2x + 6 = \text{-}2 \end{array}$$

Now, the opposite of 6 is added to each side of the equation.

$$\begin{array}{r} 2x + 6 \ = \ \text{-}2 \\ \underline{- 6 \quad - 6} \\ 2x + 0 \ = \ \text{-}8 \ \ \text{ or } 2x = \text{-}8 \text{ or } x = \text{-}4 \end{array}$$

The root of $5x + 6 = 3x - 2$ is -4.

Focus on solving equations with four terms

For the equation $4 - 5x = \text{-}3x + 6$, two terms need to be eliminated.

To generate an equation with all the x terms on the left side, the opposite of -3x is added to each side of the equation.

$$\begin{array}{r} 4 - 5x = \text{-}3x + 6 \\ \underline{+ 3x \quad + 3x} \\ 4 - 2x = \ \ 0 + 6 \end{array}$$

Now, the opposite of 4 is added to each side of the equation so that a basic equation is generated.

$$\begin{array}{r} 4 - 2x = 6 \\ \underline{- 4 \qquad - 4} \\ \text{-}2x = 2 \ \text{ or } x = \text{-}1 \end{array}$$

Therefore, the root of $4 - 5x = \text{-}3x + 6$ is -1.

Simplifying as Part of the Solving Process

The equation $7x - 4 + 3x = 5x + 16$ has three terms on its left side and two of those terms are like terms.

The first step in solving any equation is the simplification of each side of the equation.

In solving $7x - 4 + 3x = 5x + 16$, first simplify the left side of the equation.

$$7x - 4 + 3x = 5x + 16$$

$$
\begin{array}{r}
10x - 4 = 5x + 16 \\
-5x \qquad\ \ -5x \\
\hline
5x - 4 = \ \ 0 + 16
\end{array}
$$

$$
\begin{array}{r}
5x - 4 = 16 \\
+4 \ \ +4 \\
\hline
5x + 0 = 20
\end{array}
\quad \text{or } 5x = 20 \quad \text{or } x = 4
$$

Therefore, the solution of $7x - 4 + 3x = 5x + 16$ is 4.

Focus on simplifying before solving

The first step in solving
$$2(3x - 4) + 5 = 2x + 5$$
is to remove the parentheses and simplify the left side of the equation.

$$
\begin{array}{l}
2(3x - 4) + 5 = 2x + 5 \\
6x - 8 + 5 \quad = 2x + 5 \\
6x - 3 \qquad\ \ = 2x + 5
\end{array}
$$

Now the equation is solved by adding opposites. -2x is added to both sides of the equation.

$$
\begin{array}{r}
6x - 3 = 2x + 5 \\
-2x \qquad -2x \\
\hline
4x - 3 = \ \ 0 + 5
\end{array}
$$

3 is added to both sides of the equation.

$$
\begin{array}{r}
4x - 3 = 5 \\
+3 \ \ +3 \\
\hline
4x \qquad = 8 \\
x = 2
\end{array}
$$

2 is the solution of $4x = 8$ and also of $2(3x - 4) + 5 = 2x + 5$.

Simplifying Both Sides of an Equation

To solve $7 - (2x + 3) = 2(3x - 1) - 2$, first simplify both sides of the equation. Notice particularly the simplification of $7 - (2x + 3)$ below.

$$7 - (2x + 3) \quad = \quad 2(3x - 1) - 2$$
$$7 - 1(2x + 3) \qquad 2(3x - 1) - 2$$
$$7 - 2x - 3 \qquad\quad 6x - 2 - 2$$
$$4 - 2x \qquad\qquad 6x - 4$$

Now, $4 - 2x = 6x - 4$ is solved by adding opposites.

$$
\begin{array}{rcl}
4 - 2x & = & 6x - 4 \\
- 6x & & - 6x \\
\hline
4 - 8x & = & 0 - 4
\end{array}
$$

$$
\begin{array}{rcl}
4 - 8x & = & -4 \\
- 4 & & - 4 \\
\hline
-8x & = & -8 \quad \text{or} \quad x = 1
\end{array}
$$

Therefore, the root of $7 - (2x + 3) = 2(3x - 1) - 2$ is 1.

Focus on solving an equation

To solve $3(4x - 7) - 4 = 12 - (5x + 3)$, both sides of the equation need to be simplified by removing parentheses and adding like terms.

$$3(4x - 7) - 4 \quad = \quad 12 - (5x + 3)$$
$$3(4x - 7) - 4 \qquad\quad 12 - 1(5x + 3)$$
$$12x - 21 - 4 \qquad\quad 12 - 5x - 3$$
$$12x - 25 \qquad\qquad 9 - 5x$$

Now, $12x - 25 = 9 - 5x$ is solved by adding opposites. 25 is added to both sides.

$$
\begin{array}{rcl}
12x - 25 & = & 9 - 5x \\
+ 25 & & + 25 \\
\hline
12x + 0 & = & 34 - 5x
\end{array}
$$

5x is added to both sides

$$
\begin{array}{rcl}
12x & = & 34 - 5x \\
+ 5x & & + 5x \\
\hline
17x & = & 34 \qquad \text{or} \quad x = 2
\end{array}
$$

Therefore, the root of $3(4x - 7) - 4 = 12 - (5x + 3)$ is 2.

Checking a Solution for an Equation

To check -4 as the solution of $5x - 3(2x - 6) = 5 - (2x - 9)$,

1. Replace x in the $\qquad$ $5x - 3(2x - 6) \quad = \quad 5 - (2x - 9)$
   equation by -4. $\qquad$ $5 \cdot -4 - 3(2 \cdot -4 - 6) \quad 5 - (2 \cdot -4 - 9)$

2. Separately evaluate $\qquad -20 - 3(-8 - 6) \qquad 5 - (-8 - 9)$
   the two sides. $\qquad\qquad -20 - 3(-14) \qquad\qquad 5 - (-17)$
   $\qquad\qquad\qquad -20 + 42 \qquad\qquad\qquad 5 + 17$
   $\qquad\qquad\qquad\qquad 22 \qquad\qquad\qquad\qquad 22$

3. If the evaluations are the $\qquad$ The evaluations are the same.
   same, the root -4 checks. $\qquad\qquad$ The root -4 checks.

To check 2 as the root of $9 - 4(3x - 2) = 6(3x - 7) + 4x$,

Focus on checking a possible root

1. Replace x in the $\qquad$ $9 - 4(3x - 2) \quad = \quad 6(3x - 7) + 4x$
   equation by 2. $\qquad\qquad 9 - 4(3 \cdot 2 - 2) \qquad 6(3 \cdot 2 - 7) + 4 \cdot 2$

2. Separately evaluate $\qquad 9 - 4(6 - 2) \qquad\qquad 6(6 - 7) + 8$
   the two sides. $\qquad\qquad\quad 9 - 4(4) \qquad\qquad\qquad 6(-1) + 8$
   $\qquad\qquad\qquad\quad 9 - 16 \qquad\qquad\qquad\quad -6 + 8$
   $\qquad\qquad\qquad\qquad -7 \qquad\qquad\qquad\qquad\quad 2$

3. If the evaluations are the $\qquad$ The evaluations are different.
   same, then 2 checks as $\qquad$ The 2 does not check as the
   the root. $\qquad\qquad\qquad$ root.

Unit 8 Exercise

This exercise reviews the preceding unit. The exercise is divided into three parts. Part A reviews the foci of Unit 8. Part B offers opportunities to practice the skills and concepts of Unit 8. Part C contains problems that review your previous work in this text. All answers for Parts A and B are at the back of the book. Each problem of Part C is accompanied by a notation «CU» that refers to the Chapter (C) and Unit (U) in which that type of problem is studied.

Part A: Reviewing the foci of Unit 8.

1. Solve $8x = 12 + 6x$ by first adding -6x to both sides of the equation.

2. Solve $2x - 6 = 4x$ by first adding -2x to both sides of the equation.

3. Solve $9x + 2 - 7x = x + 8$ by first simplifying the left side of the equation.

4. Solve $2x - 3 = 5x + 7 - 4x$ by first simplifying the right side of the equation.

5. Solve $9 + 5(x + 4) = 2x - 2(x + 3)$ by first removing parentheses and adding like terms.

6. Solve $5 - (x - 3) = 3x - 4$ by first removing parentheses and adding like terms.

7. Solve $3x + 7 = x - 9$

8. Solve $2x - 7 = 5x + 2$

Part B: Drill and Practice

Solve.

1. $5x - (4x - 7) = -3$

13. $5x - 2 - 3x = x + 3$

25. $5x + 4 = 4 - 2x$

2. $12 + (5x - 12) = 10$

14. $5x - 7 + x = (2x + 5)$

26. $2(x - 3) = 12$

3. $6x - (3x - 2) = -4$

15. $8x - 4 + 2x = 10 + 3x$

27. $5x + 3(x + 1) = 19$

4. $13 + (3x - 13) = -12$

16. $-14 - 3x + 3 = 7(x - 3)$

28. $x - 9 = 5x + 11$

5. $4x - 7 = x + 11$

17. $x + 9 - 2x = 7 - 3x$

29. $2x + 3 - x = 2x - 8$

6. $2x - 9 = -3x + 11$

18. $5x - 2x + 7 = 4(x - 2)$

30. $8 - (2x - 3) = 17$

7. $12x - 20 = 3x + 7$

19. $2x + 3 = x - 7$

Solve and check.

8. $5x + 7 = 2x + 7$

20. $5 - 3x = -7$

31. $2x - (x - 4) = -9$

9. $-2x - 7 = 5x + 14$

21. $4x = x + 15$

32. $5 + 2(3 - 7x) = 11$

10. $3x + 8 = -14x + 8$

22. $9 - x = 5 - 2x$

33. $8x - 3(4 + 2x) = -6$

11. $11 - 2x = x + 8$

23. $4 + 3(x + 1) = 31$

34. $4 - 3x = 10 - x$

12. $5x - 3 = 2x + 3$

24. $2(3x - 7) + 4 = 2$

35. $5(2x + 4) - 17 = -7$

Part C: Review. Answers for the problems of Part C are not given. However, the notation «C,U» refers to the Chapter (C) and Unit (U) in which problems of the same type were presented.

1. $5,472 \times 86$ «2,2»

2. Write $\frac{23}{7}$ as a mixed number.   «3,2»

3. $6\frac{3}{4} + 3\frac{7}{10}$ «3,3»

4. Write $\frac{757}{10000}$ as a decimal fraction.   «4,1»

5. Evaluate $8 + [2 + 3 \cdot 5]$ «5,2»

6. Evaluate $3x + 5y$ when $x = 7$ and $y = 4$ «5,3»

7. Find the difference of 41.09 and 38.6   «4,2»

8. Find the product of 9.78 and 5.4   «2,2»

9. Find 0.63 of 438   «4,3»

10. Simplify $(9 + 3x) + 12$ «5,4»

11. Simplify $2(6x + 5) + (x + 3)$ «5,6»

12. Evaluate $-5 + 6$ «6,1»

13. Evaluate $3 - (-7)$ «6,2»

14. Evaluate $-17 \cdot 2$ «6,3»

15. Evaluate $-8 \cdot 2 - 9 \div -3$ «6,4»

16. Simplify $(4 - 3x) - 9$ «6,5»

17. Simplify $2(3x - 1) - 2(x - 4)$ «6,7»

18. Solve $4x - 5 = 19$ «6,7»

19. How many $\frac{1}{4}$ feet are there in 6 feet?.   «3,7»

20. A store advertises 0.22 discount on each item. If a chair normally sells for $84, what is the discount during this sale?   «4,3»

21. Each school lunch costs $1.65 and the lunch-room serves 285 students. What is the cost of the 285 lunches?   «4,3»

22. One number is five times a second number. If s represents the second number, write an open expression for the first number.   «5,8»

Unit 9: SOLVING WORD PROBLEMS

Writing Word Relationships in Symbols

The letter n and the open expression 3n + 2 show the relationship
of two quantities when one is 2 more than 3 times the other.

> If n represents one number, then
> 3n + 2 represents a number which is
> 2 more than 3 times the other number.

> 2 more than 3 times n

> 3n + 2

Focus on translating words into symbols

Word Problem: A rectangle has its length and width related so that the
length is 4 less than twice the width. Write an
open expression for the length in terms of the width.

The phrase "in terms of the width" indicates that a variable needs to be
chosen to represent the width.

> Let w represent the width, then
> 2w − 4 would represent the length.

The phrase "length is 4 less than twice the width" is translated.

> 4 less than twice w

> 2w − 4

Words for Addition

Words like "more," "plus," and "increase" usually are good clues that addition is involved.

Each of the following phrases could be translated as the open expression $5n + 13$.

13 plus 5 times a number

$13 + 5n$

13 more than 5 times a number

$5n + 13$

5 times a number increased by 13

$5n + 13$

13 increased by 5 times a number

$13 + 5n$

Focus on writing an open expression for addition

Word Problem: A second number is 18 more than three times a first number. Write an open expression for the second number in terms of the first number.

Select a variable to represent the first number because of the phrase "in terms of the first number."

Let f represent the first number, then $3f + 18$ represents "18 more than three times the first number."

18 more than three times f

$3f + 18$

Words for Subtraction

Words like "less," "decreased," and "diminished" usually are good clues that subtraction is involved.

Each of the following phrases could be translated as the open expression $4n - 13$.

13 less than 4 times a number

$4n - 13$

4 times a number decreased by 13

$4n - 13$

4 times a number diminished by 13

$4n - 13$

Notice that the expression $13 - 4n$ has a different meaning. "4 times a number less than 13" or "13 decreased by 4 times a number" are translated as $13 - 4n$.

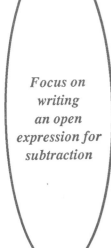

Focus on writing an open expression for subtraction

The relationship shown by the letter n and the open expression $2n - 5$ might be described in words by any of the following.

5 less than twice a number

$2n - 5$

twice a number diminished by 5

$2n - 5$

5 less than 2 times a number

$2n - 5$

decrease 2 times a number by 5

$2n - 5$

5 subtracted from twice a number

$2n - 5$

Writing Simple Equations

Many word problems can be translated into equations and
solved by algebra. In such cases there is a three-step process
that can be followed to obtain that equation.

1.   Represent an unknown quantity by a single letter.

2.   Write the other unknown quantity using an open expression
     with the letter selected in step 1.

3.   Using steps 1 or 2, or both, write an equation.

An example of this three-step process is shown below.

   Word Problem: 8 more than a number is 10.

   1.   Let the letter n represent the unknown number.

   2.   The phrase "8 more than a number" is written as the open
        expression $n + 8$.

   3.   The sentence "8 more than a number is 10" can now be
        written as the equation $n + 8 = 10$.

                8 more than a number is 10.

                    $$n + 8 = 10$$

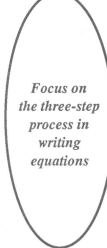

Focus on the three-step process in writing equations

As another example of the three-step process in writing equations showing word relationships, study the following.

Word Problem: 8 less than 3 times a number is 14.

1.   Let the letter n represent the unknown number.

2.   The phrase "8 less than 3 times a number" can be translated as the open expression $3n - 8$.

3.   The complete sentence can now be written as the equation $3n - 8 = 14$.

$$8 \text{ less than 3 times a number is } 14.$$
$$3n - 8 = 14$$

Systematically Attacking Word Problems

A systematic method for attacking word situations is desirable.
Such a method should follow the steps given below.

1.   Read the problem carefully.

2.   Represent an unknown quantity by a single letter.

3.   Write the other unknown quantity using an open expression with the letter selected in step 2.

4.   Using steps 2 or 3, or both, write an equation.

5.   Solve the equation.

6.   Check the solution.

Focus on solving and checking word problems

Word Problem: One number is 4 less than 3 times another number. The sum of the two numbers is 48.

1.   Read the problem carefully.

2.   Let n be "another number".

3.   Write the relationship of the number "4 less than 3 times another number."

$$(3n - 4)$$

4.   Write the equation for "the sum of the two numbers is 48."

$$(3n - 4) + n = 48$$
$$\text{or}$$
$$n + (3n - 4) = 48$$

5.   Solve the equation to find n the "another number."

$$(3n - 4) + n = 48$$

$$
\begin{aligned}
4n - 4 &= 48 \\
+4 \quad\ &\ +4 \\
\hline
4n &= 52
\end{aligned}
$$

$$n = 13$$

One number is 13. Use the expression $(3n - 4)$ to find the number that is "4 less than 3 times another number."

$$
\begin{aligned}
(3n - 4) &= 3 \cdot 13 - 4 \\
&= 39 - 4 \\
&= 35
\end{aligned}
$$

6.   Check the solution, 13 and 35, by checking "the sum of the two numbers is 48."

$$13 + 35 = 48 \text{ is true.}$$
The solution is correct.

Word Problem: Archer Company produced 17 more than 3 times the number of radios last week as Baker Company. The difference in the number of radios produced was 307.

1. Read the problem carefully.

2. Let b equal the number of radios Baker Company produces.

3. Write the relationship of "17 more than 3 times the number," to show the number of radios produced by Archer Company.

$$3b + 17$$

4. Write the equation for "the difference in the number of radios produced is 307."

$$(3b + 17) - b = 307$$

Focus on the six-step process

5. Solve the equation to find b, the number of radios produced by Baker Company.

$$(3b + 17) - b = 307$$

$$
\begin{array}{r}
2b + 17 = 307 \\
-17 \quad -17 \\
\hline
2b = 290
\end{array}
$$

$$b = 145$$

Use the expression (3b + 17) to find the number of radios produced by Archer Company.

$$
\begin{aligned}
3b + 17 &= 3 \cdot 145 + 17 \\
&= 435 + 17 \\
&= 452
\end{aligned}
$$

6. Check the solution by checking the difference in the number of radios produced.

$$452 - 145 = 307 \text{ is true.}$$
The solution is correct.

Unit 9 Exercise

This exercise reviews the preceding unit. The exercise is divided into three parts. Part A reviews the foci of Unit 9. Part B offers opportunities to practice the skills and concepts of Unit 9. Part C contains problems that review your previous work in this text. All answers for Parts A and B are at the back of the book. Each problem of Part C is accompanied by a notation «CU» that refers to the Chapter (C) and Unit (U) in which that type of problem is studied.

Part A: Reviewing the foci of Unit 9.

1. National Corporation makes 2 types of resistors. Last Tuesday the output of Type A resistors was 7 less than 8 times the number of Type B. If b represents the number of Type B resistors, what open expression would represent the number of Type A resistors?

2. A strong solution of muriatic acid had 3 more than 11 times as much hydrochloric acid in it as a weak solution. If the amount of hydrochloric acid in the weak solution is represented by the letter h, what open expression would show the amount of hydrochloric acid in the strong solution?

3. It takes a refrigerator repairman repairing an old unit 18 minutes more than twice the time required to replace the old unit with a new one. If the letter m represents the minutes spent in replacing the old unit by a new unit, what open expression would represent the repair time on an old unit?

4. If two numbers are related by the fact that the smaller is 107 less than the product of 12 and the larger, and if the letter w is used to represent the larger number, what open expression should represent the smaller number?

Write equations to represent the relationships in the following sentences.

5. 9 less than a number is 12.

6. 13 increased by a number gives a sum of 39.

7. Twice a number is 20.

8. 6 more than 5 times a number is 41.

9. One number is 18 more than another. Their sum is 64. (Let s represent the smaller number.)

10. A motor car manufacturer made 300 less sedans than sports cars. The total car production was 10,500. (Let s represent the number of sports cars.)

11. One number is 13 less than 3 times a smaller number. Their difference is 55. (Let s represent the smaller number.)

12. The fuel in a rocket weighed 3 times as much as the frame of the rocket. Together the weight was 18,000 pounds. (Let f represent the weight of the frame.)

Part B: Drill and Practice

1. A jet engine had 800 pounds less than 8 times the thrust of a turbine engine. If t represents the thrust of the turbine engine, write an open expression that shows the thrust of the jet engine.

2. A picture frame had a width represented by w. The length of the frame was 4 more than twice the width. Write an open expression that shows the length of the frame.

3. In a highway there were 18 more than 9 times as many miles of straight road as there were miles of curved highway. If c represents the miles of curved road, find an open expression for the miles of straight highway.

4. The weight of an object on earth was 100 tons less than 13 times the weight of the object on another planet. If p represents the weight of the object on the planet, what open expression represents the weight of the object on earth?

5. When he was down exploring a sunken ship, the pressure on the body of a diver was 45 pounds more than 2 times the pressure on him at the surface. Write an open expression that shows the pressure on his body down at the sunken ship when s represents the pressure at the surface.

Write equations to represent the relationships in the following sentences.

6. 8 more than a number is 21.

7. 9 less than a number is 45.

8. The sum of a number and 23 is 98.

9. The difference when 8 is subtracted from a number is 32.

10. 7 less than 3 times a number is 47.

11. 67 is 14 increased by 2 times a number.

12. 9 diminished by a number is 37.

13. One number is 5 more than another number. The sum of the numbers is 93.

14. One number is 4 times another number. Their sum is 130.

15. Decrease 14 times a number by 7 and the result is 169.

16. One number is 3 more than twice another. The total of the numbers is 105.

17. Ace Plumbing had 14 more men employed than Zero Plumbing. Together the two companies had 64 employees.

18. Mr. North had 300 less grams of sodium chloride than sodium nitrate. He had a total of 850 grams of the two chemicals.

19. A larger number is 3 increased by the product of 8 and a smaller number. The difference of the two numbers is 52.

20. A grocer sold 7 less crates of carrots than tomatoes. He sold a total of 334 crates of the two vegetables.

21. A lighthouse is 50 feet more than twice has high as a house beside it. The difference in their heights is 150 feet.

22. If a larger number is 2 more than 3 times a smaller number and the smaller is represented by s. The sum of the two numbers is 93.

23. Book A had 20 less than 3 times the number of pages in Book B. Let b represents the number of pages in Book B. The total pages in the two books is 765.

24. 12 more than a number is 69.

25. 54 is 6 times a number.

26. A number diminished by 8 is 23.

Part C: Review. Answers for the problems of Part C are not given. However, the notation «C,U» refers to the Chapter (C) and Unit (U) in which problems of the same type were presented.

1. $85,097 - 41,804$ «1,3»

2. Find the lcm for 12 and 8.   «3,2»

3. Change $2\frac{3}{5}$ to an improper fraction.   «3,2»

4. $4\frac{1}{2} \div \frac{7}{12}$ «3,6»

5. Subtract 9.76 from 14.3   «4,2»

6. Evaluate $(6 + 3 \cdot 5) + 8$ «5,2»

7. Simplify $(14 + x) + 2$ «5,4»

8. Simplify $(7x + 1) + 4(x + 2)$ «5,6»

9. Evaluate $2 + (-7)$ «6,1»

10. Evaluate $-9 - 3 - 5$ «6,2»

11. Evaluate $-11 \cdot -6$ «6,3»

12. Simplify $8(3x)$ «6,5»

13. Simplify $(2x + 3) - (x + 5)$ «6,6»

14. Solve $6x - 5 = 19$ «6,7»

15. Solve $5 + 2(3 - 7x) = 11$ «6,8»

16. A first number is the same as 4 decreased by a second number. If s represents the second number, write an open expression for the first number.   «5,8»

17. Find the sum of 5.673 and 5.43   «4,2»

18. Divide 6.97 by 0.9 and round off the answer to 2 decimal places.   «4,4»

19. A business had sales of $2,985, $3,315, and $5,412 in three successive weeks. What was the average weekly sales during that 3-week period?   «4,4»

20. Change $\frac{1}{8}$ to a decimal fraction rounded off at two decimal places.   «4,4»

CHAPTER 6 TEST

«6,U» shows the unit in which the problem is found in the chapter.

Evaluate each of the following:

Simplify each of the following:

1. $17 - 6$ «6,1»

2. $-12 - 8$ «6,1»

3. $-3 \cdot 9$ «6,3»

4. $0 \cdot 7$ «6,3»

5. $5 \cdot 9$ «6,3»

6. $-6 \cdot -4$ «6,3»

7. $7 \cdot -10$ «6,3»

8. $8 + (-13)$ «6,1»

9. $14 - (-6)$ «6,1»

10. $-9 + (-5)$ «6,1»

11. $5 + (-2)$ «6,1»

12. $-15 - (-9)$ «6,1»

13. $(5 + x) - 8$ «6,6»

14. $-6(4x)$ «6,5»

15. $-11x + x$ «6,6»

16. $-(4 - 7x)$ «6,6»

17. $5(2x - 7)$ «6,6»

18. $4(3 - 2x) - (2 - 5x)$ «6,6»

19. $2(x - 4) - 3(x + 4)$ «6,6»

20. $5(2x + 1) - (x - 1)$ «6,6»

Solve each of the following:

21. $x + 9 = 3$ «6,7»

22. $x - 7 = 11$ «6,7»

23. $6y = -54$ «6,7»

24. $-7y = -14$ «6,7»

25. $5x + 4 = 29$ «6,7»

26. $8x - 6 = x + 8$ «6,2»

27. $3x + 4 = 9 - 2x$ «6,2»

28. $7 - 4x = 3 - 2x$ «6,2»

29. $2(3x - 5) + 6 = 14$ «6,8»

30. $4(x - 6) + 3x = 25$ «6,8»

31. In a building there were 42 more than 3 times as many windows as there were doors. Write an open expression representing the number of windows letting d represent the number of doors.   «6,9»

32. A number diminished by 39 is 23. Find the number.   «6,9»

7
Algebra of
the Rational Numbers

Unit 1:   MULTIPLYING RATIONAL NUMBERS

In Chapter 6, the numbers used were primarily integers.
The set of integers is {. . . , -3, -2, -1, 0, 1, 2, 3, . . .}.

In this chapter, algebra concepts and skills are extended to
fractions (the set of rational numbers).

Definition		Ratio of Two Integers
The fraction $\frac{x}{y}$ is a **ratio of two integers**	**if and only if**	both x and y are integers and y is not zero.

$\frac{5}{7}, \frac{-7}{4}$, and $\frac{4}{-9}$ are ratios of two integers. In each case,
the numerator and denominator are integers and the
denominator is not zero.

*Focus on
the meaning
of a ratio
of two
integers*

Any fraction which has an integer as its numerator, and a non-zero integer as its denominator is a ratio of two integers.

Examples of fractions that are ratios of two integers are:

$$\frac{1}{2} \qquad \frac{8}{5} \qquad \frac{-5}{11} \qquad \frac{0}{7} \qquad \frac{0}{-4} \qquad \frac{-5}{-9} \qquad \frac{-35}{7}$$

Examples of fractions that are not ratios of two integers are:

$$\frac{1}{0} \qquad \frac{0}{0} \qquad \frac{-7}{0}$$

Definition		Set of Rational Numbers
A number is an element of the **set of rational numbers**	**if and only if**	the number is a ratio of two integers and the denominator is not zero.

*Focus on
the meaning
of rational
number*

For the open expression $\frac{x}{y}$, if x is replaced by any integer and y is replaced by any integer except zero, the result is a rational number. Consequently, numbers such as $\frac{5}{6}, \frac{-71}{9}, \frac{8}{3}$, and $\frac{4}{-7}$ are called rational numbers.

For the open expression $\frac{x}{y}$, x can be replaced by any integer and y can be replaced by any integer except zero. The result from such replacements is a rational number.

Writing Integers as Rational Numbers

The integer 5 can be written as the rational number $\frac{5}{1}$.

Similarly, the integer -12 can be written as the rational number $\frac{-12}{1}$.

To write any integer as a rational number, the integer is written as the numerator and positive one (1) is used as the denominator.

Consequently, -5 can be written as $\frac{-5}{1}$, a rational number.

Focus on writing integers as rational numbers

The integers 5, 0, and -8 can be written as rational numbers as shown below.

5 can be written as $\frac{5}{1}$

0 can be written as $\frac{0}{1}$

-8 can be written as $\frac{-8}{1}$

Simplifying Rational Numbers

Remember that $\frac{4}{6}$ can be written more simply as $\frac{2}{3}$.

$\frac{4}{6}$ can be simplified because 2 is a factor of both 4 and 6.

$$\frac{4 \div 2}{6 \div 2} = \frac{2}{3} \qquad \text{Therefore, } \frac{4}{6} = \frac{2}{3}.$$

Notice that $4 \cdot 3 = 6 \cdot 2$ is also a true equality because the product of $4 \cdot 3$ is 12 and also the product of $6 \cdot 2$ is 12.

The factors of $4 \cdot 3$ are the first numerator and second denominator of $\frac{4}{6} = \frac{2}{3}$.

The factors of $6 \cdot 2$ are the first denominator and second numerator of $\frac{4}{6} = \frac{2}{3}$.

Definition   **Equal Rational Numbers**

Two rational   if and   a • d is equal
numbers, $\frac{a}{b}$ and   only if   to b • c.
$\frac{c}{d}$, are **equal**

$\frac{a}{b} = \frac{c}{d}$ if and   ad = bc.
                only if

Focus on the meaning of equal rational numbers

To decide whether $\frac{3}{5} = \frac{6}{10}$ is a true statement, use the statement $3 \cdot 10 = 5 \cdot 6$. Since $30 = 30$ is true, then $\frac{3}{5} = \frac{6}{10}$.

To decide whether $\frac{4}{7} = \frac{12}{21}$ is a true statement, use the statement $4 \cdot 21 = 7 \cdot 12$. Since both products are 84, the rational numbers are equal.

Rational Numbers with Positive Denominators

$\frac{-2}{5}$ is equal to $\frac{2}{-5}$ because $-2 \cdot -5 = 10$ and $5 \cdot 2 = 10$.

Whenever a rational number consists of one positive and one negative integer, it can be written with a negative numerator and a positive denominator. The preferred or simplest form will have a positive denominator.

$\frac{-4}{7}$ is equal to $\frac{4}{-7}$ because $-4 \cdot -7 = 28$ and $7 \cdot 4 = 28$.

$\frac{-4}{7}$ is preferred over $\frac{4}{-7}$, because its denominator is positive.

Focus on writing rational numbers with positive denominators

$\frac{-5}{-9} = \frac{5}{9}$ is a true statement. Whenever a rational number consists of two negative integers, it can be written using two positive integers.

$\frac{2}{7}$ is the preferred simplification of $\frac{-2}{-7}$.

$\frac{9}{5} = \frac{-9}{-5}$ is a true statement because $9 \cdot -5 = 5 \cdot -9$. $\frac{9}{5}$ is preferred over $\frac{-9}{-5}$, because the denominator is positive.

Multiplying Two Rational Numbers

To evaluate the multiplication of two rationals:

1. The numerators are multiplied to give the numerator of the product.

2. The denominators are multiplied to give the denominator of the product.

To evaluate $\frac{3}{5} \cdot \frac{2}{7}$, multiply the numerators and separately multiply the denominators. The following steps are used to evaluate $\frac{3}{5} \cdot \frac{2}{7}$.

$$\frac{3}{5} \cdot \frac{2}{7} = \frac{3 \cdot 2}{5 \cdot 7} = \frac{6}{35}$$

Focus on multiplying rational numbers

To evaluate $\frac{-3}{5} \cdot \frac{4}{11}$, the following steps are used.

$$\frac{-3}{5} \cdot \frac{4}{11} = \frac{-3 \cdot 4}{5 \cdot 11} = \frac{-12}{55}$$

To evaluate $\frac{-3}{8} \cdot \frac{-1}{2}$ the following steps are used.

$$\frac{-3}{8} \cdot \frac{-1}{2} = \frac{-3 \cdot -1}{8 \cdot 2} = \frac{3}{16}$$

Multiplying a Rational Number and an Integer

To multiply a rational number by an integer, the integer should be changed to a rational number using positive one as its denominator.

For example, to multiply $\frac{-2}{7}$ by 3, write the problem as $\frac{-2}{7} \cdot \frac{3}{1}$ and complete the evaluation as shown.

$$\frac{-2}{7} \cdot 3 = \frac{-2}{7} \cdot \frac{3}{1} = \frac{-2 \cdot 3}{7 \cdot 1} = \frac{-6}{7}$$

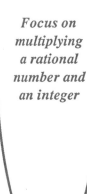

*Focus on
multiplying
a rational
number and
an integer*

To multiply a rational number by 0, we write 0 as $\frac{0}{1}$. For example, to multiply $\frac{-3}{8}$ by 0, we write $\frac{-3}{8} \cdot \frac{0}{1}$, and complete the evaluation as shown below.

$$\frac{-3}{8} \cdot 0 = \frac{-3}{8} \cdot \frac{0}{1} = \frac{-3 \cdot 0}{8 \cdot 1} = \frac{0}{8} = 0$$

Notice that zero times any rational number is zero.

To evaluate $\frac{5}{8} \cdot 1$, we write 1 as $\frac{1}{1}$ and complete the evaluation as shown below.

$$\frac{5}{8} \cdot 1 = \frac{5}{8} \cdot \frac{1}{1} = \frac{5 \cdot 1}{8 \cdot 1} = \frac{5}{8}$$

Notice that 1 times any rational number gives a product that is the same rational number.

To multiply any rational number by -1, we write -1 as $\frac{-1}{1}$. For example, to multiply $\frac{-2}{5} \cdot -1$ we write the problem as $\frac{-2}{5} \cdot \frac{-1}{1}$ and complete the evaluation as shown below.

$$\frac{-2}{5} \cdot -1 = \frac{-2}{5} \cdot \frac{-1}{1} = \frac{-2 \cdot -1}{5 \cdot 1} = \frac{2}{5}$$

Notice that -1 times a rational number gives a product that is the opposite of the rational number.

Expressing One as a Rational Number

$\frac{1}{1}$ is not the only numeral for the rational number 1. In fact, there are many different numerals for 1.
$\frac{7}{7}$ is another numeral for 1. $\frac{7}{7} = \frac{1}{1}$

$\frac{10}{10}, \frac{-11}{-11}$, and $\frac{73}{73}$ are other numerals for the rational number 1. In fact, if x is replaced by any integer other than zero in the equation $\frac{x}{x} = 1$, the resulting equality will be true.

Any rational number of the form $\frac{x}{x}$, when x is not replaced by zero, is equal to 1.

To simplify $\frac{15}{20}$, the fact that $\frac{5}{5} = 1$ is used. The simplification of $\frac{15}{20}$ is shown below.

$$\frac{15}{20} = \frac{5 \cdot 3}{5 \cdot 4} = \frac{5}{5} \cdot \frac{3}{4} = 1 \cdot \frac{3}{4} = \frac{3}{4}$$

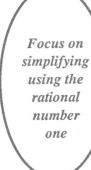

Focus on simplifying using the rational number one

To simplify $\frac{-6}{8}$, first determine that 2 is the highest common factor of 6 and 8. Since $\frac{2}{2} = 1$, the simplification is done as follows.

$$\frac{-6}{8} = \frac{2 \cdot -3}{2 \cdot 4} = \frac{2}{2} \cdot \frac{-3}{4} = 1 \cdot \frac{-3}{4} = \frac{-3}{4}$$

Simplifying Fractions with Variables

$\frac{3x}{4x}$ is an open rational expression. The highest common factor of 3x and 4x will be the number that replaces x.

To simplify the open expression $\frac{3x}{4x}$, the fact that $\frac{x}{x} = 1$ for all non-zero replacements of x is used.

$$\frac{3x}{4x} = \frac{3 \cdot x}{4 \cdot x} = \frac{3}{4} \cdot \frac{x}{x} = \frac{3}{4} \cdot 1 = \frac{3}{4}$$

$\frac{3x}{4x}$ simplifies to $\frac{3}{4}$

Focus on simplifying open rational expressions

To simplify $\frac{-5x}{10x}$, simplify $\frac{-5}{10}$ and $\frac{x}{x}$ separately.

$\frac{-5x}{10x}$ is simplified as shown below.

$$\frac{-5x}{10x} = \frac{-5}{10} \cdot \frac{x}{x} = \frac{-1}{2} \cdot 1 = \frac{-1}{2}$$

$\frac{-5x}{10x}$ simplifies to $\frac{-1}{2}$

Cancelling — A Valuable Shortcut

Recall that the multiplication of $\frac{3}{5} \cdot \frac{2}{7}$ is completed by multiplying the numerators and separately multiplying the denominators.

$$\frac{3}{5} \cdot \frac{2}{7} = \frac{3 \cdot 2}{5 \cdot 7} = \frac{6}{35}$$

The multiplication of $\frac{3}{4} \cdot \frac{7}{3}$ can be simplified by the "cancelling" process. The first step in multiplying $\frac{3}{4} \cdot \frac{7}{3}$ is to cancel like factors. The multiplication of $\frac{3}{4} \cdot \frac{7}{3}$ is simplified as follows.

$$\frac{3}{4} \cdot \frac{7}{3} = \frac{\overset{1}{\cancel{3}}}{4} \cdot \frac{7}{\underset{1}{\cancel{3}}} = \frac{1}{4} \cdot \frac{7}{1} = \frac{7}{4}$$

To evaluate $\frac{-6}{5} \cdot \frac{7}{6}$, the following steps are used.

$$\frac{-6}{5} \cdot \frac{7}{6} = \frac{\overset{-1}{\cancel{-6}}}{5} \cdot \frac{7}{\underset{1}{\cancel{6}}} = \frac{-1}{5} \cdot \frac{7}{1} = \frac{-7}{5}$$

Focus on the use of cancelling

To simplify the multiplication of $\frac{-5}{7} \cdot \frac{8}{5}$, the following steps are used.

$$\frac{-5}{7} \cdot \frac{8}{5} = \frac{\overset{-1}{\cancel{-5}}}{7} \cdot \frac{8}{\underset{1}{\cancel{5}}} = \frac{-1}{7} \cdot \frac{8}{1} = \frac{-8}{7}$$

To evaluate $\frac{12}{5} \cdot \frac{7}{6}$, use cancellation. The complete simplification of $\frac{12}{5} \cdot \frac{7}{6}$ is shown below.

$$\frac{12}{5} \cdot \frac{7}{6} = \frac{\overset{2}{\cancel{12}}}{5} \cdot \frac{7}{\underset{1}{\cancel{6}}} = \frac{2}{5} \cdot \frac{7}{1} = \frac{14}{5}$$

To simplify the evaluation of $\frac{9}{7} \cdot \frac{14}{3}$, the following steps are used.

$$\frac{9}{7} \cdot \frac{14}{3} = \frac{\overset{3}{\cancel{9}}}{\underset{1}{\cancel{7}}} \cdot \frac{\overset{2}{\cancel{14}}}{\underset{1}{\cancel{3}}} = \frac{3}{1} \cdot \frac{2}{1} = \frac{6}{1} = 6$$

Unit 1 Exercise

This exercise reviews the preceding unit. The exercise is divided into three parts. Part A reviews the foci of Unit 1. Part B offers opportunities to practice the skills and concepts of Unit 1. Part C contains problems that review your previous work in this text. All answers for Parts A and B are at the back of the book. Each problem of Part C is accompanied by a notation «CU» that refers to the Chapter (C) and Unit (U) in which that type of problem is studied.

Part A: Reviewing the foci of Unit 1.

1.   Is $\frac{46}{-3}$ a rational number?

2.   Is $\frac{-15}{0}$ a rational number?

3.   Write the integer 9 as a rational number.

4.   Write -15 as a rational number.

5.   Is $4 \cdot 3 = 6 \cdot 2$ a true statement?

6. $\frac{4}{6} = \frac{2}{3}$ is true because $4 \cdot 3 = 6 \cdot$ ___ is a true statement.

Determine if each of the following equalities is true.

7. $\frac{7}{12} = \frac{9}{20}$

8. $\frac{-4}{7} = \frac{12}{-21}$

9. $\frac{0}{8} = \frac{0}{-6}$

10. $\frac{-20}{-4} = 5$

11. Is $\frac{18}{18} = 1$ a true statement?

12. Is $\frac{-12}{-12} = 1$ a true statement?

Write each fraction with a positive denominator.

13. $\frac{4}{-9}$

14. $\frac{-9}{-8}$

15. $\frac{-7}{-15}$

16. $\frac{9}{-16}$

Evaluate.

17. $\frac{1}{2} \cdot \frac{3}{4}$

18. $\frac{2}{7} \cdot \frac{-4}{3}$

19. $\frac{-4}{5} \cdot \frac{-2}{7}$

20. $\frac{-9}{4} \cdot \frac{-3}{2}$

21. $\frac{-4}{5} \cdot 6$

22. $\frac{-7}{16} \cdot 0$

23. $\frac{-3}{4} \cdot 1$

24. $\frac{4}{11} \cdot -1$

25. $\frac{1}{1} \cdot \frac{-8}{3}$

Simplify.

26. $\frac{10}{25}$

27. $\frac{14}{49}$

28. $\frac{7}{28}$

29. $\frac{-10}{16}$

30. $\frac{-3}{-18}$

31. $\frac{14}{-7}$

32. $\frac{9x}{10x}$

33. $\frac{-8x}{40x}$

Evaluate each of the problems using cancelling wherever possible.

34. $\frac{3}{4} \cdot \frac{5}{7}$

35. $\frac{-3}{7} \cdot \frac{2}{5}$

36. $\frac{2}{9} \cdot \frac{9}{11}$

37. $\frac{-3}{11} \cdot \frac{11}{14}$

38. $\frac{14}{5} \cdot \frac{-3}{14}$

39. $\frac{4}{7} \cdot \frac{3}{4}$

40. $\frac{-4}{9} \cdot \frac{11}{4}$

41. $\frac{5}{6} \cdot \frac{-6}{7}$

42. $\frac{-9}{16} \cdot \frac{-5}{9}$

43. $\frac{-17}{3} \cdot \frac{-2}{17}$

44. $\frac{3}{5} \cdot \frac{10}{7}$

45. $\frac{8}{9} \cdot \frac{3}{4}$

46. $\frac{14}{3} \cdot \frac{6}{7}$

47. $\frac{2}{3} \cdot \frac{3}{2}$

Part B: Drill and Practice

Determine whether each of the following is a rational number.

1. $\frac{14}{25}$

2. $\frac{-16}{9}$

3. $\frac{0}{6}$

4. $\frac{0}{-4}$

5. $\frac{15}{0}$

6. $\frac{-5}{-9}$

7. $\frac{3}{0}$

Simplify.

8. $\frac{9}{15}$

9. $\frac{24}{8}$

10. $\frac{30}{18}$

11. $\frac{8}{20}$

12. $\frac{-9}{6}$

13. $\frac{-7}{35}$

14. $\frac{12}{48}$

15. $\frac{-8}{24}$

16. $\frac{12}{-14}$

17. $\frac{15}{20}$

18. $\frac{10x}{7x}$

19. $\frac{-8x}{5x}$

20. $\frac{12x}{17x}$

21. $\frac{-5x}{13x}$

22. $\frac{-9x}{-16x}$

23. $\frac{27x}{8x}$

24. $\frac{5x}{-9x}$

25. $\frac{-14x}{7x}$

26. $\frac{27x}{3x}$

27. $\frac{9x}{15x}$

28. $\frac{-20x}{15x}$

29. $\frac{10x}{2x}$

30. $\frac{-9x}{12x}$

31. $\frac{-36x}{-9x}$

32. $\frac{-32x}{16x}$

33. $\frac{12x}{40x}$

34. $\frac{-9x}{45x}$

Evaluate.

35. $\frac{3}{4} \cdot \frac{5}{7}$

36. $\frac{1}{5} \cdot \frac{2}{7}$

37. $\frac{-2}{7} \cdot \frac{1}{5}$

38. $\frac{7}{9} \cdot \frac{-2}{5}$

39. $\frac{2}{7} \cdot 4$

40. $5 \cdot \frac{-13}{14}$

41. $\frac{4}{3} \cdot \frac{-14}{5}$

42. $\frac{4}{7} \cdot \frac{-3}{5}$

43. $0 \cdot \frac{4}{7}$

44. $\frac{-9}{7} \cdot \frac{-3}{2}$

45. $\frac{-1}{2} \cdot \frac{-3}{5}$

46. $\frac{-9}{7} \cdot \frac{1}{4}$

47. $\frac{-3}{4} \cdot 1$

48. $\frac{-5}{8} \cdot 0$

49. $\frac{5}{4} \cdot \frac{7}{3}$

50. $\frac{-4}{9} \cdot 1$

51. $\frac{-3}{4} \cdot \frac{9}{5}$

52. $0 \cdot \frac{-0}{7}$

53. $\frac{4}{11} \cdot 0$

54. $\frac{-2}{3} \cdot \frac{-5}{7}$

55. $\frac{5}{3} \cdot \frac{2}{9}$

56. $\frac{-3}{5} \cdot \frac{4}{7}$

57. $\frac{2}{9} \cdot 0$

58. $\frac{-7}{3} \cdot \frac{-4}{5}$

59. $\frac{-6}{7} \cdot 3$

60. $\frac{-8}{9} \cdot \frac{9}{17}$

61. $\frac{4}{7} \cdot \frac{-9}{4}$

62. $\frac{-11}{4} \cdot \frac{4}{3}$

63. $\frac{7}{15} \cdot \frac{8}{7}$

64. $\frac{11}{3} \cdot \frac{8}{11}$

65. $\frac{7}{5} \cdot \frac{-11}{7}$

66. $\frac{6}{13} \cdot \frac{13}{7}$

67. $\frac{-10}{11} \cdot \frac{11}{19}$

68. $\frac{-7}{4} \cdot \frac{-4}{9}$

69. $\frac{-14}{3} \cdot \frac{-3}{11}$

70. $\frac{6}{13} \cdot \frac{-13}{11}$

71. $\frac{15}{4} \cdot \frac{7}{15}$

72. $\frac{-3}{11} \cdot \frac{-7}{3}$

73. $\frac{3}{13} \cdot \frac{-7}{3}$

74. $\frac{11}{14} \cdot \frac{5}{11}$

75. $\frac{7}{16} \cdot \frac{-9}{7}$

Part C: Review. Answers for the problems of Part C are not given. However, the notation «C,U» refers to the Chapter (C) and Unit (U) in which problems of the same type were presented.

1. 63,592 – 41,804   «1,3»

2. Find the lcm for 14 and 35.   «3,2»

3. Change $2\frac{7}{8}$ to an improper fraction.   «3,2»

4. $4\frac{1}{3} \div \frac{7}{12}$ «3,6»

5. Subtract 8.17 from 12.3   «4,2»

6. Evaluate $(5 + 3 \cdot 6) + 4$ «5,2»

7. Simplify $(24 + x) + 3$ «5,4»

8. Simplify $(3x + 2) + 3(x + 5)$ «5,6»

9. Evaluate $4 + (-8)$ «6,1»

10. Evaluate $-6 - 3 - 2$ «6,2»

11. Evaluate $-9 \cdot -6$ «6,3»

12. Simplify $7(4x)$ «6,5»

13. Simplify $(5x + 3) - (x - 7)$ «6,6»

14. Solve $7x - 5 = 37$ «6,7»

15. Solve $4 + 3(2 - 3x) = 28$ «6,8»

16. A first number is the same as 9 decreased by a second number. If s represents the second number, write an open expression for the first number.   «5,8»

17. Find the sum of 4.833 and 7.43   «4,2»

18. Divide 9.76 by 0.8 and round off the answer to 2 decimal places.   «4,4»

19. In three successive months, Marty's income was $1,485, $2,316, and $1,812. What was the average monthly income during that 3-month period?   «4,4»

20. Change $\frac{1}{6}$ to a decimal fraction rounded off at two decimal places.   «4,4»

21. Write an equation that shows: the difference when 8 is subtracted from a number is 32. «6,9»

Unit 2:   ADDING RATIONAL NUMBERS

The First Step in Adding Rational Numbers

Addition can only be performed on numbers with the same
labels. For rational numbers, this means the first step in addition
always involves the need for a common denominator.

To evaluate $\frac{-2}{3}$ + $\frac{4}{7}$, the following steps are used.

$$\frac{-2}{3} + \frac{4}{7} = \frac{-14}{21} + \frac{12}{21} = \frac{-2}{21}$$

*Focus on
the addition
of rational
numbers*

To evaluate $\frac{-7}{4}$ + $\frac{-2}{3}$, the following steps are used.

$$\frac{-7}{4} + \frac{-2}{3} = \frac{-21}{12} + \frac{-8}{12} = \frac{-29}{12}$$

Do not attempt to write the rational numbers of this section as
mixed numbers.

$$\frac{-29}{12} \text{ is preferred over } -2\frac{5}{12}$$

Adding an Integer and a Rational Number

To evaluate -2 + $\frac{5}{7}$, which is the sum of an integer and a
rational number, the integer should always be written first
as a rational number; –2 should be written as $\frac{-2}{1}$.

To evaluate -2 + $\frac{5}{7}$, the following steps are used.

$$-2 + \frac{5}{7} = \frac{-2}{1} + \frac{5}{7} = \frac{-14}{7} + \frac{5}{7} = \frac{-9}{7}$$

Focus on adding an integer and a rational number

The addition of 6 and $\frac{-3}{4}$ requires writing the integer as a rational number and then finding a common denominator.

$$6 + \frac{-3}{4} = \frac{6}{1} + \frac{-3}{4} = \frac{24}{4} + \frac{-3}{4} = \frac{21}{4}$$

$\frac{21}{4}$ or $5\frac{1}{4}$ is the sum of 6 and $\frac{-3}{4}$; $\frac{21}{4}$ is preferred.

Simplifying Before Adding

To add $\frac{-5}{-8}$ and $\frac{1}{-6}$ both rational numbers need to be written with positive denominators.

$$\frac{-5}{-8} + \frac{1}{-6} = \frac{5}{8} + \frac{-1}{6} = \frac{15}{24} + \frac{-4}{24} = \frac{11}{24}$$

Focus on simplifying before adding

To evaluate $\frac{-3}{-2} + \frac{5}{3}$, first change $\frac{-3}{-2}$ to $\frac{3}{2}$ as shown in the first step of the evaluation below.

$$\frac{-3}{-2} + \frac{5}{3} = \frac{3}{2} + \frac{5}{3} = \frac{9}{6} + \frac{10}{6} = \frac{19}{6}$$

Opposites of Rational Numbers

To find the opposite of any rational number:

    1. change the numerator to its opposite;

    2. do not change the denominator.

The opposite of $\frac{7}{16}$ is $\frac{-7}{16}$ because $\frac{7}{16} + \frac{-7}{16} = 0$.

Definition   **Subtraction of Rational Numbers**

$\frac{c}{d}$ is subtracted   **If and**   the opposite of $\frac{c}{d}$

from $\frac{a}{b}$ **only if**   is added to $\frac{a}{b}$.

$$\frac{a}{b} - \frac{c}{d} \quad \text{means} \quad \frac{a}{b} + \frac{-c}{d}$$

The Subtraction of Rational Numbers

Whenever the minus sign is between two rational numbers, the opposite of the second number is to be added to the first number.

$\frac{2}{7} - \frac{1}{3}$ means $\frac{2}{7} + \frac{-1}{3}$, because the minus sign between any two rational numbers indicates that the opposite of the second number is to be added to the first number.

Similarly, $\frac{3}{4} - \frac{2}{5}$ means $\frac{3}{4} + \frac{-2}{5}$.

Focus on subtracting rational numbers

The use of the minus sign only affects the number that immediately follows it. The minus sign between any two rational numbers means that the opposite of the second number is to be added to the first number.

$$\frac{-3}{9} - \frac{1}{4} \text{ means } \frac{-3}{9} + \frac{-1}{4} \text{ and } \frac{6}{7} - \frac{2}{3} \text{ means } \frac{6}{7} + \frac{-2}{3}.$$

The minus sign always means to add the opposite of the second number.

$$\frac{3}{4} - \frac{-5}{8} \text{ means } \frac{3}{4} + \frac{5}{8}, \text{ because } \frac{5}{8} \text{ is the opposite of } \frac{-5}{8}.$$

Similarly, $\frac{-7}{8} - \frac{-3}{5}$ means $\frac{-7}{8} + \frac{3}{5}$.

Evaluating Subtraction Expressions

To evaluate $\frac{3}{4} - \frac{1}{3}$, the following steps are used.

$$\frac{3}{4} - \frac{1}{3} = \frac{3}{4} + \frac{-1}{3} = \frac{9}{12} + \frac{-4}{12} = \frac{5}{12}$$

Focus on evaluating subtraction expressions

To evaluate $\frac{4}{5} - \frac{-1}{3}$, the following steps are used.

$$\frac{4}{5} - \frac{-1}{3} = \frac{4}{5} + \frac{1}{3} = \frac{12}{15} + \frac{5}{15} = \frac{17}{15}$$

Addition Expressions with More Than Two Terms

Only two rational numbers can be added at one time. The parentheses in the numerical expression, $(\frac{1}{3} + \frac{-1}{2}) + \frac{2}{5}$, indicate that $\frac{1}{3}$ and $\frac{-1}{2}$ are to be added first.

To evaluate $(\frac{1}{3} + \frac{-1}{2}) + \frac{2}{5}$, the following steps are used.

$$\begin{aligned}
(\frac{1}{3} + \frac{-1}{2}) + \frac{2}{5} &= (\frac{2}{6} + \frac{-3}{6}) + \frac{2}{5} \\
&= \frac{-1}{6} + \frac{2}{5} \\
&= \frac{-5}{30} + \frac{12}{30} \\
&= \frac{7}{30}
\end{aligned}$$

To evaluate $\frac{1}{2} + (\frac{1}{3} - \frac{8}{5})$, the following steps are used.

Focus on evaluating additions with three addends

$$\frac{1}{2} + (\frac{1}{3} - \frac{8}{5}) = \frac{1}{2} + (\frac{1}{3} + \frac{-8}{5})$$

$$= \frac{1}{2} + (\frac{5}{15} + \frac{-24}{15})$$

$$= \frac{1}{2} + \frac{-19}{15}$$

$$= \frac{15}{30} + \frac{-38}{30}$$

$$= \frac{-23}{30}$$

Notice that the evaluation of the expression in parentheses is always done first.

Unit 2 Exercise

This exercise reviews the preceding unit. The exercise is divided into three parts. Part A reviews the foci of Unit 2. Part B offers opportunities to practice the skills and concepts of Unit 2. Part C contains problems that review your previous work in this text. All answers for Parts A and B are at the back of the book. Each problem of Part C is accompanied by a notation «CU» that refers to the Chapter (C) and Unit (U) in which that type of problem is studied.

Part A: Reviewing the foci of Unit 2.

1. $\frac{3}{8} - \frac{2}{5}$ means _____.

2. $\frac{-5}{7} - \frac{-3}{8}$ means _____.

3. Evaluate $\frac{2}{-3} + \frac{-5}{8}$ by first simplifying $\frac{2}{-3}$.

4. Evaluate $\frac{2}{-7} + \frac{5}{-3}$ by first simplifying both numbers.

Evaluate each of the following by first finding a common denominator.

5. $\frac{3}{7} + \frac{2}{5}$

6. $\frac{3}{4} + 5$

12. $\frac{3}{10} + \frac{1}{5}$

Evaluate each of the following by first evaluating the expression inside the parentheses.

7. $\frac{-3}{4} + \frac{2}{5}$

13. $\frac{3}{7} + 0$

17. $(\frac{-7}{3} - \frac{-1}{2}) + \frac{1}{5}$

8. $\frac{-5}{3} + \frac{-3}{4}$

14. $\frac{4}{-3} + \frac{5}{8}$

18. $\frac{3}{4} + (\frac{-1}{3} - \frac{2}{5})$

9. $\frac{1}{6} + \frac{1}{3}$

15. $\frac{5}{6} - \frac{2}{5}$

19. $(\frac{1}{3} - \frac{2}{5}) - \frac{-1}{2}$

10. $\frac{2}{7} + \frac{3}{7}$

16. $\frac{4}{7} - \frac{-2}{5}$

20. $\frac{3}{7} - (\frac{2}{5} + \frac{1}{3})$

11. $\frac{-3}{4} + \frac{-1}{2}$

Part B: Drill and Practice

1. $\frac{4}{3} + \frac{1}{7}$

7. $\frac{2}{5} + \frac{4}{7}$

13. $\frac{-7}{3} + \frac{1}{-2}$

2. $\frac{4}{9} + \frac{2}{5}$

8. $\frac{7}{10} + \frac{1}{5}$

14. $\frac{3}{-8} + \frac{-5}{1}$

3. $\frac{2}{7} + 3$

9. $\frac{-1}{5} + \frac{5}{6}$

15. $\frac{-7}{5} + \frac{3}{4}$

4. $\frac{1}{8} + \frac{2}{3}$

10. $\frac{2}{9} + \frac{-2}{5}$

16. $\frac{5}{9} + \frac{1}{4}$

5. $\frac{5}{6} + \frac{1}{5}$

11. $\frac{-5}{8} + \frac{-3}{7}$

17. $\frac{11}{12} + \frac{-3}{5}$

6. $\frac{3}{5} + 2$

12. $\frac{-2}{7} + \frac{4}{-5}$

18. $\frac{-6}{5} + 0$

19. $\frac{-4}{5} + \frac{-1}{4}$

20. $\frac{3}{-2} + \frac{4}{-7}$

21. $\frac{-2}{-3} + \frac{7}{4}$

22. $\frac{-2}{3} + 0$

23. $\frac{5}{4} + \frac{1}{7}$

24. $\frac{2}{3} + \frac{-1}{5}$

25. $\frac{-5}{2} + \frac{1}{-4}$

26. $\frac{-1}{2} + \frac{-3}{4}$

27. $\frac{2}{3} + \frac{-1}{-6}$

28. $\frac{4}{9} + \frac{-4}{9}$

29. $\frac{-7}{3} + \frac{7}{3}$

30. $\frac{6}{7} - \frac{1}{3}$

31. $\frac{-3}{5} - \frac{1}{6}$

32. $\frac{-7}{8} - \frac{-2}{5}$

33. $\frac{5}{11} + \frac{-4}{3}$

34. $\frac{5}{8} - \frac{-2}{3}$

35. $\frac{-3}{8} + \frac{1}{3}$

36. $\frac{-2}{3} + \frac{-3}{5}$

37. $\frac{1}{3} - \frac{1}{5}$

38. $\frac{2}{5} + \frac{1}{4}$

39. $\frac{3}{7} + \frac{-1}{2}$

40. $\frac{3}{8} - \frac{-1}{5}$

41. $\frac{7}{5} - \frac{-1}{4}$

42. $\frac{-8}{5} + \frac{-2}{3}$

43. $\frac{3}{4} - \frac{5}{7}$

44. $\frac{-4}{3} - \frac{1}{8}$

45. $\frac{-8}{3} - \frac{-3}{4}$

46. $\frac{3}{5} - (\frac{2}{3} - \frac{1}{4})$

47. $(\frac{1}{2} - \frac{1}{3}) - \frac{1}{4}$

48. $(4 - \frac{1}{3}) + \frac{3}{2}$

49. $\frac{1}{5} + (\frac{2}{3} - \frac{1}{3})$

50. $\frac{2}{5} + (\frac{1}{3} - \frac{1}{2})$

51. $\frac{1}{7} - (\frac{1}{2} - \frac{1}{3})$

52. $(\frac{5}{2} - \frac{3}{4}) - \frac{2}{7}$

53. $\frac{5}{8} + (\frac{1}{3} + \frac{1}{2})$

54. $(\frac{1}{2} - \frac{3}{5}) - \frac{1}{4}$

55. $\frac{7}{8} + (2 - \frac{1}{3})$

Part C: Review. Answers for the problems of Part C are not given. However, the notation «C,U» refers to the Chapter (C) and Unit (U) in which problems of the same type were presented.

1.   48,090 + 53,099 + 4,532   «1,2»

2.   Simplify $\frac{18}{34}$ «3,1»

3.   Find the numerator of $\frac{5}{9} = \frac{?}{45}$ «3,2»

4. $8\frac{2}{3} \times \frac{4}{9}$ «3,5»

5.   Add 8.947, 13.38, and 0. 087   «4,2»

6.   Find the difference of 23.47 and 19.6   «4,2»

7.   Find the product of 4.789 and 3.5   «2,2»

8.   Simplify (6 + 2x) + 8   «5,4»

9.   Simplify 5(2x + 3) + 2(x + 3)   «5,6»

10. Evaluate 9 − (-3)   «6,2»

11. Evaluate 4 − (-10) • -2 ÷ 5   «6,4»

12. Simplify 5(2x − 1) + 7   «6,5»

13. Simplify 4(2x − 1) − 3(x − 4)   «6,6»

14. Solve 8x − 9 = 39   «6,7»

15. Solve 4 − 3x = 10 − x   «6,8»

16. Evaluate $\frac{-7}{4} \cdot \frac{-4}{7}$ «7,1»

17. How many $\frac{1}{4}$ inches are there in 8 inches?
     «3,7»

18. Mark's test scores in chemistry were 78, 82, 78, and 56. Find his average on these four tests rounded off to the nearest whole number.
     «4,4»

19. Change $\frac{5}{6}$ to a decimal fraction rounded off at two decimal places.   «4,4»

20. One number is five times a second number. If s represents the second number, write an open expression for the first number.   «5,8»

21. Write an equation that shows: 6 less than a number is 33.   «6,9»

Unit 3:   POSTULATES FOR ADDITION

Evaluating Open Expressions

$7 - 3x$ is an open expression. The letter x represents a place for a rational number. If x is replaced by $\frac{1}{2}$, the open expression $7 - 3x$ becomes the numerical expression $7 - 3 \cdot \frac{1}{2}$.

The evaluation of $7 - 3 \cdot \frac{1}{2}$ is:

$$7 - 3 \cdot \frac{1}{2} = 7 - \frac{3}{2} = \frac{14}{2} - \frac{3}{2} = \frac{11}{2}$$

Notice that there are no parentheses in the expression $7 - 3 \cdot \frac{1}{2}$. Whenever a numerical expression has no parentheses, multiplication is performed before addition.

Focus on evaluating an open expression with no parentheses

If $\frac{5}{2}$ is used as a replacement for x in the open expression -2 + 3x, the numerical expression $-2 + 3 \cdot \frac{5}{2}$ is obtained.

$-2 + 3 \cdot \frac{5}{2}$ has an evaluation of $\frac{11}{2}$ as shown.

$-2 + 3x$ has an evaluation of $\frac{11}{2}$ when x $= \frac{5}{2}$.

If $\frac{-2}{3}$ is used as a replacement for x in the open expression -2 + 3x, the numerical expression $-2 + 3 \cdot \frac{-2}{3}$ is obtained.

$-2 + 3 \cdot \frac{-2}{3}$ has an evaluation of -4 as shown.

$-2 + 3x$ has an evaluation of -4 when x $= \frac{-2}{3}$.

$-2 + 3x$ when x $= \frac{5}{2}$

$-2 + 3 \cdot \frac{5}{2}$

$-2 + \frac{15}{2}$

$\frac{-2}{1} + \frac{15}{2}$

$\frac{-4}{2} + \frac{15}{2}$

$\frac{11}{2}$

$-2 + 3x$ when x $= \frac{-2}{3}$

$-2 + 3 \cdot \frac{-2}{3}$

$-2 + (-2)$

$-2 - 2$

-4

Evaluating Open Expressions with Parentheses

The parentheses in $2(y + 4)$ indicate that when y is replaced by a rational number, the addition of the numerical expression in the parentheses should be performed first.

The evaluation of $2(y + 4)$ when $y = \frac{-1}{2}$ is shown below.

$$2(y + 4) \quad \text{when } y = \frac{-1}{2}$$

$$2(\frac{-1}{2} + 4)$$

$$2(\frac{-1}{2} + \frac{8}{2})$$

$$2 \cdot \frac{7}{2}$$

$$7$$

Focus on evaluating an open expression with parentheses

If $\frac{8}{3}$ is used as a replacement in the open expression $-3(x - 4)$, the numerical expression $-3(\frac{8}{3} - 4)$ is obtained.

$$-3(x - 4) \quad \text{when } x = \frac{8}{3}$$

$$-3(\frac{8}{3} - 4)$$

$$-3(\frac{8}{3} - \frac{12}{3})$$

$$-3 \cdot \frac{-4}{3}$$

$$4$$

$-3(\frac{8}{3} - 4)$ has an evaluation of 4 as shown.

$-3(x - 4)$ has an evaluation of 4 when $x = \frac{8}{3}$.

The evaluation of $\frac{2}{3} - (7 - x)$ when $x = \frac{17}{5}$ is shown at the right.

$$\frac{2}{3} - (7 - x) \quad \text{when } x = \frac{17}{5}$$

$$\frac{2}{3} - (7 - \frac{17}{5})$$

$$\frac{2}{3} - (\frac{35}{5} - \frac{17}{5})$$

$$\frac{2}{3} - \frac{18}{5}$$

$$\frac{10}{15} - \frac{54}{15}$$

$$\frac{-44}{15}$$

Notice the manner in which the minus sign is used. $\frac{2}{3} - (7 - x)$ has an evaluation of $\frac{-44}{15}$ when $x = \frac{17}{5}$.

The Commutative Law of Addition

The equation $x + y = y + x$ becomes a true equality when x is replaced by $\frac{1}{4}$ and y is replaced by $\frac{3}{5}$.

$$\frac{1}{4} + \frac{3}{5} = \frac{3}{5} + \frac{1}{4}$$

$x + y = y + x$ will become a true equality for all number replacements for x and y. This is the Commutative Law of Addition.

The Commutative Law of Addition states $x + y$ is equivalent to $y + x$ for all number replacements of x and y. The order of two addends does not effect the sum.

Focus on the Commutative Law of Addition

The Commutative Law of Addition states that the sum of two numbers is not affected by the order in which they are added.

By the Commutative Law of Addition the open expression $a + \frac{2}{7}$ is equivalent to $\frac{2}{7} + a$. In other words, for any number replacement of a, $a + \frac{2}{7}$ and $\frac{2}{7} + a$ will have the same evaluation.

$a + \frac{2}{7}$ and $\frac{2}{7} + a$ may be substituted for each other.

The Associative Law of Addition

The Associative Law of Addition states that $(x + y) + z$ is equivalent to $x + (y + z)$ for all number replacements of x, y, and z.

The Associative Law of Addition states that the sum of three numbers is not affected by the grouping of the numbers.

$(x + y) + z = x + (y + z)$ will become a true equality for all number replacements of x, y, and z.

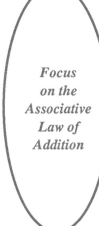

By the Associative Law of Addition $(x + \frac{2}{3}) - \frac{1}{2}$ is equivalent to $x + (\frac{2}{3} - \frac{1}{2})$. Each expression contains x, $\frac{2}{3}$, and $\frac{-1}{2}$.

Focus on the Associative Law of Addition

This means that $x + (\frac{2}{3} - \frac{1}{2})$ can be substituted for $(x + \frac{2}{3}) - \frac{1}{2}$.

Such a substitution leads to the simplification of $(x + \frac{2}{3}) - \frac{1}{2}$ shown at the right.

$$(x + \tfrac{2}{3}) - \tfrac{1}{2}$$

$$x + (\tfrac{2}{3} - \tfrac{1}{2})$$

$$x + \tfrac{1}{6}$$

$(x + \frac{2}{3}) - \frac{1}{2}$ has been simplified to $x + \frac{1}{6}$. Any number can replace x and the two expressions will have the same evaluations.

Simplifying Addition Expressions

To simplify $(\frac{5}{6} + x) - \frac{3}{8}$, both the Commutative and Associative Laws of Addition are used.

By the Commutative Law of Addition:

$(\frac{5}{6} + x) - \frac{3}{8}$ is equivalent to $(x + \frac{5}{6}) - \frac{3}{8}$

By the Associative Law of Addition:

$(x + \frac{5}{6}) - \frac{3}{8}$ is equivalent to $x + (\frac{5}{6} - \frac{3}{8})$

Therefore, $(\frac{5}{6} + x) - \frac{3}{8}$ can be simplified to $x + \frac{11}{24}$.

Focus on simplifying addition expressions

To simplify $(3x - \frac{4}{7}) + \frac{2}{3}$, first note that the expression contains 3x, $\frac{-4}{7}$, and $\frac{2}{3}$.

Using the Associative Law of Addition to regroup the terms leads to the simplification of $(3x - \frac{4}{7}) + \frac{2}{3}$.

$(3x - \frac{4}{7}) + \frac{2}{3}$ simplifies to $3x + \frac{2}{21}$.

$$(3x - \tfrac{4}{7}) + \tfrac{2}{3}$$

$$(3x + \tfrac{-4}{7}) + \tfrac{2}{3}$$

$$3x + (\tfrac{-4}{7} + \tfrac{2}{3})$$

$$3x + (\tfrac{-12}{21} + \tfrac{14}{21})$$

$$3x + \tfrac{2}{21}$$

The Addition Law of Zero

$x + 0 = x$ will become a true equality for all number replacements for x. This is the Addition Law of Zero.

The Addition Law of Zero states $x + 0$ is equivalent to x for all number replacements of x.

$$\tfrac{2}{3}x + 0 \text{ is equivalent to } \tfrac{2}{3}x$$

$$0 - \tfrac{9}{7}x \text{ is equivalent to } \tfrac{-9}{7}x$$

$$\tfrac{-5}{8}x + (\tfrac{-3}{4} + \tfrac{3}{4}) \text{ is equivalent to } \tfrac{-5}{8}x$$

Focus on the Addition Law of Zero

The Addition Law of Zero states that the sum of zero and any number is equal to that same number. For this reason, zero is sometimes called the Identity Element for Addition.

By the Addition Law of Zero the open expression $\tfrac{-2}{3}x + 0$ is equivalent to $\tfrac{-2}{3}x$. In other words, for any number replacement of x, $\tfrac{-2}{3}x + 0$ and $\tfrac{-2}{3}x$ will have the same evaluation.

$\tfrac{2}{3}x + 0$ and $\tfrac{-2}{3}x$ may be substituted for each other.

The Law of Additive Inverses

In the set of rational numbers every number has an opposite.

$$\text{The opposite of } \tfrac{-4}{3} \text{ is } \tfrac{4}{3}.$$

$$\text{The opposite of } \tfrac{5}{12} \text{ is } \tfrac{-5}{12}.$$

The introduction of signed numbers in Chapter 6 made it possible for every number to have an opposite. The Inverse Law of Addition states that every number has an opposite and the sum of two opposites is zero.

$\tfrac{x}{y} + \tfrac{-x}{y} = 0$, when x is any integer, and y is any integer except zero.

Unit 3 Exercise

This exercise reviews the preceding unit. The exercise is divided into three parts. Part A reviews the foci of Unit 3. Part B offers opportunities to practice the skills and concepts of Unit 3. Part C contains problems that review your previous work in this text. All answers for Parts A and B are at the back of the book. Each problem of Part C is accompanied by a notation «CU» that refers to the Chapter (C) and Unit (U) in which that type of problem is studied.

Part A: Reviewing the foci of Unit 3.

1. Evaluate $-2a - 3$ when $a = \frac{1}{4}$.

2. Evaluate $-5(x - \frac{1}{4})$ when $x = \frac{2}{5}$.

3. Evaluate $2a + 3b$ when $a = \frac{3}{2}$ and $b = \frac{-5}{6}$.

4. According to the Commutative Law of Addition $x - \frac{1}{4}$ is equivalent to _____.

5. According to the Commutative Law of Addition $\frac{5}{9} - 3x$ is equivalent to _____.

6. According to the Associative Law of Addition $(x - \frac{2}{3}) + \frac{3}{7}$ is equivalent to _____.

7. According to the Associative Law of Addition $\frac{8}{3} + (x - \frac{1}{5})$ is equivalent to _____.

8. Simplify $(z - \frac{1}{5}) + \frac{1}{4}$ using the Commutative and/or Associative Laws of Addition. _____.

9. Simplify $(\frac{2}{5} + x) + \frac{6}{7}$ using the Commutative and/or Associative Laws of Addition. _____.

10. According to the Addition Law of Zero, $\frac{1}{4}x + 0$ is equivalent to _____.

11. According to the Addition Law of Zero, $\frac{-3}{10}x + (\frac{-5}{9} + \frac{5}{9})$ is equivalent to _____.

12. According to the Inverse Law of Addition, there is a number that can be added to $\frac{4}{9}$ to give 0. What is this number?

13. According to the Inverse Law of Addition, there is an expression that can be added to $\frac{3}{4}x$ to give 0. What is this expression?

Part B: Drill and Practice

Evaluate.

1. $3x + \frac{1}{2}$ when $x = \frac{2}{5}$

2. $\frac{2}{3}x - \frac{1}{4}$ when $x = \frac{3}{8}$

3. $a + \frac{5}{2}$ when $a = \frac{5}{8}$

4. $\frac{4}{7}z - \frac{1}{3}$ when $z = \frac{1}{2}$

5. $\frac{1}{3} - 4r$ when $r = \frac{1}{2}$

6. $\frac{1}{7} + \frac{1}{2}c$ when $c = -3$

7. $\frac{1}{3}x + \frac{2}{3}$ when $x = \frac{3}{4}$

8. $\frac{2}{3}x + 5y + \frac{1}{4}z$ when $x = \frac{3}{10}$, $y = \frac{4}{5}$, and $z = 8$

9. $a - 3b$ when $a = \frac{7}{3}$ and $b = \frac{1}{4}$

10. $4x + 3c$ when $x = \frac{1}{5}$ and $c = \frac{1}{4}$

11. $2ab$ when $a = \frac{7}{4}$ and $b = \frac{1}{6}$

12. $\frac{1}{4}x + 3y$ when $x = \frac{2}{3}$ and $y = \frac{5}{12}$

Simplify.

13. $(8x - \frac{7}{8}) + \frac{5}{4}$

14. $(3x - \frac{2}{3}) + \frac{5}{9}$

15. $(\frac{1}{5}x - \frac{5}{7}) + \frac{5}{7}$

16. $(\frac{2}{5} + 4x) - \frac{4}{5}$

17. $\frac{3}{5} + (\frac{5}{12} - 3x)$

18. $(\frac{9}{8} + x) + \frac{5}{6}$

19. $(\frac{4}{9}x - \frac{4}{9}) + \frac{3}{4}$

20. $(\frac{6}{5}x + \frac{4}{3}) - \frac{8}{5}$

21. $(\frac{1}{4}x + 3) - \frac{7}{2}$

22. $(2x - \frac{5}{6}) + \frac{2}{9}$

23. $(-3x + \frac{9}{10}) - \frac{7}{15}$

24. $(3x - \frac{2}{3}) + \frac{8}{15}$

25. $(\frac{9}{8} + x) + \frac{2}{5}$

Part C: Review. Answers for the problems of Part C are not given. However, the notation «C,U» refers to the Chapter (C) and Unit (U) in which problems of the same type were presented.

1. What is the tens digit of 6,248,271?   «1,1»

2. $19 \overline{)\, 51548}$ «2,4»

3. Select the smaller fraction from: $\frac{5}{6}, \frac{9}{13}$ «3,2»

4. $7\frac{4}{5} - 3\frac{2}{3}$ «3,4»

5. Write 0.247 as a fraction.   «4,1»

6. Simplify   6(7b)   «5,5»

7. Solve   8x = 40   «5,7»

8. Evaluate   11 + (-8)   «6,1»

9. Evaluate   -8 • -6   «6,3»

10. Simplify   -7(9x)   «6,5»

11. Simplify   4(2x − 7) − 3(x − 2)   «6,6»

12. Solve   3x − 7 = 23   «6,7»

13. Solve   8x − 3(4 + 2x) = -6   «6,8»

14. Evaluate $\frac{5}{8} \cdot \frac{8}{5}$ «7,1»

15. Evaluate $\frac{1}{7} - (\frac{1}{2} - \frac{1}{3})$ «7,2»

16. A salesman has completed $\frac{3}{4}$ of his appointments after 6 hours work. If he has 24 appointments for the day, how many appointments has he completed?   «3,7»

17. A first number is the same as 11 increased by a second number. If s represents the second number, write an open expression for the first number.   «5,8»

18. What is the average, to two decimal places, of 312, 406, and 579?   «4,4»

19. Write an equation that shows: the sum of a number and 19 is 46.   «6,9»

Unit 4:   POSTULATES FOR MULTIPLICATION

The Commutative Law of Multiplication

The equation xy = yx becomes a true equality when x is replaced by $\frac{5}{8}$ and y is replaced by $\frac{1}{4}$.

$$\frac{5}{8} \cdot \frac{1}{4} = \frac{1}{4} \cdot \frac{5}{8}$$

xy = yx will become a true equality for all number replacements for x and y. This is the Commutative Law of Multiplication.

The Commutative Law of Multiplication states xy is equivalent to yx for all number replacements of x and y. The order of two factors does not effect the product.

Focus on the Commutative Law of Multiplication

The Commutative Law of Multiplication states that the product of two numbers is not affected by the order in which they are multiplied.

By the Commutative Law of Multiplication the open expression $\frac{2}{3}x \cdot \frac{7}{9}$ is equivalent to $\frac{7}{9} \cdot \frac{2}{3}x$. In other words, for any number replacement of x, $\frac{2}{3}x \cdot \frac{7}{9}$ and $\frac{7}{9} \cdot \frac{2}{3}x$ will have the same evaluation.

$\frac{2}{3}x \cdot \frac{7}{9}$ and $\frac{7}{9} \cdot \frac{2}{3}x$ may be substituted for each other.

The Associative Law of Multiplication

The Associative Law of Multiplication states that (xy)z is equivalent to x(yz) for all number replacements of x, y, and z.

The Associative Law of Multiplication states that the product of three numbers is not affected by the grouping of the numbers.

(xy)z = x(yz) will become a true equality for all number replacements of x, y, and z.

Simplifying Multiplication Expressions

The Associative Law of Multiplication allows the simplification of open expressions such as $5 \cdot (\frac{3}{4}x)$, as shown below.

$$5 \cdot (\tfrac{3}{4}x) = (5 \cdot \tfrac{3}{4})x = \tfrac{15}{4}x$$

To simplify $-3 \cdot (\frac{4}{7}x)$, the following steps are used.

$$-3 \cdot (\tfrac{4}{7}x) = (-3 \cdot \tfrac{4}{7})x = \tfrac{-12}{7}x$$

Focus on simplifying multiplication expressions

To simplify $(\frac{7}{8}x) \cdot 8$, the Commutative Law of Multiplication and the Associative Law of Multiplication are used. The Commutative Law of Multiplication is used to substitute $8 \cdot (\frac{7}{8}x)$ for $(\frac{7}{8}x) \cdot 8$.

The simplification of $(\frac{7}{8}x) \cdot 8$ is shown below.

$$(\tfrac{7}{8}x) \cdot 8 = 8 \cdot (\tfrac{7}{8}x) = (8 \cdot \tfrac{7}{8})x = 7x$$

To simplify $(\frac{-5}{4}x) \cdot \frac{-4}{5}$ the following steps are used.

$$(\tfrac{-5}{4}x) \cdot \tfrac{-4}{5} = \tfrac{-4}{5} \cdot (\tfrac{-5}{4}x) = (\tfrac{-4}{5} \cdot \tfrac{-5}{4})x = 1 \cdot x = x$$

The Multiplication Laws of 1 and -1

The rational numbers 1 and -1 have special properties in multiplication.

$1x = x$ is a true equality for any number replacement of x. This is the statement of the Multiplication Law of One.

$1x = x$ for all number replacements of x.

When any rational number is multiplied by -1, the product is the opposite of the number. $-1x = -x$ for any number replacement of x.

$-1x = -x$ for all number replacements of x.

$-1 \cdot \frac{5}{6}$ is $\frac{-5}{6}$ and the opposite of $\frac{5}{6}$ is also $\frac{-5}{6}$

$-1 \cdot \frac{-3}{4}$ is $\frac{3}{4}$ and the opposite of $\frac{-3}{4}$ is also $\frac{3}{4}$

Focus on multiplication properties of 1 and -1

The use of 1 and -1 as factors arises in a number of open expressions.

For example, by the Multiplication Law of One $\frac{1}{5}x + x$ is equivalent to $\frac{1}{5}x + 1x$.

-1 is often important as a factor in an open expression. $\frac{4}{9}x - x$ is equivalent to $\frac{4}{9}x - 1x$.

$\frac{6}{11} + (\frac{3}{5}x - \frac{2}{7})$ is equivalent to $\frac{6}{11} + 1(\frac{3}{5}x - \frac{2}{7})$.

$\frac{7}{8} - (\frac{1}{6}x - \frac{4}{7})$ is equivalent to $\frac{7}{8} - 1(\frac{1}{6}x - \frac{4}{7})$.

A New Postulate for Multiplication

The rational numbers gives us one new postulate for algebra because every rational number,except 0, can be multiplied by another rational number to give a product of 1.

> **The Law of Multiplicative Inverses: For every rational number except zero there is another rational number such that the product of the two numbers is 1.**

For example, $\frac{7}{12}$ is a rational number and $\frac{12}{7}$ is a second rational number such that their product is 1.

$$\frac{7}{12} \cdot \frac{12}{7} = 1$$

Definition	Reciprocals	
Two rational numbers are **reciprocals**	**If and only if**	their product is one.

$\frac{3}{8}$ and $\frac{8}{3}$ are **reciprocals** because $\frac{3}{8} \cdot \frac{8}{3} = 1$

$\frac{-5}{11}$ and $\frac{-11}{5}$ are **reciprocals** because $\frac{-5}{11} \cdot \frac{-11}{5} = 1$

Focus on pairs of reciprocals

To find the reciprocal of any positive rational number, interchange the numerator and denominator.

$\frac{5}{7}$ has numerator 5 and denominator 7. Its reciprocal

$\frac{7}{5}$ has numerator 7 and denominator 5.

$$\frac{5}{7} \cdot \frac{7}{5} = 1 \quad \text{and} \quad \frac{7}{5} \cdot \frac{5}{7} = 1$$

To find the reciprocal of any negative rational number, interchange the numerator and denominator, but simplify the result.

$\frac{-3}{8}$ has numerator -3 and denominator 8. Its reciprocal

$\frac{-8}{3}$ has numerator -8 and denominator 3. Note: $\frac{8}{-3} = \frac{-8}{3}$

$$\frac{-3}{8} \cdot \frac{-8}{3} = 1 \quad \text{and} \quad \frac{-8}{3} \cdot \frac{-3}{8} = 1$$

Unit 4 Exercise

This exercise reviews the preceding unit. The exercise is divided into three parts. Part A reviews the foci of Unit 4. Part B offers opportunities to practice the skills and concepts of Unit 4. Part C contains problems that review your previous work in this text. All answers for Parts A and B are at the back of the book. Each problem of Part C is accompanied by a notation «CU» that refers to the Chapter (C) and Unit (U) in which that type of problem is studied.

Part A: Reviewing the foci of Unit 4.

1. By the Commutative Law of Multiplication $y \cdot \frac{5}{9}$ is equivalent to _____.

2. By the Commutative Law of Multiplication $(\frac{-1}{6}z) \cdot \frac{4}{3}$ is equivalent to _____.

3. By the Associative Law of Multiplication $\frac{5}{13}(\frac{-7}{10}x)$ is equivalent to _____.

4. By the Associative Law of Multiplication $\frac{3}{5}(\frac{5}{3}x)$ is equivalent to _____.

5. Simplify $(\frac{4}{9}x) \cdot \frac{3}{4}$ using the Commutative and/or Associative Laws of Multiplication.

6. Simplify $(\frac{-5}{8}x) \cdot \frac{6}{7}$ using the Commutative and/or Associative Laws of Multiplication.

7. Is 1x equivalent to x for any number replacement for x?

8. Is $x + \frac{2}{7}x$ equivalent to $1x + \frac{2}{7}x$?

9. Is -1x equivalent to -x for any number replacement for x?

10. Is $-1(\frac{-4}{5} - x)$ equivalent to $-(\frac{-4}{5} - x)$?

11. Is $\frac{2}{7} - \frac{3}{5}x$ equivalent to $1(\frac{2}{7} - \frac{3}{5}x)$?

12. The product of two reciprocals is 1. Are $\frac{15}{17}$ and $\frac{17}{15}$ reciprocals?

13. The product of two reciprocals is 1. Are $\frac{2}{9}$ and $\frac{-2}{9}$ reciprocals?

14. Interchange the numerator and denominator of $\frac{7}{12}$ to find its reciprocal.

15. Interchange the numerator and denominator of $\frac{-3}{10}$ and then simplify it to find its reciprocal.

Part B: Drill and Practice

Simplify.

1. $\frac{2}{7} \cdot (\frac{3}{5}x)$

2. $\frac{5}{6} \cdot (6x)$

3. $(\frac{9}{14}x) \cdot 14$

4. $\frac{3}{5} (\frac{-5}{3}x)$

5. $\frac{3}{4} \cdot (\frac{6}{7}x)$

6. $\frac{2}{7} (\frac{7}{2}x)$

7. $-3 \cdot (\frac{4}{7}x)$

8. $8 (\frac{3}{8}x)$

9. $(\frac{-6}{5}x) \cdot \frac{-1}{7}$

10. $\frac{2}{3} (\frac{3}{2}x)$

11. $\frac{3}{4} (\frac{4}{3}x)$

12. $\frac{-9}{8} (\frac{-8}{9}x)$

13. $\frac{5}{13} (\frac{13}{5}x)$

14. $(\frac{2}{3}x) \cdot \frac{3}{2}$

15. $\frac{1}{5} (5x)$

16. $\frac{-1}{9} (-9x)$

17. $\frac{-2}{5} (\frac{-5}{2}x)$

18. $\frac{-6}{7} (\frac{-7}{6}x)$

19. $-8 (\frac{-3}{8}x)$

20. $(\frac{5}{6}x) \cdot 6$

21. $\frac{3}{4} (\frac{-5}{9}x)$

22. $\frac{-11}{8} (\frac{4}{9}x)$

23. $\frac{2}{5} (\frac{10}{7}b)$

24. $\frac{4}{7} (\frac{-14}{5}r)$

25. $\frac{5}{6} (\frac{8}{7}x)$

26. $\frac{8}{13} (\frac{13}{8}x)$

Part C: Review. Answers for the problems of Part C are not given. However, the notation «C,U» refers to the Chapter (C) and Unit (U) in which problems of the same type were presented.

1. 6,172 x 8   «2,1»

2. Write $\frac{28}{5}$ as a mixed number.   «3,2»

3. $6\frac{7}{8} + 3\frac{7}{10}$ «3,3»

4. Write $\frac{7}{100}$ as a decimal fraction.   «4,1»

5. Evaluate 7 + [5 + 3 • 2]   «5,2»

6. Evaluate 4x + 3y when x = 7 and y = 4   «5,3»

7. Find the difference of 14.09 and 7.6   «4,2»

8. Find the product of 4.38 and 5.9 «2,2»

9. Find 0.32 of 673   «4,3»

10. Simplify (12 + 3x) + 5   «5,4»

11. Simplify 3(2x + 5) + (x + 7)   «5,6»

12. Evaluate 6 − (-8)   «6,2»

13. Evaluate -5 • 2 − 12 ÷ -4   «6,4»

14. Simplify 2(3x − 1) − 2(x − 4)   «6,6»

15. Solve 5(2x + 4) − 17 = -7   «6,8»

16. Evaluate $\frac{9}{11} \cdot \frac{11}{9}$ «7,1»

17. Evaluate $(\frac{15}{2} - \frac{3}{4}) - \frac{2}{7}$ «7,2»

18. Simplify $(\frac{6}{5}x + \frac{4}{3}) - \frac{8}{5}$ «7,3»

19. How many $\frac{1}{8}$ feet are there in 5 feet?   «3,7»

20. Clarice estimates that 0.3 of her time is spent sleeping? In a 24 hour day, how much does Clarice estimate she sleeps?   «4,4»

21. Each box car in a freight train contains 80 refrigerators. If the train has 52 box cars, how many refrigerators are on board?   «4,3»

22. One number is seven times a second number. If s represents the second number, write an open expression for the first number.   «5,8»

23. Write an equation that shows: 15 more than a number is 31.   «6,9»

Unit 5: THE DISTRIBUTIVE LAW OF MULTIPLICATION OVER ADDITION

Evaluating Numerical Expressions

The expressions $10(\frac{3}{5} - \frac{2}{3})$ and $10 \cdot \frac{3}{5} + 10 \cdot \frac{-2}{3}$ involve the same numbers, but are evaluated differently.

$$10(\tfrac{3}{5} - \tfrac{2}{3})$$
$$10(\tfrac{9}{15} - \tfrac{10}{15})$$
$$10 \cdot \tfrac{-1}{15}$$
$$2 \cdot \tfrac{-1}{3}$$
$$\tfrac{-2}{3}$$

$$10 \cdot \tfrac{3}{5} + 10 \cdot \tfrac{-2}{3}$$
$$2 \cdot \tfrac{3}{1} + \tfrac{-20}{3}$$
$$6 + \tfrac{-20}{3}$$
$$\tfrac{18}{3} + \tfrac{-20}{3}$$
$$\tfrac{-2}{3}$$

The evaluations of $10(\frac{3}{5} - \frac{2}{3})$ and $10 \cdot \frac{3}{5} + 10 \cdot \frac{-2}{3}$ are the same, $\frac{-2}{3}$. This is no coincidence. It is an example of the Distributive Law of Multiplication Over Addition.

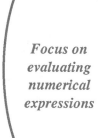

Focus on evaluating numerical expressions

The evaluations of $\frac{2}{3}(8 + \frac{1}{4})$ and $\frac{2}{3} \cdot 8 + \frac{2}{3} \cdot \frac{1}{4}$ are shown below. Notice that the steps in the evaluations are different, but the results are the same.

$$\tfrac{2}{3}(8 + \tfrac{1}{4})$$
$$\tfrac{2}{3}(\tfrac{32}{4} + \tfrac{1}{4})$$
$$\tfrac{2}{3} \cdot \tfrac{33}{4}$$
$$\tfrac{1}{3} \cdot \tfrac{33}{2}$$
$$\tfrac{1}{1} \cdot \tfrac{11}{2}$$
$$\tfrac{11}{2}$$

$$\tfrac{2}{3} \cdot 8 + \tfrac{2}{3} \cdot \tfrac{1}{4}$$
$$\tfrac{2}{3} \cdot 8 + \tfrac{1}{3} \cdot \tfrac{1}{2}$$
$$\tfrac{16}{3} + \tfrac{1}{6}$$
$$\tfrac{32}{6} + \tfrac{1}{6}$$
$$\tfrac{33}{6}$$
$$\tfrac{11}{2}$$

The Distributive Law of Multiplication Over Addition

The equation $x(y + z) = xy + xz$ becomes a true equality for all rational number replacements of x, y, and z.

The equivalence of $x(y + z)$ with $xy + xz$ is the Distributive Law of Multiplication over Addition.

Because of the Distributive Law, the expression $x(\frac{7}{8} - \frac{3}{5})$ can be substituted for $\frac{7}{8}x - \frac{3}{5}x$.

Also by the Distributive Law, the expression $\frac{3}{4} \cdot \frac{1}{2} + \frac{3}{4} \cdot 8x$ can be written in place of $\frac{3}{4}(\frac{1}{2} + 8x)$.

Focus on equivalent open expressions

The Distributive Law of Multiplication over Addition is a powerful property in algebra. It makes it possible to simplify many open expressions.

For example, $15x - 9x$ is, by the Distributive Law, equivalent to $x(15 - 9)$ which in turn is equivalent to $x \cdot 6$ or $6x$.

Similarly, the parentheses of $6(5 - \frac{1}{2}x)$ may be removed using the Distributive Law because the expression is equivalent to $6 \cdot 5 - 6 \cdot \frac{1}{2}x$ or $30 - 3x$.

Adding Like Terms

Recall from Chapters 5 and 6 that expressions such as $2x + 5x$ can be added because the terms are **like terms** (each term contains exactly the same variable factor).

$2x + 5x$ is simplified to 7x because:

   $2x + 5x$ is equivalent to $(2 + 5)x$ which is 7x.

$\frac{3}{4}x$ and $\frac{-2}{3}x$ are also like terms and $\frac{3}{4}x - \frac{2}{3}x$ is simplified by combining **like terms**.

   $\frac{3}{4}x - \frac{2}{3}x$ is equivalent to $(\frac{3}{4} - \frac{2}{3})x$ which is $\frac{1}{12}x$.

Focus on combining like terms

To simplify $\frac{4}{5}x + \frac{3}{8}x$ the like terms are added

using the Distributive Law of Multiplication

over Addition.

The final result is $\frac{47}{40}x$.

$$\frac{4}{5}x + \frac{3}{8}x$$
$$(\frac{4}{5} + \frac{3}{8})x$$
$$(\frac{32}{40} + \frac{15}{40})x$$
$$\frac{47}{40}x$$

Removing Parentheses from Open Expressions

Recall that parentheses are removed from $5(2x + 3)$ by separately multiplying 2x and 3 by 5.

$$5(2x + 3) = 5 \cdot 2x + 5 \cdot 3$$
$$= 10x + 15$$

Similarly, parentheses are removed from $5(x + \frac{2}{3})$ by

separately multiplying x and $\frac{2}{3}$ by 5.

$$5(x + \frac{2}{3}) = 5 \cdot x + 5 \cdot \frac{2}{3}$$
$$= 5x + \frac{10}{3}$$

Focus on removing parentheses

The parentheses of $\frac{1}{4}(\frac{1}{2}x - 6)$ are

removed by multiplying each term

of $(\frac{1}{2}x - 6)$ by $\frac{1}{4}$.

$$\frac{1}{4}(\frac{1}{2}x - 6)$$
$$\frac{1}{4} \cdot \frac{1}{2}x - \frac{1}{4} \cdot 6$$
$$\frac{1}{8}x - \frac{3}{2}$$

Simplifying Open Expressions with Parentheses

To simplify $2(x - 1) + 3(4 - 2x)$ both parentheses are removed and then like terms are grouped.

$$2(x - 1) + 3(4 - 2x)$$
$$2x - 2 + 12 - 6x$$
$$(2x - 6x) + (-2 + 12)$$
$$-4x + 10$$

Focus on parentheses preceded by 1 or -1

To simplify $(3x - 4) - (2x - 6)$ the parentheses must be removed. Notice that 1 is the multiplier of $(3x - 4)$ and -1 is the multiplier of $(2x - 6)$.

$$(3x - 4) - (2x - 6)$$
$$1(3x - 4) - 1(2x - 6)$$
$$3x - 4 - 2x + 6$$
$$x + 2$$

Simplifying Open Expressions with Rational Terms

To simplify $3(\frac{1}{4}x + \frac{1}{5}) - 2(\frac{1}{3}x + \frac{1}{5})$ the following steps are used.

$$3(\tfrac{1}{4}x + \tfrac{1}{5}) - 2(\tfrac{1}{3}x + \tfrac{1}{5})$$
$$3 \cdot \tfrac{1}{4}x + 3 \cdot \tfrac{1}{5} - 2 \cdot \tfrac{1}{3}x - 2 \cdot \tfrac{1}{5}$$
$$\tfrac{3}{4}x + \tfrac{3}{5} - \tfrac{2}{3}x - \tfrac{2}{5}$$
$$\tfrac{3}{4}x - \tfrac{2}{3}x + \tfrac{3}{5} - \tfrac{2}{5}$$
$$\tfrac{1}{12}x + \tfrac{1}{5}$$

Focus on adding terms with like denominators

The fractions $\frac{5}{x}$ and $\frac{7}{x}$ are like terms. They are added by adding the numerators.

$$\frac{5}{x} + \frac{7}{x} = \frac{12}{x}$$

Similarly, $\frac{4}{x}$ and $\frac{9}{x}$ are like terms.

$$\frac{4}{x} - \frac{9}{x} = \frac{-5}{x}$$

Unit 5 Exercise

This exercise reviews the preceding unit. The exercise is divided into three parts. Part A reviews the foci of Unit 5. Part B offers opportunities to practice the skills and concepts of Unit 5. Part C contains problems that review your previous work in this text. All answers for Parts A and B are at the back of the book. Each problem of Part C is accompanied by a notation «CU» that refers to the Chapter (C) and Unit (U) in which that type of problem is studied.

Part A: Reviewing the foci of Unit 5.

1. Evaluate $8(\frac{-3}{2} + \frac{1}{4})$ and $8 \cdot \frac{-3}{2} + 8 \cdot \frac{1}{4}$. Are the evaluations the same?

Simplify.

6. $\frac{7}{8}x + \frac{-1}{5}x$

2. By using the Distributive Law of Multiplication over Addition, $8y + 8 \cdot \frac{-3}{8}$ can be substituted for _____.

7. $(5 + \frac{1}{2}x) + (\frac{1}{3}x + 6)$

8. $x + 5 + \frac{2}{3}x$

3. By using the Distributive Law of Multiplication over Addition, $x(3 + \frac{-5}{6})$ can be substituted for _____.

9. $3(2x - \frac{1}{7}) + (3x + \frac{5}{6})$

4. By using the Distributive Law of Multiplication over Addition, $-5w + -5 \cdot \frac{1}{3}$ can be substituted for _____.

10. $\frac{1}{3}(\frac{1}{4} - \frac{2}{5}x) + \frac{5}{8}x$

11. $6 - 4(1 - 3x)$

5. Use the Distributive Law of Multiplication over Addition to simplify $7y + 5y$.

12. $2x - (4 - 3x)$

13. $5x + 2(x - 7)$

14. $4 + (7x - 3)$

Part B: Drill and Practice

Simplify the following.

1. 5x + 3x

2. 9x + 4x

3. 2x + 4x

4. 9x − 7x

5. 3x − 11x

6. -2x − 5x

7. $2x + \frac{-5}{6}x$

8. $\frac{2}{3}x - \frac{1}{4}x$

9. $\frac{7}{6}y - \frac{1}{2}y$

Remove the parentheses from the following.

10. $-5(\frac{3}{5}x - 2)$

11. $\frac{1}{4}(8x - 3)$

12. -2(7x + 5)

13. $(8 - \frac{17}{5}x)$

14. 7(5 − 2x)

15. $\frac{2}{5}(10 - 5x)$

16. -(3x − 4)

17. -3(4x + 1)

18. -6(5 − x)

19. -4(5 − x)

20. 2(5 − 3x)

21. -(7 − x)

22. -8(4 − 2x)

23. (6 − 9x)

Simplify the following.

24. 3x − 2(x + 5)

25. 4 + 7(x − 2)

26. -3 − 5(1 − 2x)

27. 9x − 3(2x + 1)

28. 2x − (5x − 1)

29. (5x − 7) − (3x + 5)

30. 4(2x − 3) + 3(x + 4)

31. 2x − 7(x + 3) + 5

32. 3(2x − 5) + 2(5x − 2)

33. $(\frac{2}{5}x + 4) + (7 + \frac{1}{3}x)$

34. $(\frac{7}{8}x + 3) + (\frac{2}{3} + \frac{1}{8}x)$

35. $3(2x - \frac{1}{4}) - 2(x + 3)$

36. $\frac{3}{x} + \frac{7}{x}$

37. $\frac{6}{x} - \frac{9}{x}$

Part C: Review. Answers for the problems of Part C are not given. However, the notation «C,U» refers to the Chapter (C) and Unit (U) in which problems of the same type were presented.

1. $57,092 - 41,804$ «1,3»

2. Find the lcm for 12 and 20.   «3,2»

3. Change $2\frac{8}{9}$ to an improper fraction.   «3,2»

4. $3\frac{3}{4} \div \frac{5}{12}$ «3,6»

5. Simplify $(5x + 7) + 2(x + 5)$ «5,6»

6. Evaluate $9 + (-4)$ «6,1»

7. Evaluate $-7 \cdot -6$ «6,3»

8. Simplify $8(4x)$ «6,5»

9. Simplify $(3x + 5) - (x - 7)$ «6,6»

10. Solve $6x - 5 = 37$ «6,7»

11. Solve $6 - 5(2 - 3x) = 26$ «6,8»

12. Evaluate $\frac{14}{3} \cdot \frac{18}{7}$ «7,1»

13. Evaluate $\frac{1}{5} + (\frac{2}{3} - \frac{1}{3})$ «7,2»

14. Simplify $(\frac{4}{9}x - \frac{4}{9}) + \frac{3}{4}$ «7,3»

15. Simplify $\frac{3}{4}(\frac{-5}{9}x)$ «7,4»

16. A first number is the same as 12 decreased by a second number. If s represents the second number, write an open expression for the first number.   «5,8»

17. Find the sum of 5.813 and 9.47   «4,2»

18. Divide 4.03 by 0.7 and round off the answer to 2 decimal places.   «4,4»

19. In three successive days the temperature at noon was 78, 82, and 86 degrees. What was the average noon temperature during that 3-day period?   «4,4»

20. Change $\frac{5}{11}$ to a decimal fraction rounded off at two decimal places.   «4,4»

21. Write an equation that shows: the difference when 13 is subtracted from a number is 19.   «6,9»

Unit 6:   SOLVING BASIC EQUATIONS

Recall that in Chapter 6 equations were solved by first
generating basic equations. In this unit and in Unit 7 the
same process is followed. Once again, the solving of an
equation will involve generating an equivalent basic equation.

Many basic equations have integer solutions that are easy to
see. For example, $5x = 10$ has 2 as its solution because $5 \cdot 2$
is equal to 10.

Other basic equations do not have obvious solutions and a
method for solving them will be taught later in this unit. For
example, $7x = -13$ and $\frac{3}{4}x = \frac{2}{3}$ are basic equations which do
not have integer solutions.

Reviewing Basic Equations

A basic equation is one like

$$5x = 10 \quad \text{or} \quad 7x = -13 \quad \text{or} \quad \tfrac{3}{4}x = \tfrac{2}{3}$$

- In each case, a basic equation consists of two terms – one on
the left side of the equal sign and one on the right side of the
equal sign. The term on the left side of the equal sign must
contain a variable and the term on the right side of the equal
sign must be a number (sometimes called a constant).

*Focus on
the meaning
of a basic
equation*

The equation $8x = -48$ is a basic equation with -6 as its root.

The equation $12 = -4x$ needs to be rewritten as the basic equation
$-4x = 12$. Its root is -3.

The equation $5x + 3 = 38$ is not a basic equation. It has three terms.
To solve $5x + 3 = 38$, a new, equivalent basic equation must be
generated.

Generating Equivalent Equations

$5x + 3 = 38$ and $5x = 35$ are **equivalent** equations because both have 7 as their root.

In this text there are two ways taught for generating equivalent equations. The first method is:

> **Any number or open expression can be added to both sides of an equation to generate a new equivalent equation.**

For example, the equation $5x + 3 = 38$ can be used to generate another equivalent equation by adding -3 to both sides of $5x + 3 = 38$.

$$
\begin{array}{c|c}
\text{left side} & \text{right side} \\[4pt]
5x + 3 & = \quad 38 \\
\underline{\quad -\ 3\quad} & \underline{\quad -\ 3\quad} \\
5x & = \quad 35 \quad \text{or } x = 7
\end{array}
$$

$5x + 3 = 38$ and its generated equivalent $5x = 35$ have the same root which is 7.

Focus on generating equivalent equations

Any number can be added to both sides of an equation to generate a new equivalent equation. Below are shown three examples of generating equivalent, basic equations.

$$
\begin{array}{lll}
7x + 13 = 6 & 3x + 8 = 20 & 2x + 7 = 18 \\
\underline{\quad -13 \quad -13} & \underline{\quad -8 \quad -8} & \underline{\quad -7 \quad -7} \\
7x + 0 = -7 & 3x + 0 = 12 & 2x + 0 = 11
\end{array}
$$

The basic equations $7x = -7$ and $3x = 12$ have obvious roots, -1 and 4 respectively. The root of $2x = 11$ is not obvious. Later in this unit a method for finding its root will be shown.

Solving Equations of the Form ax + b = c

To solve $9x + 6 = -4$, first generate an equivalent basic
equation by adding something to both sides of $9x + 6 = -4$.

To decide what should be added to both sides of $9x + 6 = -4$,
ask:

> **What term of $9x + 6 = -4$ must be
> eliminated to make a basic equation?**

The answer to that question is that 6 needs to be eliminated to
generate a basic equation from $9x + 6 = -4$. Consequently, -6
is added to both sides to eliminate the 6.

$$9x + 6 = -4$$
$$\underline{\quad - 6 \quad - 6}$$
$$9x + 0 = -10 \quad \text{or} \quad 9x = -10$$

The basic equation is $9x = -10$. Its root is $\frac{-10}{9}$ and
a method for finding it will be shown shortly.

*Focus on
generating
a basic
equation*

The equation $3x + 5 = 19$ is
used to generate an equivalent
basic equation.

$$3x + 5 = 19$$
$$\underline{\quad - 5 \quad - 5}$$
$$3x + 0 = 14$$

Since $3x = 14$ is equivalent to $3x + 5 = 19$, the equations have the same
root. The root is $\frac{14}{3}$ and a method for finding it will be shown in the
next explanation.

Solving Basic Equations

The first method for generating equivalent equations allowed any number to be added to both sides of the equation. A basic equation has no addition – only multiplication. The second method for generating equivalent equations deals with eliminating any non-zero multiplication factor.

Any number except zero may be multiplied by both sides of an equation to generate a new equivalent equation.

If both sides of the equation $3x = 6$ are multiplied by $\frac{1}{3}$, the equivalent equation $1x = 2$ is obtained.

$$3x = 6$$
$$\frac{1}{3} \cdot 3x = \frac{1}{3} \cdot 6$$
$$1x = 2$$

Since 1x is equivalent to x, the only number that can replace x in $1x = 2$ to give a true statement is 2.

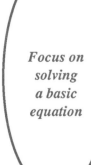

Focus on solving a basic equation

To solve $7x = 15$, both sides of the equation are multiplied by $\frac{1}{7}$.

$$7x = 15$$
$$\frac{1}{7} \cdot 7x = \frac{1}{7} \cdot 15$$
$$1x = \frac{15}{7}$$

The root of $7x = 15$ is $\frac{15}{7}$.

To solve $-5x = 11$, both sides of the equation are multiplied by $\frac{-1}{5}$.

$$-5x = 11$$
$$\frac{-1}{5} \cdot -5x = \frac{-1}{5} \cdot 11$$
$$1x = \frac{-11}{5}$$

The root of $-5x = 11$ is $\frac{-11}{5}$.

Solving Equations by Using Reciprocals

The product of any number and its reciprocal is 1.

In the equation $\frac{3}{5}x = 7$, the number $\frac{3}{5}$ is called the **coefficient** of x. To solve $\frac{3}{5}x = 7$, both sides of the equation are multiplied by the reciprocal of $\frac{3}{5}$, the **coefficient** of x. The reciprocal of $\frac{3}{5}$ is $\frac{5}{3}$, because $\frac{3}{5} \cdot \frac{5}{3} = 1$.

$$\frac{3}{5}x = 7$$
$$\frac{5}{3} \cdot \frac{3}{5}x = \frac{5}{3} \cdot 7$$
$$1x = \frac{35}{3}$$

The only replacement for x that will make $1x = \frac{35}{3}$ a true statement is its root, $\frac{35}{3}$.

Focus on solving basic equations

To solve $\frac{-5}{9}x = 2$, first multiply both sides of the equation by the reciprocal of the coefficient of x, $\frac{-5}{9}$.

$$\frac{-5}{9}x = 2$$
$$\frac{-9}{5} \cdot \frac{-5}{9}x = \frac{-9}{5} \cdot 2$$

The equation $1x = \frac{-18}{5}$ is obtained.

Therefore, the root of $\frac{-5}{9}x = 2$ is $\frac{-18}{5}$.

$$1x = \frac{-18}{5}$$
$$x = \frac{-18}{5}$$

The coefficient of x in $7x = 10$ is 7. The reciprocal of 7 is $\frac{1}{7}$. The root of $7x = 10$ is found by first multiplying both sides of the equation by the reciprocal of 7.

$$7x = 10$$
$$\frac{1}{7} \cdot 7x = \frac{1}{7} \cdot 10$$
$$1x = \frac{10}{7}$$
$$x = \frac{10}{7}$$

The root of $7x = 10$ is $\frac{10}{7}$.

Focus on solving an equation with coefficient -1

The reciprocal of -1 is -1, because -1 • -1 = 1.

To solve -x = 4, both sides of the equation are multiplied by -1.

Therefore, the root of -x = 4 is -4.

$$
\begin{aligned}
-x &= 4 \\
-1 \bullet x &= 4 \\
-1 \bullet -1x &= -1 \bullet 4 \\
1x &= -4 \\
x &= -4
\end{aligned}
$$

Solving Equations using Opposites and Reciprocals

To solve the equation 5x + 7 = 13, first generate a basic equation by adding the opposite of 7 to both sides of the equation.

$$
\begin{aligned}
5x + 7 &= 13 \\
-7 \quad &-7 \\
\hline
5x + 0 &= 6
\end{aligned}
$$

After the basic equation 5x = 6 is generated, the second step is to multiply both sides by the reciprocal of 5.

$$
\begin{aligned}
5x &= 6 \\
\tfrac{1}{5} \bullet 5x &= \tfrac{1}{5} \bullet 6 \\
1x &= \tfrac{6}{5} \\
x &= \tfrac{6}{5}
\end{aligned}
$$

The root of 5x + 7 = 13 is $\tfrac{6}{5}$.

Focus on solving equations

To solve the equation 4x + 3 = 21, the first step is to add the opposite of 3 to both sides of the equation. The new equation is 4x = 18.

$$
\begin{aligned}
4x + 3 &= 21 \\
-3 \quad &-3 \\
\hline
4x + 0 &= 18
\end{aligned}
$$

The second step is to multiply both sides of the basic equation by the reciprocal of 4.

$$
\begin{aligned}
4x &= 18 \\
\tfrac{1}{4} \bullet 4x &= \tfrac{1}{4} \bullet 18
\end{aligned}
$$

After simplifying both sides, it can be seen that the root of 4x + 3 = 21 is $\tfrac{9}{2}$.

$$
\begin{aligned}
1x &= \tfrac{9}{2} \\
x &= \tfrac{9}{2}
\end{aligned}
$$

Removing Parentheses from an Equation

To solve $7 + 3(x + 1) = x + 4$,
first remove the parentheses
and simplify the left side.

$$
\begin{aligned}
7 + 3(x + 1) &= x + 4 \\
7 + 3x + 3 &= x + 4 \\
3x + 10 &= x + 4
\end{aligned}
$$

The equation $3x + 10 = x + 4$
is used to generate a basic
equation by adding the opposite
of 10 and the opposite of x.

$$
\begin{array}{rcr}
3x + 10 &=& x + 4 \\
- 10 & & - 10 \\
\hline
3x &=& x - 6 \\
- x & & - x \\
\hline
2x &=& -6
\end{array}
$$

Both sides of the basic equation
are multiplied by the reciprocal
of 2.

$$
\begin{aligned}
2x &= -6 \\
\tfrac{1}{2} \cdot 2x &= \tfrac{1}{2} \cdot -6 \\
1x &= -3 \\
x &= -3
\end{aligned}
$$

The root of the original equation and each of its generated equivalent
equations is -3.

*Focus on
solving
equations with
parentheses*

To solve $3(4x - 5) - 8 = 10 - (x - 6)$,
first simplify both sides of the equation.

$$
\begin{aligned}
3(4x - 5) - 8 &= 10 - (x - 6) \\
12x - 15 - 8 &= 10 - x + 6 \\
12x - 23 &= 16 - x
\end{aligned}
$$

The equation $12x - 23 = 16 - x$ is used
to generate a basic equation by adding the
opposite of -23 and the opposite of -x to both
sides of the equation.

$$
\begin{array}{rcr}
12x - 23 &=& 16 - x \\
+ 23 & & + 23 \\
\hline
12x &=& 39 - x \\
+ x & & + x \\
\hline
13x &=& 39
\end{array}
$$

Both sides of the basic equation are
multiplied by the reciprocal of 13.

$$
\begin{aligned}
13x &= 39 \\
\tfrac{1}{13} \cdot 13x &= \tfrac{1}{13} \cdot 39 \\
1x &= 3 \\
x &= 3
\end{aligned}
$$

The root of the original equation,
$3(4x - 5) - 8 = 10 - (x - 6)$ is 3.

Unit 6 Exercise

This exercise reviews the preceding unit. The exercise is divided into three parts. Part A reviews the foci of Unit 6. Part B offers opportunities to practice the skills and concepts of Unit 6. Part C contains problems that review your previous work in this text. All answers for Parts A and B are at the back of the book. Each problem of Part C is accompanied by a notation «CU» that refers to the Chapter (C) and Unit (U) in which that type of problem is studied.

Part A: Reviewing the foci of Unit 6.

1. In the basic equation $8x = 33$, what is the coefficient of x?

2. What number should be multiplied by both sides of $8x = 33$ to solve the equation?

3. In the basic equation $\frac{3}{4}x = 5$, what is the coefficient of x?

4. What number should be multiplied by both sides of $\frac{3}{4}x = 5$ to solve the equation?

5. In the basic equation $-3x = 25$, what is the coefficient of x?

6. What number should be multiplied by both sides of $-3x = 25$ to solve the equation?

7. In the basic equation $\frac{-7}{5}x = 4$, what is the coefficient of x?

8. What number should be multiplied by both sides of $\frac{-7}{5}x = 4$ to solve the equation?

9. Solve $-9x - 7 = 4$ by first adding 7 to both sides and then multiplying by the reciprocal of -9.

10. Solve $-2(x + 5) = -7$ by first removing its parentheses.

11. Solve the equation $9x + 3 = 22 + 4x$ by adding opposites and multiplying by reciprocals.

12. Solve $2x + 3(x + 2) = 2(x - 5) + 8$ by first removing the parentheses.

Part B: Drill and Practice

Solve the following.

1. $2x - 5 = 27$

2. $5x - 3 = 27$

3. $\frac{9}{4}x = 3$

4. $\frac{7}{9}x = 7$

5. $\frac{-4}{5}x = 2$

6. $\frac{-2}{5}x = -3$

7. $\frac{5}{13}x = -1$

8. $\frac{9}{5}x = -3$

9. $\frac{2}{7}x = \frac{5}{3}$

10. $\frac{-3}{8}x = \frac{1}{4}$

11. $-4x = 7$

12. $-3x = 8$

13. $9x = 7$

14. $-5x = -7$

15. $4x = 15$

16. $-5x = 21$

17. $-3x = -17$

18. $\frac{1}{8}x = 4$

19. $\frac{-1}{3}x = 5$

20. $\frac{1}{5}x = -4$

21. $\frac{-1}{6}x = -3$

22. $6x + 9 = 5$

23. $-3x + 2 = -8$

24. $11x + 4 = -3$

25. $4x - 9 = 14$

26. $-3x + 4 = -4$

27. $7x + 2 = -8$

28. $15x + 4 = 14$

29. $-8x + 9 = 5$

30. $15 = 4x - 1$

31. $23 = 5x + 4$

32. $4 = 9x + 3$

33. $2(3x - 4) = 9$

34. $4x - 3(x + 5) = 5 - 2x$

35. $6x - 4 = 5 - (x + 2)$

36. $3(5x - 4) - 8 = 3 - x$

37. $(4x - 7) = 2(x - 3) + 5$

38. $2(3x + 4) + 4 = 3(x - 5)$

Part C: Review. Answers for the problems of Part C are not given. However, the notation «C,U» refers to the Chapter (C) and Unit (U) in which problems of the same type were presented.

1. 56,990 + 23,899 + 6,732   «1,2»

2. Simplify $\frac{19}{38}$ «3,1»

3. Find the numerator of $\frac{3}{5} = \frac{?}{45}$ «3,2»

4. $7\frac{1}{3}$ x $\frac{9}{10}$ «3,5»

5. Add 4.827, 16.38, and 0. 708   «4,2»

6. Find the difference of 92.47 and 76.6   «4,2»

7. Simplify (9 + 3x) + 7   «5,4»

8. Simplify 4(3x + 1) + 5(x + 3)   «5,6»

9. Evaluate 2 − (-5)   «6,2»

10. Evaluate 9 − (-12) • -4 ÷ 6   «6,4»

11. Simplify 3(4x − 1) − 2(x − 4)   «6,6»

12. Solve 6 − 5x = 10 − x   «6,8»

13. Evaluate $\frac{-7}{4} \cdot \frac{-4}{7}$ «7,1»

14. Evaluate $\frac{2}{5}$ + $(\frac{1}{3} - \frac{1}{2})$ «7,2»

15. Simplify $(\frac{1}{4}x + 3) - \frac{7}{2}$ «7,3»

16. Simplify $\frac{-11}{8}(\frac{4}{9}x)$ «7,4»

17. Simplify $\frac{4}{x} - \frac{9}{x}$ «7,5»

18. How many $\frac{1}{4}$ feet are there in 9 feet?.   «3,7»

19. Mary's test scores in Spanish were 56, 95, 93, and 88. Find her average on these four tests rounded off to the nearest whole number. «4,4»

20. Change $\frac{5}{17}$ to a decimal fraction rounded off at two decimal places.   «4,4»

21. One number is nine times a second number. If s represents the second number, write an open expression for the first number.   «5,8»

22. Write an equation that shows: 3 less than a number is 52.   «6,9»

Unit 7:   EQUATIONS WITH RATIONAL ROOTS

In Unit 6 of this Chapter, equations such as:

$$7x + 19 = 31 \text{ and } 8x + 3 = -41$$

where a, b, and c are integers were solved.

In this unit, equations such as $\frac{2}{5}x + \frac{1}{4} = \frac{1}{6}$ are solved.

Solving Equations Involving Rational Numbers

Any equation involving rational numbers can be changed to an equivalent equation involving only integers. This is accomplished by multiplying both sides of the equation by the common denominator of its terms.

This process is demonstrated in the example below.

The common denominator of the fractions is 24.
$$\frac{7}{8}x + \frac{5}{6} = \frac{1}{12}$$

24 is multiplied by both sides of the equation.
$$24\left(\frac{7}{8}x + \frac{5}{6}\right) = 24 \cdot \frac{1}{12}$$

24 is multiplied by each term of the equation.
$$24 \cdot \frac{7}{8}x + 24 \cdot \frac{5}{6} = 24 \cdot \frac{1}{12}$$

Cancelling makes each denominator a 1.
$$\overset{3}{\cancel{24}} \cdot \frac{7}{\underset{1}{\cancel{8}}}x + \overset{4}{\cancel{24}} \cdot \frac{5}{\underset{1}{\cancel{6}}} = \overset{2}{\cancel{24}} \cdot \frac{1}{\underset{1}{\cancel{12}}}$$

The equation now involves only integers and is solved by adding the opposite of 20 and multiplying by the reciprocal of 21.
$$21x + 20 = 2$$
$$21x = -18$$
$$x = \frac{-6}{7}$$

$\frac{-6}{7}$ is the solution of $\frac{7}{8}x + \frac{5}{6} = \frac{1}{12}$. Notice that $\frac{7}{8} \cdot \frac{-6}{7} + \frac{5}{6} = \frac{1}{12}$ is true.

To solve $\frac{5}{3}x + \frac{1}{4} = \frac{7}{9}$ the following steps are used.

1. The common denominator of the terms is 36.

$$\frac{5}{3}x + \frac{1}{4} = \frac{7}{9}$$

2. 36 is multiplied by both sides of the equation.

$$36(\frac{5}{3}x + \frac{1}{4}) = 36 \cdot \frac{7}{9}$$

3. 36 is multiplied separately by each term of the equation.

$$36 \cdot \frac{5}{3}x + 36 \cdot \frac{1}{4} = 36 \cdot \frac{7}{9}$$

4. Cancelling makes each denominator equal to 1.

$$\overset{12}{\cancel{36}} \cdot \frac{5}{\cancel{3}}x + \overset{9}{\cancel{36}} \cdot \frac{1}{\cancel{4}} = \overset{4}{\cancel{36}} \cdot \frac{7}{\cancel{9}}$$
$$\phantom{36 \cdot \frac{5}{3}x}111$$

5. The resulting equation only involves integers.

$$60x + 9 = 28$$

6. -9 is added to both sides.

$$60x = 19$$

7. $\frac{1}{60}$ is multiplied by both sides.

$$x = \frac{19}{60}$$

Focus on solving a fractional equation

$\frac{19}{60}$ is the solution of $\frac{5}{3}x + \frac{1}{4} = \frac{7}{9}$.

The root, $\frac{19}{60}$, should be checked to show the equation $\frac{5}{3}x + \frac{1}{4} = \frac{7}{9}$ is a true equality.

When $x = \frac{19}{60}$, the equation $\frac{5}{3}x + \frac{1}{4} = \frac{7}{9}$ becomes:

$$\frac{5}{3} \cdot \frac{19}{60} + \frac{1}{4} = \frac{7}{9}$$

$$\frac{19}{36} + \frac{1}{4} = \frac{7}{9}$$

$$\frac{19}{36} + \frac{9}{36} = \frac{7}{9}$$

$$\frac{28}{36} = \frac{7}{9}$$

$\frac{28}{36} = \frac{7}{9}$ is true and this checks the root $\frac{19}{60}$.

Solving Equations with Four Terms

To solve $\frac{3}{8}x - \frac{7}{5} = \frac{9}{10} - \frac{3}{4}x$ the four terms of the equation are first multiplied by their common denominator, 40.

$$\frac{3}{8}x - \frac{7}{5} = \frac{9}{10} - \frac{3}{4}x$$

$$40 \cdot \frac{3}{8}x - 40 \cdot \frac{7}{5} = 40 \cdot \frac{9}{10} - 40 \cdot \frac{3}{4}x$$

$$\overset{5}{\cancel{40}} \cdot \frac{3}{\cancel{8}}x - \overset{8}{\cancel{40}} \cdot \frac{7}{\cancel{5}} = \overset{4}{\cancel{40}} \cdot \frac{9}{\cancel{10}} - \overset{10}{\cancel{40}} \cdot \frac{3}{\cancel{4}}x$$

$$15x - 56 = 36 - 30x$$

$$45x - 56 = 36$$

$$45x = 92$$

$$x = \frac{92}{45}$$

The solution of $\frac{3}{8}x - \frac{7}{5} = \frac{9}{10} - \frac{3}{4}x$ is $\frac{92}{45}$.

The root, $\frac{92}{45}$, should be checked to show the equation $\frac{3}{8}x - \frac{7}{5} = \frac{9}{10} - \frac{3}{4}x$ is a true equality.

When $x = \frac{92}{45}$, the equation $\frac{3}{8}x - \frac{7}{5} = \frac{9}{10} - \frac{3}{4}x$ becomes:

$$\frac{3}{8} \cdot \frac{92}{45} - \frac{7}{5} = \frac{9}{10} - \frac{3}{4} \cdot \frac{92}{45}$$

$$\frac{23}{30} - \frac{7}{5} = \frac{9}{10} - \frac{23}{15}$$

$$\frac{23}{30} - \frac{42}{30} = \frac{27}{30} - \frac{46}{30}$$

$$\frac{-19}{30} = \frac{-19}{30}$$

$\frac{-19}{30} = \frac{-19}{30}$ is true and this checks the root $\frac{92}{45}$.

The solution of $\frac{2}{5}x + \frac{1}{4} = \frac{1}{6}x - \frac{1}{2}$ is shown at the right.

$$\frac{2}{5}x + \frac{1}{4} = \frac{1}{6}x - \frac{1}{2}$$

1. First, the common denominator is multiplied by both sides.

$$60(\frac{2}{5}x + \frac{1}{4}) = 60(\frac{1}{6}x - \frac{1}{2})$$

2. 60 is multiplied by each term.

$$60 \cdot \frac{2}{5}x + 60 \cdot \frac{1}{4} = 60 \cdot \frac{1}{6}x - 60 \cdot \frac{1}{2}$$

3. Cancellation makes each denominator 1.

$$\overset{12}{\cancel{60}} \cdot \frac{2}{\underset{1}{\cancel{5}}}x + \overset{15}{\cancel{60}} \cdot \frac{1}{\underset{1}{\cancel{4}}} = \overset{10}{\cancel{60}} \cdot \frac{1}{\underset{1}{\cancel{6}}}x - \overset{30}{\cancel{60}} \cdot \frac{1}{\underset{1}{\cancel{2}}}$$

4. The new equation has no fractions.

$$24x + 15 = 10x - 30$$

5. The opposite of 10x is added to both sides.

$$14x + 15 = -30$$

6. The opposite of 15 is added to both sides.

$$14x = -45$$

7. The reciprocal of 14 is multiplied by both sides.

$$x = \frac{-45}{14}$$

Focus on solving a four-term fractional equation

$\frac{-45}{14}$ is the solution of $\frac{2}{5}x + \frac{1}{4} = \frac{1}{6}x - \frac{1}{2}$.

The root, $\frac{-45}{14}$, should be checked to show the equation $\frac{2}{5}x + \frac{1}{4} = \frac{1}{6}x - \frac{1}{2}$ is a true equality.

When $x = \frac{-45}{14}$, the equation $\frac{2}{5}x + \frac{1}{4} = \frac{1}{6}x - \frac{1}{2}$ becomes:

$$\frac{2}{5} \cdot \frac{-45}{14} + \frac{1}{4} = \frac{1}{6} \cdot \frac{-45}{14} - \frac{1}{2}$$

$$\frac{-9}{7} + \frac{1}{4} = \frac{-15}{28} - \frac{1}{2}$$

$$\frac{-36}{28} + \frac{7}{28} = \frac{-15}{28} - \frac{14}{28}$$

$$\frac{-29}{28} = \frac{-29}{28}$$

$\frac{-29}{28} = \frac{-29}{28}$ is true and this checks the root $\frac{-45}{14}$.

Solving an Equation with a Variable Denominator

To solve $\frac{5}{x} + \frac{2}{3} = \frac{4}{x}$, the following steps are used.

1. The common denominator is 3x.

    $$\frac{5}{x} + \frac{2}{3} = \frac{4}{x}$$

2. 3x is multiplied by each term.

    $$3x \cdot \frac{5}{x} + 3x \cdot \frac{2}{3} = 3x \cdot \frac{4}{x}$$

3. Cancelling makes each denominator 1.

    $$\overset{3}{\cancel{3x}} \cdot \frac{5}{\cancel{x}} + \cancel{3x} \cdot \frac{2}{\cancel{3}} = \overset{3}{\cancel{3x}} \cdot \frac{4}{\cancel{x}}$$
    $$\quad 1 \qquad\quad 1 \qquad\quad 1$$

4. The new equation has no fractions.

    $$15 + 2x = 12$$

5. -15 is added to both sides.

    $$2x = -3$$

6. $\frac{1}{2}$ is multiplied by both sides.

    $$x = \frac{-3}{2}$$

$\frac{-3}{2}$ is the root of $\frac{5}{x} + \frac{2}{3} = \frac{4}{x}$.

Focus on solving an equation with a variable denominator

Study carefully the procedure used in solving the equation $\frac{3}{5} - \frac{2}{x} = \frac{1}{10}$.

$$\frac{3}{5} - \frac{2}{x} = \frac{1}{10}$$

$$10x \cdot \frac{3}{5} - 10x \cdot \frac{2}{x} = 10x \cdot \frac{1}{10}$$

$$6x - 20 = x$$

$$5x = 20$$

$$x = 4$$

4 is the solution of $\frac{3}{5} - \frac{2}{x} = \frac{1}{10}$.

Checking Roots of Fractional Equations

To check $\frac{-6}{7}$ as the solution of $\frac{7}{8}x + \frac{5}{6} = \frac{1}{12}$ the x must be

replaced by $\frac{-6}{7}$ and the equation shown to be an equality.

$$\frac{7}{8}x + \frac{5}{6} = \frac{1}{12}$$

Replace x by $\frac{-6}{7}$. $\frac{7}{8} \cdot \frac{-6}{7} + \frac{5}{6}$? $\frac{1}{12}$

Multiply $\frac{7}{8} \cdot \frac{-6}{7}$. $\frac{-3}{4} + \frac{5}{6}$? $\frac{1}{12}$

Add $\frac{-3}{4} + \frac{5}{6}$. $\frac{-9}{12} + \frac{10}{12}$? $\frac{1}{12}$

Since both sides are equal, $\frac{1}{12} = \frac{1}{12}$
the root checks.

$\frac{-6}{7}$ is the solution of $\frac{7}{8}x + \frac{5}{6} = \frac{1}{12}$.

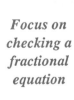

*Focus on
checking a
fractional
equation*

To check 4 as the solution of $\frac{3}{5} - \frac{2}{x} = \frac{1}{10}$ the x must be replaced by 4 and
the equation shown to be an equality.

$$\frac{3}{5} - \frac{2}{x} = \frac{1}{10}$$

Replace x by 4. $\frac{3}{5} - \frac{2}{4}$? $\frac{1}{10}$

Simplify $\frac{2}{4}$. $\frac{3}{5} - \frac{1}{2}$? $\frac{1}{10}$

Add $\frac{3}{5} - \frac{1}{2}$. $\frac{6}{10} - \frac{5}{10}$? $\frac{1}{10}$

Since both sides are equal, $\frac{1}{10} = \frac{1}{10}$
the root checks.

4 is the solution of $\frac{3}{5} - \frac{2}{x} = \frac{1}{10}$.

Unit 7 Exercise

This exercise reviews the preceding unit. The exercise is divided into three parts. Part A reviews the foci of Unit 7. Part B offers opportunities to practice the skills and concepts of Unit 7. Part C contains problems that review your previous work in this text. All answers for Parts A and B are at the back of the book. Each problem of Part C is accompanied by a notation «CU» that refers to the Chapter (C) and Unit (U) in which that type of problem is studied.

Part A: Reviewing the foci of Unit 7.

1. What is the common denominator of the terms of $\frac{5}{6}x + \frac{1}{3} = \frac{2}{3}$?

2. Solve $\frac{5}{6}x + \frac{1}{3} = \frac{2}{3}$ by multiplying each term by 6.

3. What is the common denominator of the terms of $\frac{2}{5}x + \frac{1}{4} = \frac{7}{8}$?

4. Solve $\frac{2}{5}x + \frac{1}{4} = \frac{7}{8}$ by multiplying each term by 40.

5. What is the common denominator of the terms of $\frac{3}{4}x - \frac{2}{3} = \frac{1}{5}$?

6. Solve $\frac{3}{4}x - \frac{2}{3} = \frac{1}{5}$ by multiplying each term by 60.

7. What is the common denominator of the terms of $\frac{4}{3}x + \frac{1}{2} = \frac{3}{4}$?

8. Solve $\frac{4}{3}x + \frac{1}{2} = \frac{3}{4}$ by multiplying each term by 12.

9. What is the common denominator of the terms of $\frac{2}{7} - \frac{3}{x} = \frac{-5}{7}$?

10. Solve $\frac{2}{7} - \frac{3}{x} = \frac{-5}{7}$ by multiplying each term by 7x.

11. What is the common denominator of the terms of $\frac{5}{x} - \frac{9}{4} = \frac{1}{2x}$?

12. Solve $\frac{5}{x} - \frac{9}{4} = \frac{1}{2x}$ by multiplying each term by 4x.

13. What is the common denominator of the terms of $\frac{5}{x} - \frac{2}{3} = \frac{3}{2x}$?

14. Solve $\frac{5}{x} - \frac{2}{3} = \frac{3}{2x}$ by multiplying each term by 6x.

15. What is the common denominator of the terms of $\frac{1}{x} - \frac{3}{5} = \frac{3}{x} - \frac{2}{5}$?

16. Solve $\frac{1}{x} - \frac{3}{5} = \frac{3}{x} - \frac{2}{5}$ by multiplying each term by 5x.

Part B: Drill and Practice

Solve.

1. $2x - \frac{2}{5} = \frac{3}{4}$

2. $5x - \frac{1}{2} = \frac{5}{8}$

3. $\frac{2}{5}x - 3 = \frac{1}{5}$

4. $\frac{7}{8}x + \frac{1}{2} = 4$

5. $\frac{2}{5}x - \frac{3}{4} = \frac{7}{10}$

6. $\frac{7}{8}x + \frac{5}{6} = \frac{1}{12}$

7. $\frac{5}{6}x + \frac{7}{12} = \frac{2}{3}$

8. $\frac{2}{3}x + \frac{1}{4} = \frac{5}{12}$

9. $\frac{3}{8}x - \frac{5}{12} = \frac{1}{6}$

10. $\frac{2}{7}x + \frac{3}{14} = \frac{1}{4}$

11. $\frac{1}{4}x - \frac{3}{8} = \frac{1}{2}$

12. $\frac{5}{12}x + \frac{2}{3} = \frac{1}{4}$

13. $\frac{2}{3}x - \frac{2}{5} = \frac{3}{5}x - \frac{1}{3}$

14. $\frac{5}{6}x - \frac{1}{3} = \frac{1}{4}x + \frac{1}{2}$

15. $\frac{3}{4}x - \frac{1}{2} = \frac{1}{4}x + \frac{11}{2}$

16. $\frac{2}{3} - \frac{3}{5}x = \frac{1}{3}x + \frac{1}{2}$

17. $\frac{-1}{2} + \frac{5}{4}x = \frac{7}{8} - \frac{1}{2}x$

18. $\frac{-3}{2}x - \frac{1}{3} = \frac{-2}{5}x + \frac{1}{6}$

19. $\frac{4}{3}x + \frac{1}{2} = \frac{5}{6}x - \frac{1}{2}$

20. $\frac{1}{6} - \frac{2}{x} = \frac{1}{2}$

21. $\frac{5}{8} - \frac{3}{2x} = \frac{1}{4}$

22. $\frac{2}{5} - \frac{4}{x} = \frac{3}{10}$

23. $\frac{2}{5} - \frac{1}{x} = \frac{3}{5x}$

24. $\frac{6}{x} - \frac{1}{2} = \frac{1}{4}$

25. $\frac{4}{x} - \frac{2}{3} = \frac{5}{6x}$

26. $\frac{3}{x} - \frac{1}{2} = \frac{5}{6}$

27. $\frac{3}{4x} - \frac{1}{2} = \frac{2}{3} + \frac{3}{x}$

28. $\frac{-2}{5x} - \frac{1}{3} = \frac{2}{5} + \frac{4}{x}$

29. $\frac{1}{x} + \frac{2}{7} = \frac{2}{x} + \frac{3}{14}$

30. $\frac{-5}{6} - \frac{2}{3x} = \frac{5}{6} + \frac{1}{x}$

Solve and check.

31. $\frac{2}{3}x + \frac{5}{6} = \frac{1}{12}$

32. $\frac{3}{4}x - \frac{1}{5} = \frac{1}{10}$

33. $\frac{2}{3}x - \frac{3}{8} = \frac{1}{2}x + \frac{17}{8}$

34. $\frac{5}{x} + \frac{1}{2} = \frac{3}{x} + \frac{7}{10}$

35. $\frac{4}{3x} + \frac{1}{2} = \frac{2}{x}$

Part C: Review. Answers for the problems of Part C are not given. However, the notation «C,U» refers to the Chapter (C) and Unit (U) in which problems of the same type were presented.

1. What is the ten thousands digit of 6,248,271? «1,1»

2. Select the smaller fraction from: $\frac{5}{6}, \frac{14}{17}$ «3,2»

3. $9\frac{3}{7} - 3\frac{2}{3}$ «3,4»

4. Write 0.697 as a fraction.   «4,1»

5. Solve $9x = 36$ «5,7»

6. Evaluate $4 + (-7)$ «6,1»

7. Evaluate $-7 \cdot -3$ «6,3»

8. Simplify $-5(4x)$ «6,5»

9. Solve $7x - 4 = 17$ «6,7»

10. Evaluate $\frac{2}{3} \cdot \frac{3}{5}$ «7,1»

11. Evaluate $\frac{1}{8} - (\frac{1}{2} - \frac{1}{6})$ «7,2»

12. Simplify $(2x - \frac{5}{6}) + \frac{2}{9}$ «7,3»

13. Simplify $\frac{2}{5}(\frac{10}{7}b)$ «7,4»

14. Simplify $3x + 2(x - 5)$ «7,5»

15. Solve $6x + 3 = x - 7$ «7,6»

16. A runner has completed $\frac{5}{6}$ of his run after 2 hours. If he has a total of 15 miles to run, how many miles has he completed?   «3,7»

17. A first number is the same as 24 increased by a second number. If s represents the second number, write an open expression for the first number.   «5,8»

18. What is the average, to two decimal places, of 35, 41, and 43?   «4,4»

19. Write an equation that shows the sum of a number and 33 is 56.   «6,9»

Chapter 7 Test

«7,U» shows the unit in which the problem is found in the chapter.

Evaluate.

1. $\frac{5}{3} \cdot \frac{-6}{7}$ «7,1»

2. $\frac{-4}{5} \cdot \frac{3}{-10}$ «7,1»

3. $\frac{-3}{8} \cdot \frac{0}{5}$ «7,1»

4. $\frac{4}{3} \cdot \frac{4}{3}$ «7,1»

5. $\frac{1}{3} + \frac{3}{7}$ «7,2»

6. $\frac{-2}{5} + \frac{1}{10}$ «7,2»

7. $\frac{-3}{7} - \frac{3}{4}$ «7,2»

8. $\frac{5}{3} + \frac{-2}{5}$ «7,2»

9. $\frac{-2}{9} + \frac{3}{4}$ «7,2»

10. $2 + \frac{-5}{9}$ «7,2»

Simplify.

11. $\frac{1}{3}(2x + 6)$ «7,5»

12. $\frac{5}{6}x - \frac{3}{4}x$ «7,5»

13. $\frac{8}{x} + \frac{3}{x}$ «7,5»

14. $\frac{5}{x} - \frac{7}{x}$ «7,5»

Solve.

15. $5x + 3 = 2$ «7,6»

16. $4x - 7 = 16$ «7,6»

17. $-6x + 12 = 31$ «7,6»

18. $2x - 5 = 3x + 7$ «7,6»

19. $x + 7 = 3 - 5x$ «7,6»

20. $x - \frac{5}{8} = \frac{1}{4}$ «7,7»

21. $\frac{3}{5}x = \frac{2}{7}$ «7,6»

22. $\frac{-5}{2}x = \frac{1}{5}$ «7,6»

23. $4(2x + 3) - 6 = x + 1$
       «7,6»

24. $5 - (6x + 7) = 7$ «7,6»

25. $\frac{4}{x} + \frac{3}{x} = 5$ «7,7»

26. $\frac{2}{x} - 3 = \frac{1}{x} + 2$ «7,7»

27. $\frac{3}{x} + 4 = \frac{5}{x} - 3$ «7,7»

8
Ratios, Proportions, and Percents

Unit 1: THE MEANING OF A RATIO

Ratios

A ratio is a comparison by division of two numbers. Three
ways to show the same ratio are listed below.

$$2 \text{ to } 3 \qquad 2{:}3 \qquad \frac{2}{3}$$

This unit will be concerned with the first and third method
of showing a ratio.

*Focus on
expressing
ratios as
fractions*

The ratio of 4 to 5 can be expressed as the fraction $\frac{4}{5}$. 4 is the numerator and 5 is the denominator.

The ratio of 7 to 4 expressed as a fraction is $\frac{7}{4}$.

The ratio of 12 sophomores to 8 freshmen is $\frac{12}{8}$ which simplifies to $\frac{3}{2}$.

The ratio of 4 books to $48 is $\frac{4}{48}$ and is simplified to $\frac{1}{12}$.

Simplifying Ratios of Decimal Numerals

The ratio $\frac{0.1}{3.2}$ can be simplified by first multiplying both numerator and denominator by 10. The multiplier 10 is used because each of the numerals, 0.1 and 3.2, has one digit right of the decimal.

$$\frac{0.1}{3.2} = \frac{0.1 \times 10}{3.2 \times 10} = \frac{1.}{32.} = \frac{1}{32}$$

Notice that all decimal points are removed from both the numerator and denominator of a fraction.

*Focus on
simplifying
a ratio*

To simplify the ratio of 2.8 to 5.79, first note that 2.8 has one decimal place and 5.79 has two decimal places.

To make 2.8 a whole number it would have be be multiplied by 10.

$$2.8 \cdot 10 = 28$$

To make 5.79 a whole number it would have be be multiplied by 100.

$$5.79 \cdot 100 = 579$$

Therefore, the ratio of 2.8 to 5.79 needs to be multiplied by $\frac{100}{100}$ so that both 2.8 and 5.79 will become whole numbers.

$$\frac{2.8}{5.79} = \frac{2.8 \cdot 100}{5.79 \cdot 100} = \frac{280}{579}$$

The ratio of 2.8 to 5.79 in its simplest form is $\frac{280}{579}$.

Definition	Proportion	
An equation or an equality is a **proportion**	**If and only If**	each side of the equality (equation) is a ratio.

Proportions

A **proportion** is an equality of two ratios.

Three examples of **proportions** are given below.

$$\frac{2}{3} = \frac{4}{6} \qquad \frac{6}{4} = \frac{18}{12} \qquad \frac{x}{5} = \frac{3}{8}$$

The first two examples above are true equalities. The third example is an equation.

Focus on proportions

Two ratios with an equal sign between them form a proportion.

$\frac{8}{10} = \frac{5}{4}$ is a proportion; a proportion can be true or false.

$\frac{3}{4} = \frac{6}{10}$ is a proportion, but it is also a false equality.

$\frac{5}{8} = \frac{15}{24}$ is a proportion, and a true equality.

$\frac{2}{3} = \frac{x}{17}$ is a proportion, and an equation. For most number replacements of x it would become a false equality, but if x is replaced by $\frac{34}{3}$ it becomes a true equality. Later in this Chapter, it will be shown how to find the $\frac{34}{3}$.

The Means and Extremes of a Proportion

For the proportion $\frac{3}{8} = \frac{12}{32}$ the first numerator, 3, and the second denominator, 32, are called **extremes**.

The first denominator, 8, and the second numerator, 12, are called **means**.

> **In a proportion that is a true equality,**
> **the product of the extremes is equal to**
> **the product of the means.**

$\frac{3}{8} = \frac{12}{32}$ is a true equality because $3 \cdot 32$ equals $8 \cdot 12$. Both equal 96.

Focus on the extremes and means of proportions

For the proportion $\frac{1}{2} = \frac{2}{4}$ the extremes are 1 and 4, and the means are 2 and 2. The product of the extremes is $1 \cdot 4$ or 4. The product of the means is $2 \cdot 2$ or 4. Since the products in both cases are 4, the proportion $\frac{1}{2} = \frac{2}{4}$ is a true equality.

Proportions that Are Equations

The proportions

$$\frac{n}{8} = \frac{3}{5} \qquad \frac{7}{n} = \frac{3}{11} \qquad \frac{12}{17} = \frac{n}{6}$$

are equations.

The following equivalences are found using the idea that the product of the extremes must be equal to the product of the means.

$\frac{n}{8} = \frac{3}{5}$ is equivalent to $n \cdot 5 = 8 \cdot 3$

$\frac{7}{n} = \frac{3}{11}$ is equivalent to $7 \cdot 11 = n \cdot 3$

$\frac{12}{17} = \frac{n}{6}$ is equivalent to $12 \cdot 6 = 17 \cdot n$

Unit 1 Exercise

This exercise reviews the preceding unit. The exercise is divided into three parts. Part A reviews the foci of Unit 1. Part B offers opportunities to practice the skills and concepts of Unit 1. Part C contains problems that review your previous work in this text. All answers for Parts A and B are at the back of the book. Each problem of Part C is accompanied by a notation «CU» that refers to the Chapter (C) and Unit (U) in which that type of problem is studied.

Part A: Reviewing the foci of Unit 1.

1. The ratio of 7 to 6 is the fraction with the numerator 7 and the denominator 6. Write the ratio of 7 to 6 as a fraction.

2. The ratio of 3 to 11 is the fraction $\frac{3}{11}$. Which fraction, $\frac{8}{7}$ or $\frac{7}{8}$, is the ratio of 8 to 7?

3. Does $\frac{9}{6}$ show the ratio of 6 to 9?

4. When n and d represent numbers, the ratio of n to d is shown by the fraction $\frac{n}{d}$. $\frac{6}{9}$ shows the ratio of _____ to _____.

5. $\frac{7}{11}$ shows the ratio of 7 to 11. $\frac{9}{23}$ shows the ratio of _____ to _____.

6. Give the ratio of 25 drops to 20 liters in simplest terms.

7. Give the ratio of 180 kilometers to 6 hours in simplest terms.

8. Simplify the ratio of 3.8 grams to 1.25 seconds.
$$\frac{3.8}{1.25} = \frac{3.8 \cdot 100}{1.25 \cdot 100}$$

9. Simplify the ratio of 4 eggs to 1.25 cups of milk.
$$\frac{4}{1.25} = \frac{4 \quad \cdot 100}{1.25 \cdot 100}$$

10. Simplify the ratio of 1.362 mg to 4 liters.
$$\frac{1.362}{4} = \frac{1.362 \cdot 1000}{4 \quad \cdot 1000}$$

11. Is $\frac{3}{5} = \frac{4}{6}$ a true equality?

12. Is $\frac{20}{80} = \frac{2}{8}$ a true equality?

13. Is $\frac{1}{3} = \frac{1}{2}$ a true equality?

14. Is $\frac{6}{51} = \frac{2}{17}$ a true equality?

15. Is $\frac{4.5}{9} = \frac{3}{6}$ a true equality?

16. Is $\frac{3}{9} = \frac{6.6}{20}$ a true equality?

17. For the proportion $\frac{3}{4} = \frac{12}{16}$ the extremes are 3 and 16 and the means are 4 and 12. Find the product of the extremes and the product of the means.

18. In the proportion $\frac{3}{12} = \frac{1}{4}$ the extremes are 3 and 4 and the means are _____ and _____.

19. What are the means of the proportion below?
$$\frac{15}{18} = \frac{5}{6}$$

20. $\frac{6}{n} = \frac{13}{14}$ is equivalent to $6 \cdot 14 = n \cdot 13$.
$\frac{8}{13} = \frac{9}{n}$ is equivalent to _____ = _____.

21. $\frac{47}{n} = \frac{28}{33}$ is equivalent to _____ = _____.

22. $\frac{2}{9} = \frac{n}{18}$ is equivalent to _____ = _____.

23. $\frac{7}{8} = \frac{14}{n}$ is equivalent to _____ = _____.

Part B: Drill and Practice

Give the following ratios in simplest form.

1. Give the ratio of 10 to 32 as a fraction.

2. The ratio of 28 to 18 is shown by the fraction _____.

3. The ratio of 2.7 to 0.8 is shown by the fraction _____.

4. The ratio of 18 to 7.8 is the fraction _____.

5. Give the ratio of 40 units to 80 kilograms in simplest terms.

6. Give the ratio of 210 cycles to 60 minutes in simplest terms.

7. Give the ratio of $20 to 5 liters of whisky in simplest terms.

8. Give the ratio of 4,000 particles to 8,000 particles of pollution in simplest terms.

9. Simplify the ratio of 0.3 cm to 5 kilometers.

10. Simplify the ratio of 7.5 kilometers to 0.5 centimeters.

11. Simplify the ratio of 4.5 grams to 5.5 milliliters.

12. Simplify the ratio of $2.61 to $2.

13. Simplify the ratio of 0.76 mg to 1.8 liters.

14. Does $\frac{7}{13}$ show the ratio of 7 to 13?

15. Give the ratio of 6 to 13.

16. Give the ratio of 1.3 to 7.

17. Give the ratio of 0.8 to 0.3.

18. 8 to 10

19. 240 km to 2 hrs.

20. 18 problems correct to 20 problems total.

21. $1,260 total price to 4 refrigerators

22. 3.4 grams to 0.5 cu. cc

23. 403 grams to 5.51 liters

24. 0.58 kilograms to 23 cu. cc

25. 4,756 kilometers to 2.3 hrs.

26. Determine if $\frac{9}{18} = \frac{13}{26}$ is a true equality using the products of the extremes and means.

27. Determine if $\frac{11}{22} = \frac{15}{32}$ is a true equality using the products of the extremes and means.

28. Determine if $\frac{2.5}{5} = \frac{2.4}{4.8}$ is a true equality using the products of the extremes and means.

29. Determine if $\frac{4}{8} = \frac{0.2}{4}$ is a true equality using the products of the extremes and means.

30. $\frac{n}{8} = \frac{2}{16}$ is equivalent to _____ = _____

31. $\frac{37}{18} = \frac{n}{10}$ is equivalent to _____ = _____

32. $\frac{16}{20} = \frac{5}{n}$ is equivalent to _____ = _____

Part C: Review. Answers for the problems of Part C are not given. However, the notation «C,U» refers to the Chapter (C) and Unit (U) in which problems of the same type were presented.

1. 49,302 – 23,675   «1,3»

2. Find the lcm for 10 and 16.   «3,2»

3. $2\frac{1}{2} \div \frac{5}{12}$ «3,6»

4. Simplify $(3x + 5) + 2(x + 3)$ «5,6»

5. Evaluate $11 + (-6)$ «6,1»

6. Evaluate $-8 \cdot -12$ «6,3»

7. Simplify $-8(-x)$ «6,5»

8. Simplify $(2x + 3) - (x - 8)$ «6,6»

9. Solve $4x - 5 = 27$ «6,7»

10. Evaluate $\frac{14}{3} \cdot \frac{15}{8}$ «7,1»

11. Simplify $(\frac{2}{3}x - \frac{1}{2}) + \frac{3}{4}$ «7,3»

12. Simplify $\frac{3}{4}(\frac{-5}{6}x)$ «7,4»

13. Simplify $7(\frac{3}{7}x + 4)$ «7,5»

14. Solve $3(2x - 4) + 8 = 5 - (x + 2)$ «7,6»

15. Solve $\frac{1}{x} + \frac{2}{7} = \frac{2}{x} + \frac{3}{14}$ «7,7»

16. A first number is the same as 7 decreased by a second number. If s represents the second number, write an open expression for the first number.   «5,8»

17. Find the sum of 8.513 and 7.47   «4,2»

18. Divide 9.53 by 0.6 and round off the answer to 2 decimal places.   «4,4»

19. On three successive days Henry read 84 pages, 54 pages, and 93 pages. What was the average pages read during that 3-day period?   «4,4»

20. Change $\frac{7}{15}$ to a decimal fraction rounded off at two decimal places.   «4,4»

21. Write an equation that shows the difference when 23 is subtracted from a number is 49.   «6,9»

Unit 2:   SOLVING SIMPLE MULTIPLICATION EQUATIONS

Solving Statements of the Form 5 • n = 75

In Chapter 7, the equation $5 \cdot n = 3 \cdot 25$ was solved by:

1. Multiplying both sides of the equation by the reciprocal of 5.

$$\frac{1}{5} \cdot (5 \cdot n) = \frac{1}{5} \cdot (3 \cdot 25)$$

2. Simplifying, by cancelling, both sides of the equation.

$$\frac{1}{\cancel{5}} \cdot \frac{\cancel{5} \cdot n}{1} = \frac{1}{\cancel{5}} \cdot \frac{3 \cdot \cancel{25}^{5}}{1}$$

3. Therefore, n = 15.

$$n = 15$$

15 is the solution of $5 \cdot n = 75$.

Focus on solving by using the reciprocal

To solve $13 \cdot n = 6 \cdot 7$, it is necessary to find a number that is multiplied by 13 to give 42. It will not be a whole number because 42 is not a multiple of 13.

To solve $13 \cdot n = 6 \cdot 7$, multiply both sides of the equation by the reciprocal of 13.

$$13 \cdot n = 6 \cdot 7$$

$$\frac{1}{13} \cdot (13 \cdot n) = \frac{1}{13} \cdot (6 \cdot 7)$$

After simplifying as much as possible, the result is $n = \frac{42}{13}$.

$$\frac{1}{\cancel{13}} \cdot \frac{\cancel{13} \cdot n}{1} = \frac{1}{13} \cdot \frac{6 \cdot 7}{1}$$

$\frac{42}{13}$ or its equivalent mixed number, $3\frac{3}{13}$, is the solution of $13 \cdot n = 6 \cdot 7$.

$$n = \frac{42}{13}$$

Equivalent Equations

Two equations are equivalent when they have exactly the
same solution or solutions.

The two equations shown below are equivalent because the
two sides of an equation may be interchanged.

$$5 \cdot n = 6 \cdot 9 \text{ is equivalent to } 6 \cdot 9 = 5 \cdot n$$

The two equations shown below are equivalent because the
two factors of a multiplication problem may be commuted.

$$n \cdot 11 = 4 \cdot 7 \text{ is equivalent to } 11 \cdot n = 4 \cdot 7$$

Focus on equivalent equations

To solve $8 \cdot 13 = 9 \cdot n$,
the equation may be
written as $9 \cdot n = 8 \cdot 13$.

$8 \cdot 13 = 9 \cdot n$ is equivalent to $9 \cdot n = 8 \cdot 13$
because
two sides of an equation may be interchanged.

To solve $n \cdot 41 = 6 \cdot 54$,
the equation may be
written as $41 \cdot n = 6 \cdot 54$.

$n \cdot 41 = 6 \cdot 54$ is equivalent to $41 \cdot n = 6 \cdot 54$
because
any two factors may be commuted.

The Steps in Solving a Proportion Equation

To solve the proportion $\frac{7}{n} = \frac{9}{10}$

1. Use the products of the means and
   extremes to write a multiplication
   equation.

$$\frac{7}{n} = \frac{9}{10}$$

$$7 \cdot 10 = n \cdot 9$$

2. Solve $7 \cdot 10 = n \cdot 9$ using its
   equivalent, $9 \cdot n = 7 \cdot 10$.

$$9 \cdot n = 7 \cdot 10$$

3. Both sides are multiplied by $\frac{1}{9}$.

$$\frac{1}{9} \cdot (9 \cdot n) = \frac{1}{9} \cdot (7 \cdot 10)$$

4. Therefore, $\frac{70}{9}$ or $7\frac{7}{9}$ is the
   solution of the proportion.

$$n = \frac{70}{9}$$

Focus on solving for "n"

To solve the proportion $\frac{n}{5} = \frac{11}{6}$ first multiply the means and the extremes.

$\frac{n}{5} = \frac{11}{6}$ gives $n \cdot 6 = 5 \cdot 11$ which is equivalent to $6n = 55$.

$6n = 55$ is solved as a basic equation.

The solution is $\frac{55}{6}$ or $9\frac{1}{6}$.

$$6n = 55$$

$$\frac{1}{6} \cdot 6n = \frac{1}{6} \cdot 55$$

$$n = \frac{55}{6}$$

Unit 2 Exercise

This exercise reviews the preceding unit. The exercise is divided into three parts. Part A reviews the foci of Unit 2. Part B offers opportunities to practice the skills and concepts of Unit 2. Part C contains problems that review your previous work in this text. All answers for Parts A and B are at the back of the book. Each problem of Part C is accompanied by a notation «CU» that refers to the Chapter (C) and Unit (U) in which that type of problem is studied.

Part A: Reviewing the foci of Unit 2.

1. Solve $6 \cdot n = 7 \cdot 12$ by multiplying both sides of the equation by the reciprocal of 6 and cancelling wherever possible.

2. Solve $4 \cdot n = 3 \cdot 9$ by multiplying both sides of the equation by the reciprocal of 4 and cancelling wherever possible.

3. Solve $9 \cdot n = 7 \cdot 3$ by multiplying both sides of the equation by the reciprocal of 9 and cancelling wherever possible.

4. Solve $11 \cdot n = 6 \cdot 9$ by multiplying both sides of the equation by the reciprocal of 11 and cancelling wherever possible.

5. Write $\frac{8}{n} = \frac{10}{13}$ as a multiplication equation using the products of the means and the extremes. _____ = _____

6. Write $\frac{7}{11} = \frac{5}{n}$ as a multiplication equation using the products of the means and the extremes. _____ = _____

7. Solve $\frac{n}{8} = \frac{3}{5}$ by first writing a multiplication equation using the products of the means and the extremes.

8. Solve $\frac{17}{20} = \frac{n}{4}$ by first writing a multiplication equation using the products of the means and the extremes.

Part B: Drill and Practice

Solve.

1. $8 \cdot n = 3 \cdot 11$

2. $6 \cdot n = 8 \cdot 7$

3. $2 \cdot n = 7 \cdot 5$

4. $7 \cdot n = 2 \cdot 2$

5. $4 \cdot n = 19 \cdot 1$

6. $8 \cdot n = 11 \cdot 4$

7. $8 \cdot n = 3 \cdot 2$

8. $6 \cdot n = 2 \cdot 25$

9. $3 \cdot 4 = 2 \cdot n$

10. $17 \cdot n = 3 \cdot 7$

11. $7 \cdot n = 9 \cdot 7$

12. $6 \cdot n = 4 \cdot 5$

13. $37 \cdot 1 = 5 \cdot n$

14. $n \cdot 9 = 3 \cdot 5$

15. $8 \cdot 2 = n \cdot 7$

16. $8 \cdot n = 7 \cdot 4$

17. $\frac{6}{11} = \frac{9}{n}$

18. $\frac{4}{7} = \frac{1}{n}$

19. $\frac{6}{5} = \frac{1}{n}$

20. $\frac{4}{3} = \frac{n}{7}$

21. $\frac{12}{5} = \frac{2}{n}$

22. $\frac{5}{n} = \frac{8}{10}$

23. $\frac{9}{14} = \frac{n}{7}$

24. $\frac{6}{7} = \frac{5}{n}$

25. $\frac{3}{n} = \frac{11}{2}$

26. $\frac{n}{7} = \frac{3}{5}$

27. $\frac{n}{8} = \frac{1}{15}$

Part C: Review. Answers for the problems of Part C are not given. However, the notation «C,U» refers to the Chapter (C) and Unit (U) in which problems of the same type were presented.

1. 43,790 + 23,690 + 6,523   «1,2»

2. Simplify $\frac{13}{39}$ «3,1»

3. Find the numerator of $\frac{2}{15} = \frac{?}{45}$ «3,2»

4. Find the difference of 48.47 and 39.6   «4,2»

5. Simplify $(7 + 5x) + 6$ «5,4»

6. Simplify $5(2x + 9) + 3(x + 2)$ «5,6»

7. Evaluate $11 - (-2)$ «6,2»

8. Evaluate $6 - (-12) \cdot -2 \div 8$ «6,4»

9. Simplify $6(2x - 1) - (x - 3)$ «6,6»

10. Solve $9 - 4x = 23 - 2x$ «6,8»

11. Evaluate $\frac{1}{4} + (\frac{2}{3} - \frac{1}{2})$ «7,2»

12. Simplify $\frac{-3}{8}(\frac{4}{9}x)$ «7,4»

13. Simplify $\frac{3}{x} - \frac{7}{x}$ «7,5»

14. Solve $x + 9 = 3x - 3$ «7,6»

15. Solve $\frac{2}{3}x - \frac{1}{2} = \frac{1}{10}$ «7,7»

16. $\frac{37}{18} = \frac{n}{10}$ is equivalent to _____ = _____
    «8,1»

17. How many $\frac{1}{4}$ ounces are there in 7 ounces?.
    «3,7»

18. Lucille's test scores in Government were 76, 87, 75, and 88. Find her average on these four tests rounded off to the nearest whole number.
    «4,4»

19. Change $\frac{3}{7}$ to a decimal fraction rounded off at two decimal places.   «4,4»

20. One number is eight times a second number. If s represents the second number, write an open expression for the first number.   «5,8»

21. Write an equation that shows: 17 less than a number is 43.   «6,9»

Unit 3:   SOLVING WORD PROBLEMS WITH PROPORTIONS

Steps in Solving Word Problems With Proportions

In this unit word problems are solved using proportions.
Each problem involves a three-step approach.

1.   Write two ratios showing similar comparisons.

2.   Write a proportion between the two ratios.

3.   Solve the proportion.

Focus on writing proportions to show comparisons

Word Problem: The ratio of boys to girls in the sophomore class is $\frac{6}{5}$ and there are 45 girls in the class. How many boys are in the sophomore class?

The proportion $\frac{6}{5} = \frac{b}{45}$ contains two ratios, $\frac{6}{5}$ and $\frac{b}{45}$, which are comparisons of boys to girls. The letter b represents the number of boys in the sophomore class.

Using b to represent the number of boys, then $\frac{6}{5}$ and $\frac{b}{45}$ both are ratios of boys to girls.

$$\frac{6 \text{ boys}}{5 \text{ girls}} \qquad\qquad \frac{b \text{ boys}}{45 \text{ girls}}$$

Notice that each ratio has "boys" as its numerator and "girls" as its denominator. In writing ratios for a proportion, be certain that the numerators have the same labels and the denominators have the same labels.

The original problem is solved using the proportion $\frac{6}{5} = \frac{b}{45}$.

Since $b = 54$, the proportion's solution means that there are 54 boys in the sophomore class.

$$\frac{6}{5} = \frac{b}{45}$$
$$6 \cdot 45 = 5 \cdot b$$
$$5 \cdot b = 6 \cdot 45$$
$$\frac{1}{5} \cdot 5 \cdot b = \frac{1}{5} \cdot 6 \cdot 45$$
$$1 \cdot b = 1 \cdot 6 \cdot 9$$
$$b = 54$$

Solving a Word Problem Using a Proportion

The following problem is solved using a three-step approach:

> One pound of hamburger makes spaghetti for three people. How much hamburger is needed to make spaghetti for eight people?

1. The ratio $\frac{1}{3}$ compares hamburger amount to people.
    If n represents the hamburger needed for eight people,
    the ratio is $\frac{n}{8}$.

$$\frac{1 \text{ hamburger}}{3 \text{ people}} \qquad \frac{n \text{ hamburger}}{8 \text{ people}}$$

2. The two ratios comparing hamburger amounts to people
    are $\frac{1}{3}$ and $\frac{n}{8}$.

    The proportion to be solved is: $\frac{1}{3} = \frac{n}{8}$

3.
$$1 \cdot 8 = 3 \cdot n$$
$$3 \cdot n = 1 \cdot 8$$
$$\frac{1}{3} \cdot 3 \cdot n = \frac{1}{3} \cdot 1 \cdot 8$$
$$n = \frac{8}{3}$$

$2\frac{2}{3}$ pounds of hamburger are needed for eight people.

Focus on solving problems with proportions

Word Problem: Two inches on a map represents 15 miles. How many miles are represented by 11 inches?

The two ratios comparing inches to miles are $\frac{2}{15}$ and $\frac{11}{n}$.

$$\frac{2 \text{ inches}}{15 \text{ miles}} \qquad \frac{11 \text{ inches}}{n \text{ miles}}$$

The proportion to be solved is $\frac{2}{15} = \frac{11}{n}$.

11 inches on the map represents $82\frac{1}{2}$ miles.

$$\frac{2}{15} = \frac{11}{n}$$
$$2 \cdot n = 15 \cdot 11$$
$$\frac{1}{2} \cdot 2 \cdot n = \frac{1}{2} \cdot 15 \cdot 11$$
$$n = \frac{165}{2}$$

Unit 3 Exercise

This exercise reviews the preceding unit. The exercise is divided into three parts. Part A reviews the foci of Unit 3. Part B offers opportunities to practice the skills and concepts of Unit 3. Part C contains problems that review your previous work in this text. All answers for Parts A and B are at the back of the book. Each problem of Part C is accompanied by a notation «CU» that refers to the Chapter (C) and Unit (U) in which that type of problem is studied.

Part A: Reviewing the foci of Unit 3.

1. If 4 inches of insulation cost $2, what is the cost of 10 inches of insulation?
    a. Write two ratios comparing inches of insulation to cost.
    b. Write a proportion.
    c. Solve the proportion.

2. If 6 yards of cloth cost $9, how much do 20 yards cost?
    a. Write two ratios comparing yards of cloth to dollars.
    b. Write a proportion.
    c. Solve the proportion.

3. If 3 long distance runners cover 51 miles, how many runners would be needed to cover 85 miles?
    a. Write two ratios comparing runners to miles.
    b. Write a proportion.
    c. Solve the proportion.

4. If a mixture of nuts contains 7 pounds of pecans for every 4 pounds of cashews, how many pounds of pecans should be used with 11 pounds of cashews?
    a. Write two ratios comparing pecans to cashews.
    b. Write a proportion.
    c. Solve the proportion.

5. A solution of 3 parts medicine to 20 parts alcohol is to be made. The total amount of alcohol will be 240 milliliters (ml). How much medicine is needed?
    a. Write the proportion comparing medicine to alcohol.
    b. Solve the proportion.

6. The ratio of profit to cost for a product is 1 to 9. If the profit is $16,200 what is the cost?
    a. Write the proportion comparing profit to cost.
    b. Solve the proportion.

7. 5 grains of a drug are mixed with 160 milliliters (ml) of a solution. In a sample containing 2 grains of the drug, how much solution will be used?
    a. Write the proportion comparing grains to solution.
    b. Solve the proportion.

8. A recipe calls for 1 cup of sugar with 3 cups of flour. How much sugar is needed with 12 cups of flour?
    a. Write the proportion comparing sugar to flour.
    b. Solve the proportion.

9. An intravenous solution needs 2 liters (l) of glucose mixed with 7 units of blood. How much glucose is needed for 35 units of blood?
   a. Write the proportion comparing glucose to blood.
   b. Solve the proportion.

10. A doctor's order calls for 8 drams of drug to be used with 100 milliliters of water. How much water will be needed for 18 drams of drug?
   a. Write the proportion comparing drug to water.
   b. Solve the proportion.

Part B: Drill and Practice

1. 2 centimeters (cm) on a map represent 10 kilometers (km). How many centimeters are needed to show 37 kilometers?

2. A car used 13 liters (l) of fuel traveling 340 kilometers (km). How many liters of fuel will be used traveling 85 kilometers?

3. If 20 liters (l) of paint will cover 210 square meters of wall, how many square meters will 15 liters of paint cover?

4. A store decided to give a discount of 12 cents for every $1 of price. A refrigerator was priced at $300. How much is the discount?

5. If 1500 20-year-olds die each year per every 1,000,000 20-year-olds, how many will die out of 2,000 20-year-olds?

6. If 2 nuclear accidents can occur every 360 days, how many may occur in 900 days?

7. Ace Packing Co. makes 3 cents profit for every $200 in sales. What is the profit on a $1000 sale?

8. A nurse is ordered to make a mixture containing 7 tablespoons of drug for each 10 liters of solvent. If 2 liters of solvent are used, how much drug is needed?

9. A student gets 5 out of every 6 problems correct on a test. If there are 180 total problems on the test, how many did the student answer correctly?

10. A business found that there were 2 cents advertising cost for every 9 cents in sales. On an item selling for $18 what was the advertising cost?

11. A photograph was 3 inches high and 5 inches wide. If it is enlarged to be 7 inches high, how wide will it be?

12. In a 250 kilogram block of alloy, 2 grams out of every 5 grams of alloy are gold. How much gold is in the block?

13. The National Bank charges interest of 13 cents for every $1 loaned for one year. How much interest will be charged for a loan of $150 for one year?

14. The ratio of the width to length of a book is to be 8 to 11. How wide should a page be if it is 16 cm long?

Part C: Review. Answers for the problems of Part C are not given. However, the notation «C,U» refers to the Chapter (C) and Unit (U) in which problems of the same type were presented.

1. What is the thousands digit of 6,248,271? «1,1»

2. Select the smaller fraction from: $\frac{9}{10}, \frac{19}{20}$ «3,2»

3. $9\frac{1}{3} - 3\frac{3}{4}$ «3,4»

4. Write 0.517 as a fraction.   «4,1»

5. Solve $4x = 36$ «5,7»

6. Evaluate $3 + (-9)$ «6,1»

7. Evaluate $-7 \cdot -5$ «6,3»

8. Simplify $-5(-3x)$ «6,5»

9. Solve $4x - 9 = 19$ «6,7»

10. Evaluate $\frac{5}{6} \cdot \frac{3}{5}$ «7,1»

11. Simplify $(2x - \frac{1}{2}) + \frac{1}{8}$ «7,3»

12. Simplify $8x - 2(x - 5)$ «7,5»

13. Solve $\frac{-5}{6} - \frac{2}{3x} = \frac{5}{6} + \frac{1}{x}$ «7,7»

14. Solve $\frac{6}{7} = \frac{5}{n}$ «8,2»

15. A manufacturer has completed $\frac{3}{4}$ of an order after days work. If the order requires 52 items , how many items has he completed?   «3,7»

16. A first number is the same as 17 increased by a second number. If s represents the second number, write an open expression for the first number.   «5,8»

17. What is the average, to two decimal places, of 53, 48, and 59?   «4,4»

18. Write an equation that shows: the sum of a number and 27 is 75.   «6,9»

19. Determine if $\frac{4}{8} = \frac{0.2}{4}$ is a true equality using the products of the extremes and means. «8,1»

Unit 4:   THE MEANING OF PERCENT

Percent is a special type of ratio and all percent problems
can be solved as proportion problems. In this unit, the
study of percent begins.

Definition	Percent	
A ratio is a **percent**	**if and only if**	its denominator is 100.

Focus on the meaning of percent

The ratio $\frac{41}{100}$ is read as 41 percent.

The ratio $\frac{7}{100}$ is read as 7 percent.

The ratio $\frac{23}{100}$ is read as 23 percent.

17 percent is read as the ratio $\frac{17}{100}$.

4.7 percent is read as the ratio $\frac{4.7}{100}$.

357 percent is read as the ratio $\frac{357}{100}$.

The Symbol for Percent

The symbol for percent is "%".

67 percent is written as 67% and means $\frac{67}{100}$.

245 percent is written as 245% and means $\frac{245}{100}$.

0.8 percent is written as 0.8% and means $\frac{0.8}{100}$.

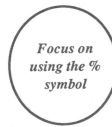

Focus on using the % symbol

% is the symbol for percent.

96 percent is 96% or the ratio $\frac{96}{100}$.

1000 percent is 1000% or the ratio $\frac{1000}{100}$.

5.25 percent is 5.25% or the ratio $\frac{5.25}{100}$.

Writing Percents as Decimals

To change 52% to a decimal first think of it as a ratio with a denominator of 100.

$$52\% = \frac{52}{100}$$

The ratio $\frac{52}{100}$ means $52 \div 100$ and division by 100 requires the movement of the decimal point two places to the left.

$$100 \overline{)\,52} \quad \text{is equivalent to} \quad 100 \overline{)\,5\,2.0\,0}^{\,0\,0.5\,2}$$

52% written as a decimal is 0.52.

Focus on writing percents as decimal fractions

To change 9% to a decimal fraction use the ratio $\frac{9}{100}$ to write the division problem $9 \div 100$. 9% is 0.09

$$100 \overline{)\,9}$$

$$100 \overline{)\,9.0\,0}^{\,0.0\,9}$$

To change 5.6% to a decimal fraction use the ratio $\frac{5.6}{100}$ to write the division problem $5.6 \div 100$. 5.6% is 0.056

$$100 \overline{)\,5.6}$$

$$100 \overline{)\,5.6\,0\,0}^{\,0.0\,5\,6}$$

To change 147% to a decimal fraction use the ratio $\frac{147}{100}$ to write the division problem $147 \div 100$. 147% is 1.47

$$100 \overline{)\,147}$$

$$100 \overline{)\,1\,4\,7.0\,0}^{\,0\,0\,1.4\,7}$$

Dividing by 100 and Placing the Decimal Point

Whenever a number is divided by 100, the quotient will be the original dividend with its decimal point moved two places to the left. This fact is helpful in changing any percent to a decimal.

9% or 9.0% is equal to 0.09

5.6% is equal to 0.056

147% or 147.0% is equal to 1.47

Writing Decimal Fractions as Percents

To write any decimal fraction as a percent first write it as a fraction and then, if necessary, change the denominator of the fraction to 100.

$$0.57 = \frac{57}{100} = 57\%$$

$$0.052 = \frac{52}{1000} = \frac{5.2}{100} = 5.2\%$$

$$6.4 = \frac{64}{10} = \frac{640}{100} = 640\%$$

Changing a percent to a decimal is accomplished by moving the decimal point two places to the left. The reverse process of changing a decimal to a percent requires moving the decimal point two places to the right.

Focus on changing a decimal to a percent

$$0.63 = 0.6\,3.\% = 63\%$$

$$0.45 = 0.4\,5.\% = 45\%$$

$$0.091 = 0.0\,9.1\% = 9.1\%$$

$$2.19 = 2.1\,9.\% = 219\%$$

Note that the decimal point in the decimal fraction has been moved two places to the right when adding the percent symbol (%).

Changing Fractions to Percents

Every percent is a ratio with denominator 100. To write a fraction like $\frac{5}{7}$ as a percent means to solve the proportion

$\frac{5}{7} = \frac{r}{100}$ where r is the percent.

$5 \cdot 100 = 7 \cdot r$

$7 \cdot r = 5 \cdot 100$

$\frac{1}{7} \cdot 7 \cdot r = \frac{1}{7} \cdot 5 \cdot 100$

$r = \frac{500}{7}$

$r = \frac{500}{7} = 71\frac{3}{7}$

In most work with percents, an answer like $71\frac{3}{7}\%$ is rounded off to the nearest one percent. Since $\frac{3}{7}$ is less than $\frac{1}{2}$ the answer $71\frac{3}{7}\%$ is rounded off to 71%.

$\frac{5}{7} \doteq 71\%$. The symbol $\doteq$ means "approximately equals."

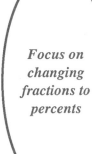

Focus on changing fractions to percents

To change $\frac{5}{8}$ to its approximate percent the first step is to write the proportion

$\frac{5}{8} = \frac{r}{100}$

The proportion is solved to find r, the percent.

$r = \frac{125}{2} = 62\frac{1}{2}$

Since $\frac{1}{2}$ is greater than or equal to $\frac{1}{2}$, $62\frac{1}{2}\%$ is rounded up to 63%.

$62\frac{1}{2}\%$ is approximately equal to 63%. $\frac{5}{8} \doteq 63\%$.

$\frac{5}{8} = \frac{r}{100}$

$5 \cdot 100 = 8 \cdot r$

$8 \cdot r = 5 \cdot 100$

$\frac{1}{8} \cdot (8 \cdot r) = \frac{1}{8} \cdot (5 \cdot 100)$

$\frac{1}{1} \cdot 1 \cdot r = \frac{1}{2} \cdot 5 \cdot 25$

$r = \frac{125}{2} = 62\frac{1}{2}$

Common Fraction – Percent Equivalents

Any fraction can be changed to a percent approximation.
However, some fractions occur frequently enough that
their equivalents are often memorized.

$\frac{1}{2} = 50\%$ $\frac{1}{3} = 33\frac{1}{3}\%$ $\frac{2}{3} = 66\frac{2}{3}\%$

$\frac{1}{4} = 25\%$ $\frac{3}{4} = 75\%$ $\frac{1}{5} = 20\%$

$\frac{1}{6} = 16\frac{2}{3}\%$ $\frac{1}{7} = 14\frac{2}{7}\%$ $\frac{1}{8} = 12\frac{1}{2}\%$

$\frac{1}{10} = 10\%$ $\frac{3}{10} = 30\%$ $\frac{9}{10} = 90\%$

Focus on improper and proper fractions

$$\frac{5}{6} = 5 \cdot \frac{1}{6} = 5 \cdot 16\frac{2}{3}\% = 83\frac{1}{3}\% \qquad \frac{5}{6} \doteq 83\%$$

$$\frac{6}{7} = 6 \cdot \frac{1}{7} = 6 \cdot 14\frac{2}{7}\% = 85\frac{5}{7}\% \qquad \frac{6}{7} \doteq 86\%$$

$\frac{5}{6}$ and $\frac{6}{7}$ are proper fractions. Proper fractions are less than 100%.

$$\frac{4}{3} = 4 \cdot \frac{1}{3} = 4 \cdot 33\frac{1}{3}\% = 133\frac{1}{3}\% \qquad \frac{4}{3} \doteq 133\%$$

$$\frac{11}{8} = 11 \cdot \frac{1}{8} = 11 \cdot 12\frac{1}{2}\% = 137\frac{1}{2}\% \qquad \frac{11}{8} \doteq 138\%$$

$\frac{4}{3}$ and $\frac{11}{8}$ are improper fractions. Improper fractions are greater than 100%.

Unit 4 Exercise

This exercise reviews the preceding unit. The exercise is divided into three parts. Part A reviews the foci of Unit 4. Part B offers opportunities to practice the skills and concepts of Unit 4. Part C contains problems that review your previous work in this text. All answers for Parts A and B are at the back of the book. Each problem of Part C is accompanied by a notation «CU» that refers to the Chapter (C) and Unit (U) in which that type of problem is studied.

Part A: Reviewing the foci of Unit 4.

1. Percent means a ratio with a denominator of 100. 5 percent is the same as 5 one-hundredths. 8 percent is the same as 8 _____-_____.

2. Percent means one-hundredths. 63 percent means $\frac{63}{100}$ or 63 _____-_____.

3. Percent means one-hundredths. 91 percent means $\frac{91}{100}$ or 91 _____-_____.

4. Percent means _____-_____.

5. 47 one-hundredths is 47 percent. 53 one-hundredths is 53 _____.

6. The terms "percent" and "one-hundredths" are interchangeable. 69 one-hundredths is the same as 69 _____.

7. 59 one-hundredths and 59 percent have the _____ (same, different) meaning.

8. Use the symbol % to write 9 percent.

9. The symbol for percent is _____.

10. Write 56 percent in symbols and as a ratio.

11. Write .3 percent in symbols and as a ratio.

12. Write 133 percent in symbols and as a ratio.

13. To change 34% to a decimal divide 34 by 100. 34% is the decimal _____.

14. To change 91% to a decimal divide 91 by 100. 91% is the decimal _____.

15. To change 413% to a decimal divide 413 by 100. 413% is the decimal _____.

16. To change 4.9% to a decimal divide 4.9 by 100. 4.9% is the decimal _____.

17. To change a percent to a decimal move the decimal point two places to the left.

   $92\% = 92.\% = .92$

   $49\% = 49.\% = $ _____

18. To change 5% to a decimal move the decimal point two places to the left.

   $5\% = 5.\% = $ _____

19. Change 2% to a decimal by moving the decimal point two places to the left.

20. To change any percent to a decimal move the decimal point _____ (how many) places left.

21. To change a percent to a decimal move the decimal point two places _____ (right, left).

22. Write 0.73 as a percent by first writing it as a fraction with denominator 100.

23. Write 0.14 as a percent by first writing it as a fraction with denominator 100.

24. Write 0.039 as a percent by first writing it as a fraction with denominator 100.

25. Write 3.7 as a percent by first writing it as a fraction with denominator 100.

26. Write .18 as a percent by moving its decimal point two places right.

27. Write .41 as a percent by moving its decimal point two places right.

28. Write .06 as a percent by moving its decimal point two places right.

29. When converting a decimal numeral to a percent, the point is moved _____ places to the _____.

30. Complete the first step in changing $\frac{3}{11}$ to its approximate percent by writing the proportion associated with $\frac{3}{11}$.

31. Solve $\frac{3}{11} = \frac{r}{100}$.

32. $\frac{3}{11} = 27\frac{3}{11}\%$. Round off $27\frac{3}{11}\%$ to the nearest one percent.

33. Complete the first step in changing $\frac{9}{46}$ to its approximate percent by writing the proportion associated with $\frac{9}{46}$.

34. Solve $\frac{9}{46} = \frac{r}{100}$.

35. $\frac{9}{46} = 19\frac{13}{23}\%$. Round off $19\frac{13}{23}\%$ to the nearest one percent.

36. $\frac{4}{11} = 36\frac{4}{11}\%$. Round off $36\frac{4}{11}\%$ to the nearest one percent.

37. $\frac{8}{15} = 53\frac{1}{3}\%$. Round off $53\frac{1}{3}\%$ to the nearest one percent.

38. $\frac{23}{32} = 71\frac{7}{8}\%$. Round off $71\frac{7}{8}\%$ to the nearest one percent.

39. $\frac{9}{5}$ is _____ (more, less) than 100%.

40. Change $\frac{9}{5}$ to a percent.

41. Change $\frac{15}{8}$ to a percent.

Part B: Drill and Practice

1. Write $\frac{150}{100}$ as a percent.

2. Write $\frac{4}{100}$ as a percent.

3. Write $\frac{1}{100}$ as a percent.

Write each of the following as a ratio.

4. 17%

5. 49%

6. 57%

7. 143%

8. 7%

9. 4.7%

10. 3195%

Write each of the following as a percent.

11. $\frac{19}{100}$

12. $\frac{81}{100}$

13. $\frac{3}{100}$

14. $\frac{181}{100}$

15. $\frac{67}{100}$

16. $\frac{5.9}{100}$

17. $\frac{615}{100}$

Write each of the following as a decimal fraction.

18. 95%

19. 57%

20. 64%

21. 8%

22. 3%

23. 140%

24. 107%

25. 2.6%

26. 57.9%

27. 78%

28. 2%

29. 110%

30. .1%

31. 1.6%

32. .03%

33. 1%

34. 27%

35. 85%

36. 56%

37. 4%

38. 37%

Write each of the following as a percent.

39. .09

40. .71

41. 1.56

42. .5

43. .09

44. 6.2

45. .01

46. .007

47. 1.01

48. .08

49. .001

50. .009

51. 1.62

Round off each of the following to the nearest one percent

52. $\frac{9}{10}$

53. $\frac{3}{8}$

54. $\frac{3}{4}$

55. $\frac{57}{89}$

56. $\frac{15}{47}$

57. $\frac{7}{104}$

58. $\frac{15}{8}$

59. $\frac{23}{9}$

60. $\frac{3}{8}$

61. $\frac{5}{9}$

62. $\frac{4}{5}$

63. $\frac{1}{3}$

64. $\frac{3}{10}$

Part C: Review. Answers for the problems of Part C are not given. However, the notation «C,U» refers to the Chapter (C) and Unit (U) in which problems of the same type were presented.

1. Evaluate $6 + [8 + 7 \cdot 2]$ «5,2»

2. Evaluate $2x + 5y$ when $x = 7$ and $y = 4$
    «5,3»

3. Find the difference of 16.39 and 9.4   «4,2»

4. Find the product of 6.78 and 9.6   «2,2»

5. Find 0.45 of 237.   «4,3»

6. Simplify $(13 + 2x) + 7$ «5,4»

7. Simplify $6(x + 3) + (x + 8)$ «5,6»

8. Evaluate $9 - (-4)$ «6,2»

9. Evaluate $-6 \cdot 2 - 12 \div -2$ «6,4»

10. Simplify $2(5x - 2) - 2(x - 3)$ «6,6»

11. Solve $2(3x + 5) - 17 = -7$ «6,8»

12. Evaluate $(\frac{1}{2} - \frac{3}{4}) - \frac{1}{3}$ «7,2»

13. Simplify $(\frac{3}{4}x + \frac{1}{2}) - \frac{3}{8}$ «7,3»

14. Simplify $\frac{4}{5}(\frac{-15}{8}r)$ «7,4»

15. Simplify $3(2x + \frac{2}{3}) + 2(\frac{3}{4}x - 5)$ «7,5»

16. Solve $6x - 3(2x + 7) = 5 - 2x$ «7,6»

17. Solve $\frac{2}{3}x + \frac{3}{4} = \frac{7}{12}$ «7,7»

18. $\frac{n}{8} = \frac{2}{16}$ is equivalent to _____ = _____
    «8,1»

19. Solve $\frac{n}{7} = \frac{3}{5}$ «8,2»

20. Alice spends 0.15 of her money on entertainment. With a monthly income of $1,500, how much does she spend on entertainment each month?   «4,4»

21. Write an equation that shows: 15 less than a number is 31.   «6,9»

22. A photograph was 3 inches high and 5 inches wide. If it is enlarged to be 7 inches high, how wide will it be?   «8,3»

Unit 5:   THE THREE NUMBERS IN A PERCENT PROBLEM

Base, Rate, and Part

Every percent problem has three numbers. They are the **base** (b), the **rate** (r), and the **part** (p).

In this unit, simple percent problems are solved.

Simple Statements Involving Percent

The statement

$$15\% \text{ of } 200 \text{ is } 30$$

illustrates the common form of most percent problems. Notice there are three numbers, 15, 200, and 30, in the statement. These numbers are called the **rate**, the **base**, and the **part**, respectively.

The percent statements

21 is 30% of 70   and   30% of 70 is 21

mean exactly the same thing. Both statements contain the numbers 30, 70, and 21 which are, respectively, the **rate**, the **base**, and the **part**.

Focus on percent statements

The percent statements

7% of 45 is 3.15   50% of 812 is 406 $33\frac{1}{3}\%$ of 600 is 200

all involve three numbers: the rate, the base, and the part.

Selecting the Rate Number in a Percent Statement

In the percent statement

$$7\% \text{ of } 45 \text{ is } 3.15$$

the number which accompanies the % symbol is the rate (r).
7 is the rate number in the statement.

In the percent statement

$$50\% \text{ of } 812 \text{ is } 406$$

the rate number r is 50 because it accompanies the % symbol.

Focus on selecting the rate number r

To identify the rate number in a percent statement, find the number with the % symbol.

In the statement "30% of 70 is 21" the rate number (r) is 30 because it has the symbol %.

Selecting the Base Number in a Percent Statement

In the percent statement

$$7\% \text{ of } 45 \text{ is } 3.15$$

the number which follows the word "of" is the base (b).
45 is the base number in the statement.

In the percent statement

$$50\% \text{ of } 812 \text{ is } 406$$

the base number b is 812 because it follows "of."

Focus on selecting the base number b

To identify the base number in a simple percent statement, find the number which follows the word "of."

In the statement "30% of 70 is 21" the base number (b) is 70 because it follows "of."

Selecting the Part Number in a Percent Statement

In the percent statement

$$7\% \ \text{of} \ 45 \ \text{is} \ 3.15$$

the rate (r) is 7 and the base (b) is 45. The only other number in the problem is 3.15 and it is the part (p).

In the percent statement

$$50\% \ \text{of} \ 812 \ \text{is} \ 406$$

the rate number r is 50 and the base number b is 812. The part number is 406.

Focus on selecting the part in a percent problem

A percent problem always involves three numbers:

1. The rate (r) is the number with the % symbol.
2. The base (b) is the number which follows "of."
3. The part (p) is the third number in the problem.

In the simple percent statement   "8% of 900 is 72"

1. The rate (r) is 8.
2. The base (b) is 900.
3. The part (p) is 72.

The Relationship Between Rate, Base, and Part

In every percent problem the rate (r), base (b), and part (b)
are related by the proportion:

$$\frac{r}{100} = \frac{p}{b}$$

The simple percent statement "7% of 45 is 3.15" has r = 7,
b = 45, and p = 3.15 and these numbers form the proportion

$$\frac{7}{100} = \frac{3.15}{45}$$

*Focus
on the
proportion
used to solve
percent
problems*

Every percent problem can be solved
with the proportion shown at the
right.

$$\frac{r}{100} = \frac{p}{b}$$

Recall that every percent is a ratio with 100 as its denominator. This
means that the proportion is a translation of the statement:

r %   is   the ratio: part to base

$$\frac{r}{100} \qquad = \qquad \frac{p}{b}$$

Finding the Part in a Percent Problem

To solve the percent problem   "68% of 95 is _____"

1. First note that r = 68 and b = 95

2. Use the proportion $$\frac{r}{100} = \frac{p}{b}$$

$$\frac{68}{100} = \frac{p}{95}$$

3. Solve the proportion $$68 \cdot 95 = 100 \cdot p$$

$$6460 = 100p$$

$$64.6 = p$$

The part (p) is 64.6

Focus on finding the part in a percent problem

In the percent statement   "43% of 56 is _____"

1. The rate (r) is 43
2. The base (b) is 56
3. The part (p) is unknown

The proportion $\frac{r}{100} = \frac{p}{b}$ is used replacing r by 43 and b by 56.

$$\frac{43}{100} = \frac{p}{56}$$

$$43 \cdot 56 = 100 \cdot p$$
$$2408 = 100p$$
$$24.08 = p$$

The part (p) is 24.08

Finding the Rate in a Percent Problem

In a percent problem where the base (b) and part (p) are known, the rate (r) is found using the proportion $\frac{r}{100} = \frac{p}{b}$.

For example, in the percent statement "___% of 308 is 46" the rate (r) is unknown, the base (b) is 308, and the part (p) is 46.

The proportion is written: $$\frac{r}{100} = \frac{46}{308}$$

Then it is solved. $$r \cdot 308 = 100 \cdot 46$$

$$308r = 4600$$

$$r = \frac{4600}{308} = 14\frac{288}{308} \doteq 15 \qquad\qquad r = \frac{4600}{308}$$

The rate is 15% because $14\frac{288}{308}$ is rounded off to 15. In those cases where the rate does not come out evenly, it will be rounded off to the nearest one percent.

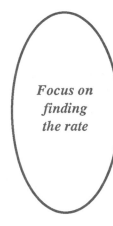

Focus on finding the rate

To find the rate in a percent problem, use the proportion $\frac{r}{100} = \frac{p}{b}$
For example, in the problem
          ___% of 812 is 56
the rate (r) is unknown,
the base (b) is 812, and
the part (p) is 56.

$$\frac{r}{100} = \frac{56}{812}$$

$$r \cdot 812 = 100 \cdot 56$$

$$812r = 5600$$

$$r = \frac{5600}{812} = 6\frac{728}{812} \doteq 7 \qquad r = \frac{5600}{812}$$

The rate to the nearest one percent is 7%.

Finding the Base in a Percent Problem

When the rate (r) and the part (p) of a percent problem are known
the base is found using the proportion $\frac{r}{100} = \frac{p}{b}$.

For example, in the percent statement "6% of ____ is 2.4"
the rate (r) is 6, the base (b) is unknown, and the part (p)
is 2.4.

The proportion is: $\frac{6}{100} = \frac{2.4}{b}$

The proportion is solved. $6 \cdot b = 100 \cdot 2.4$

                                    $6b = 240$

The base number is 40. $b = 40$

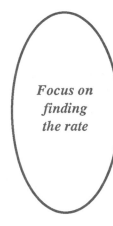

Focus on finding the base in a percent problem

To find the missing number in   "252.2 is 52% of _____"
first note that the rate is 52, the base is unknown, and the part is 252.2.

The proportion $\frac{r}{100} = \frac{p}{b}$ is used. $\frac{52}{100} = \frac{252.2}{b}$

                                              $52 \cdot b = 100 \cdot 252.2$

                                              $52b = 25{,}220$

The base is 485. $b = 485$

Unit 5 Exercise

This exercise reviews the preceding unit. The exercise is divided into three parts. Part A reviews the foci of Unit 5. Part B offers opportunities to practice the skills and concepts of Unit 5. Part C contains problems that review your previous work in this text. All answers for Parts A and B are at the back. Each problem of Part C is accompanied by a notation «CU» that refers to the Chapter (C) and Unit (U) in which that type of problem is studied.

Part A: Reviewing the foci of Unit 5.

1. The three numbers in a percent problem are the _____ (b), the rate (r), and the part (p).

2. The three numbers in a percent problem are the base (b), the _____ (r), and the part (p).

3. The three numbers in a percent problem are the base (b), the rate (r), and the _____ (p).

4. The three numbers in a percent problem are the _____ (b), the _____ (r), and the _____ (p).

5. The word "of" in a percent problem is a good clue for identifying the base (b). To find the base, look for the word "_____."

6. Find the base (b) of the following percent statement by choosing the number directly after the word "of."
    25% of 44 is 11.

7. Find the base (b) of the following percent statement.
    30 is 75% of 40.

8. Find the base (b) of the following percent statement.
    125% of 40 is 50.

9. Find the base (b) of the following percent statement.
    3 is 6% of 50.

10. Find the rate number in the following statement.
    16 is 50% of 32.

11. Find the rate number in: 18% of 300 is 54.

12. Find the rate number in: 45 is 150% of 30.

13. In a percent statement, the number that comes directly after "of" is the base (b) and the number with the percent symbol (%) is the _____.

14. Find the base, rate, and part for:
    55% of 80 is 44.
    b = _____ r = _____ p = _____

15. Find the b, r, and p for:
     85 is 17% of 500.
       b = _____ r = _____ p = _____

16. Find the b, r, and p for:
     150% of 90 is 135.
       b = _____ r = _____ p = _____

Part B: Drill and Practice

1. The three numbers in a percent statement are
   called _____ (b), _____ (r), and _____ (p).

2. To find the b in a percent statement, look
   for the word "_____."

3. To find the r in a percent statement, look
   for the symbol _____.

Find the b, r, and p, for the following:

4. 4 is 200% of 2.
     b = _____ r = _____ p = _____

5. 3% of 8 is 0.24.
     b = _____ r = _____ p = _____

6. 14 is 20% of 70.
     b = _____ r = _____ p = _____

7. 53% of 500 is 265
     b = _____ r = _____ p = _____

8. 17 is 85% of 20.
     b = _____ r = _____ p = _____

9. 125% of 60 is 75.
     b = _____ r = _____ p = _____

10. 144 is 80% of 180
     b = _____ r = _____ p = _____

Use the proportion $\frac{r}{100} = \frac{p}{b}$ to find the missing
number in each of the following percent statements.

11. 12% of 46 is _____.

12. 56% of 334 is _____.

13. _____ is 48% of 700.

14. _____ is 125% of 140.

15. 6% of 542 is _____.

16. _____ is 19% of 413.

17. 84% of 894 is _____.

18. _____ is 9% of 560.

19. _____ is 143% of 65.

20. 36% of 491 is _____.

21. _____% of 438 is 97.

22. _____% of 96 is 8.64.

23. 46 is _____% of 64.

24. 31 is _____% of 468.

25. _____% of 46 is 76.

26. _____% of 108 is 59.

27. 36 is _____% of 248.

28. 58 is _____% of 248.

29. _____% of 85 is 33.

30. _____% of 104 is 115.

31. 7% of _____ is 21

32. 43% of _____ is 258

33. 35% of _____ is 28

34. 60% of _____ is 90

35. 63 is 45% of _____

36. 5% of _____ is 43.6

37. 18% of _____ is 97.2

38. 214.62 is 49% of _____

39. 275.6 is 130% of _____

40. 16% of _____ is 5.6

Part C: Review. Answers for the problems of Part C are not given. However, the notation «C,U» refers to the Chapter (C) and Unit (U) in which problems of the same type were presented.

1. $65,802 - 43,675$ «1,3»

2. Find the lcm for 8 and 20.   «3,2»

3. Evaluate $14 + (-5)$ «6,1»

4. Evaluate $-4 \cdot -11$ «6,3»

5. Simplify $-7(-2x)$ «6,5»

6. Solve $9 - 7x = 30$ «6,7»

7. Evaluate $\frac{15}{7} \cdot \frac{21}{5}$ «7,1»

8. Simplify $(\frac{3}{4}x - \frac{1}{3}) + \frac{1}{2}$ «7,3»

9. Simplify $\frac{1}{2}(\frac{-5}{6}x)$ «7,4»

10. Simplify $5(\frac{3}{5}x + 2)$ «7,5»

11. Solve $\frac{1}{x} + \frac{2}{7} = \frac{2}{x} + \frac{3}{14}$ «7,7»

12. Solve $\frac{9}{14} = \frac{n}{7}$ «8,2»

13. Round off $\frac{9}{10}$ to the nearest one percent.
    «8,4»

14. A first number is the same as 12 increased by a second number. If s represents the second number, write an open expression for the first number.   «5,8»

15. Find the sum of 5.183 and 4.77   «4,2»

16. Divide 8.35 by 0.8 and round off the answer to 2 decimal places.   «4,4»

17. A business had sales of $5,400 in January and $7,300 in February. What was the average sales for those two months?   «4,4»

18. Change $\frac{1}{7}$ to a decimal fraction rounded off at two decimal places.   «4,4»

19. Write an equation that shows: the difference when 32 is subtracted from a number is 42.   «6,9»

20. Determine if $\frac{7}{9} = \frac{3}{4}$ is a true equality using the products of the extremes and means.   «8,1»

21. A student gets 7 out of every 8 problems correct on a test. If there are 160 total problems on the test, how many did the student answer correctly?   «8,3»

Unit 6:   WORD PROBLEMS INVOLVING PERCENTS

Solving a Percent Problem

The proportion $\frac{r}{100} = \frac{p}{b}$ can be used to solve any percent problem of the form

$$r\% \ \text{of} \ b \ \text{is} \ p$$

where r is the rate, b is the base, and p is the part.

When two of the three variables are known, the proportion results in an equation. The solution of the equation provides the answer for the third variable.

Focus on solving percent problems

To find the missing number in

18% of 82 is _____

solve the proportion
as shown at the right to find p.

$$\frac{18}{100} = \frac{p}{82}$$

$$18 \cdot 82 = 100p$$
$$100p = 1476$$
$$p = 14.76$$

To find the missing number in

___% of 24 is 6

solve the proportion
as shown at the right to find r.

$$\frac{r}{100} = \frac{6}{24}$$

$$r \cdot 24 = 100 \cdot 6$$
$$24r = 600$$
$$r = 25$$

To find the missing number in

6% of ____ is 9

solve the proportion
as shown at the right to find b.

$$\frac{6}{100} = \frac{9}{b}$$

$$6b = 100 \cdot 9$$
$$6b = 900$$
$$b = 150$$

Translating Word Problems

Word problems involving percent should be restated in the form

r % of b is p

The following problem illustrates this translation.

A class of 45 students has 19 boys. What percent of the class is boys?

Make a list of the numbers given in the problem and include the labels for those numbers.

45 students
19 boys

The question of the problem is:

What percent   of   the class   is   boys?

_____%   of   _____   is   ___

? %   of   45   is   19

r is unknown, b is 45, and p is 19. The word problem is solved using the proportion

$$\frac{r}{100} = \frac{19}{45}$$

The proportion is solved.

$$r \cdot 45 = 100 \cdot 19$$

$$45r = 1900$$

$$r = 42\frac{2}{9}$$

To the nearest one percent r is 42%.

Word Problem: 75% of a test with 60 questions are true-false
questions. How many are true-false questions?

Make a list of the numbers in the problem and
include the labels for those numbers.

  75 %
  60 questions
  __ true-false questions

Focus on translating a percent problem

In the word problem, r = 75 because of the percent symbol. This
means the problem is

$$75\% \text{ of } 60 \text{ is true-false}$$

This leads to the proportion $\dfrac{75}{100} = \dfrac{p}{60}$

Solving the proportion gives p = 45. Therefore, there are 45
true-false questions on the test.

Word Problem: 22% of a solution of rubbing alcohol is water. There
were 3.96 liters of water in the solution. How many
liters were in the solution?

Make a list of the numbers given in the problem and
include the labels for those numbers.

  22 %
  3.96 liters of water
  ___ liters of solution

Focus on solving a percent problem

In the word problem, r = 22 because of the percent symbol. This
means the problem is

$$22\% \text{ of solution is } 3.96$$

This leads to the proportion $\dfrac{22}{100} = \dfrac{3.96}{b}$

Solving the proportion gives b = 18. Therefore, there are 18
liters in the solution.

Unit 6 Exercise

This exercise reviews the preceding unit. The exercise is divided into three parts. Part A reviews the foci of Unit 6. Part B offers opportunities to practice the skills and concepts of Unit 6. Part C contains problems that review your previous work in this text. All answers for Parts A and B are at the back of the book. Each problem of Part C is accompanied by a notation «CU» that refers to the Chapter (C) and Unit (U) in which that type of problem is studied.

Part A: Reviewing the foci of Unit 6.

1. Last year Mr. Gottza paid $27,000 for his income taxes which was 38% of his income. 38% of _____ is $27,000. What was his income?

2. 2% of the students in a school have a hearing impairment. There are 1500 students in the school. Find the number of students with a hearing impairment.   2% of 1500 is _____.

3. Does the question below ask for base, rate, or part?
What percent of a ring weighing 20 grams is silver if there are 8 grams of silver in the ring?

4. Fill in the blanks below for the following question.
What percent of a ring weighing 20 grams is silver if there are 8 grams of silver in the ring?
_____% of _____ is _____.

5. Find the percent of silver in a ring of the previous question by finding the r in: r% of 20 is 8.

6. What is the weight of a rock if 3% of its weight is 36 kilograms? Does the question ask for the base, rate or part?

7. What is the weight of a rock if 3% of its weight is 36 kilograms? Fill in the blanks below. Place the numbers and the letter b where they belong. _____% of _____ is _____.

8. Find the weight of the rock in the previous problem by finding b in: 3% of b is 36.

9. How many girls are in a class of 450 students if 52% of the class is girls? Does the question ask for the base, rate, or part?

10. Complete the solution of: How many girls are in a class of 450 students if 52% of the class is girls?
52% of 450 is _____.

11. Complete the solution of: The interest on a loan of $2000 is $360 for one year. Find the interest rate charged. r% of $2000 is $360.

12. Complete the solution: 30% of the patients in a hospital had heart problems. If 210 patients had heart problems, how many patients are in the hospital? 30% of _____ is 210.

13. A store makes a 6% profit on its total sales of $2,000,000. Find the profit by first identifying 6% and $2,000,000 as the base, rate, or part.

14. 87% of milk is water. If a bottle of milk contains 2.61 liters of water, how much milk is in the bottle? Answer the question by first identifying the numbers 2.61 and 87% as the base, rate, or part.

Part B: Drill and Practice

1. 58% of 938 is _____.

2. 16 is _____% of 298.

3. 50.66 is 34% of _____.

4. _____% of 38 is 10.

5. _____ is 6% of 47.

6. 155% of _____ is 93.

7. A bottle of medicine contains 12 liters. 2.6% of the contents is active ingredients. How much of the medicine is active ingredients?

8. 92% of the body weight of a 160 lb. male is water. How much of the 160 lbs. is water?

9. At a sale, slacks were sold for $15 which was 80% of their original price. What was the original price?

10. A three-year-old house is worth 125% of its original price of $36,000. How much is the house worth?

11. About 22,000,000 American workers are union members. If 25% of the total labor force is union, what is the size of the total labor force?

12. A $600 refrigerator was sold for $360 during a sale, a reduction of $240. The reduction is what percent of the original price?

13. The pollution count in a bay was 72 parts per million. It increased by 9.3%. How much did the pollution in the bay increase?

14. A microwave oven uses 35% less power to cook a roast than a conventional oven which uses 800 watts (w.). How much power would the microwave oven save for the same roast?

15. A 6% commission is paid on sales of $40,000. What is the commission?

16. A team won 30 games which was 40% of the games it played. How many games were played?

17. 18 girls are in a class of 42 students. What percent of the class is girls?

18. 80% of a 40 member biology class passed the course. How many passed the course?

19. A sample of uranium ore weighs 90 kilograms. How much of it is uranium if 2.5% of the ore is uranium?

20. A lawyer's fee for a lawsuit was 35% of the award for the case. The fee was $10,500. Find the size of the award.

Part C: Review. Answers for the problems of Part C are not given. However, the notation «C,U» refers to the Chapter (C) and Unit (U) in which problems of the same type were presented.

1. $43,790 + 23,690 + 6,523$ «1,2»

2. Simplify $\frac{14}{26}$ «3,1»

3. Find the numerator of $\frac{4}{5} = \frac{?}{45}$ «3,2»

4. Find the difference of 84.67 and 73.6   «4,2»

5. Simplify $(6 + 7x) + 5$ «5,4»

6. Evaluate $11 - (-6) \cdot -2 \div 4$ «6,4»

7. Simplify $3(2x - 1) - 2(x - 3)$ «6,6»

8. Solve $7 - 4x = 22 - x$ «6,8»

9. Evaluate $\frac{1}{6} + (\frac{2}{3} - \frac{3}{4})$ «7,2»

10. Simplify $\frac{-3}{8}(\frac{2}{5}x)$ «7,4»

11. Simplify $\frac{8}{x} - \frac{2}{x}$ «7,5»

12. Solve $x + 13 = 4x - 8$ «7,6»

13. Solve $\frac{3}{4}x - \frac{1}{2} = \frac{2}{5}$ «7,7»

14. $\frac{36}{25} = \frac{n}{13}$ is equivalent to _____ = _____
    «8,1»

15. How many $\frac{2}{3}$ quarts are there in 6 quarts?
    «3,7»

16. Art's test scores in Economics were 88, 94, 97, and 78. Find his average on these four tests rounded off to the nearest whole number.
    «4,4»

17. Change $\frac{2}{9}$ to a decimal fraction rounded off at two decimal places.   «4,4»

18. One number is twice a second number. If s represents the second number, write an open expression for the first number.   «5,8»

19. Write an equation that shows: 21 less than a number is 38.   «6,9»

20. Solve $\frac{3}{n} = \frac{11}{2}$ «8,2»

21. A business found that there was a 1 cent advertising cost for every 9 cents in sales. On an item selling for $27 what was the advertising cost?   «8,3»

22. Round off $\frac{3}{8}$ to the nearest one percent.
    «8,4»

23. _____% of 104 is 115.   «8,5»

Chapter 8 Test

«8,U» shows the unit in which the problem is found in the chapter.

Solve.

1. $\frac{n}{8} = \frac{8}{16}$ «8,2»

2. $\frac{9}{n} = \frac{3}{5}$ «8,2»

3. $\frac{18}{20} = \frac{27}{n}$ «8,2»

4. $\frac{12}{5} = \frac{n}{20}$ «8,2»

5. If the F.C.C. ruled that the ratio of advertising time to program time is to be 2 to 5, how much program time is needed to air 16 minutes of advertising?   «8,3»

6. A weather forecaster found that 3 rainy days occur for every 8 cloudy days. In 20 cloudy days, about how many rainy days should occur?   «8,3»

7. It is found that 2 malfunctions occur for every 45 days of operating a nuclear plant. How many days may pass for 3 malfuctions to occur?   «8,3»

8. A bank charges $11 interest on a $100 loan for one year. How much could be borrowed for a year for $220 interest?   «8,3»

9. A 2 liter sample of river water contained 13 particles of pollution. How many particles would be in 50 liters of the water?   «8,3»

10. 3 drams of a drug is to be given for 12 kilograms of body weight. Find the size of a person that would be given 14 drams of the drug.   «8,3»

Write the following percents as both fractions and decimals.

11. 2%   «8,4»

12. 110%   «8,4»

13. 55%   «8,4»

14. 9%   «8,4»

In problems 15-18 write the given number as a percent.

15. $\frac{1}{2}$ «8,4»

16. 0.2   «8,4»

17. 1.01   «8,4»

18. 0.58   «8,4»

19. In the statement   "r% of s is t,"
    the base is ____, the rate is _____,
    and the part is ____.   «8,5»

20. ____% of 80 is 40.   «8,5»

21. 3% of ____ is 9.   «8,5»

22. 10% of 200 is ____.   «8,5»

23. The interest rate on a $50,000 loan for one
    year is 12%. Find the interest.   «8,6»

24. A 20 liter sample of water contains 1 liter
    of pollution. What percent of the sample is
    pollution?   «8,6»

25. An inspector found that 2% of a shipment
    of meat was spoiled. If 46 kilograms of the
    meat was spoiled, how much meat was in
    the shipment?   «8,6»

26. Find 152% of 46.   «8,6»

27. 10 is 125% of what number?   «8,6»

9
Geometry

Unit 1:   POINTS, LINES, PLANES

Centuries ago Egyptians were surveying plots of land along the
Nile. They were using ideas of shape and size which were the
basis of the mathematics called geometry. The surveying origins
of this type of mathematics is the basis for its name because
geometry means "earth measure."

Despite its beginnings, geometry today encompasses far more
than surveying skills. Many practical applications of geometry
are found in many occupations and avocations, but the esthetic
qualities of geometry involved in designing buildings, automobiles,
etc. are of almost equal importance.

Your study of geometry needs to recognize it as a practical,
functional body of knowledge which is combined with eye-
pleasing qualities that have a psychological effect.

Symmetry

One of the eye-pleasing qualities of some geometric figures is symmetry. Symmetry is a quality of balance.

A heart-shape is a symmetrical figure in which the left and right sides of the heart are identical in shape and size. A straight line from the cleft of the heart to its point separates the heart into two equal mirror-image halves. The straight line is called the **line of symmetry.**

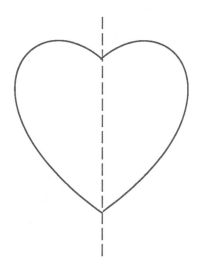

If a symmetrical figure is drawn on a sheet of paper and the paper is folded along the line of symmetry then the two halves must coincide exactly.

Focus on a square as a symmetrical figure

The four-sided figure (square) shown at the right is a symmetrical figure.

The square has four lines of symmetry shown by dashed lines in the figure.

A heart-shape is symmetrical, but has only one line of symmetry.

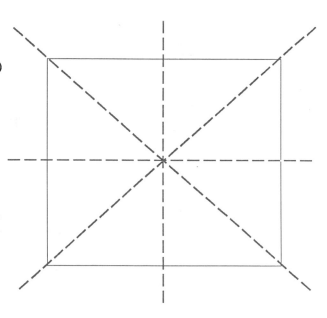

Points

The most basic of all geometric figures is the point. A point
has no size. It has no length, no width, and no depth. A
thousand points can be placed end-to end, side-by-side, stacked
one-upon-another without ever having a length, a width, or a
height.

Specific points in geometry are
commonly named by capital letters.
Two points, named by the capital
letters A and B are shown at the
right.

A point is a marvelous theoretical concept and it is the
building block for all other elements in geometry. Every other
object in geometry is a set of points.

*Focus on
points in
geometry*

The idea that every geometric figure
is a collection of points would describe
a square as a set of points arranged in
the position shown at the right. Such
a view of a square involves the idea of
infinity because there is a never ending
number of points comprising the square.

Another view of the square is that it is the
path of a single moving point. This is a
very different view. It is the idea that a
point, like a child in a field of freshly fallen
snow, has marked its travels for all to see.

Planes

Almost as basic as the point is the geometric figure called the plane. The point needs an environment in which to operate and one of the most interesting environments, historically and practically, is a surface along which the points may group or travel.

Such a surface, called a plane, is usually pictured as a flat, table-top shape that continues lengthwise and sidewise without end. A plane is a two-dimensional object. It has length and width, but it has no height or depth. This means that a moving point on a plane may move along it in a forward, backwards, or sideways motion, but cannot hop about or dig below the surface.

Focus on the idea of a plane

A plane will be shown as a table-top surface like the one shown at the right.

Any figure shown on a piece of paper or a movie screen is a plane figure. Special drawings or movie effects are used to give plane figures the appearance of having depth, but plane figures can have only two dimensions, length and width.

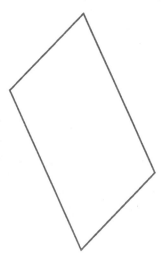

Lines — Straight Lines

The third, and last, basic geometric figure to be presented is the straight line — usually referred to simply as a line. The idea of "straight" is closely related to distance or length. In a plane, the straight line distance between two points is the shortest distance between those two points. All other, longer paths, between the two points are not straight lines.

For any two points located in a plane there is a unique, one-of-a-kind, straight line joining them. Consequently, a line is often named by two of its points.

Focus on straight lines and their symbols

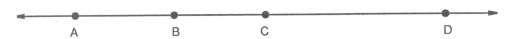

The line above may be symbolized as $\overleftrightarrow{AB}$, $\overleftrightarrow{AC}$, $\overleftrightarrow{CA}$, or $\overleftrightarrow{DB}$. The symbolization uses any two points of the line, in any order. The double-headed arrow is placed above the two points.

A line is a one-dimensional object. It has length, but no width or depth.

Rays and Line Segments

A ray is only part of a line. The ray with endpoint A and extending through B is shown at the right.

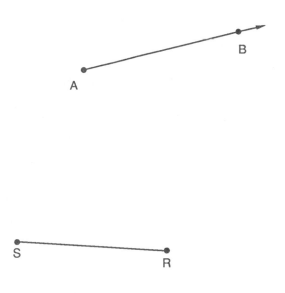

The ray consists of its endpoint, A, and that portion of line $\overleftrightarrow{AB}$ which includes point B. The symbol for the ray shown is $\overrightarrow{AB}$. Note that $\overrightarrow{AB}$ is not the same as $\overrightarrow{BA}$.

A line segment is also only part of a line. The line segment with endpoints R and S is shown at the right.

The line segment consists of its endpoints, R and S, and that portion of line $\overleftrightarrow{RS}$ between R and S. The symbol for the line segment shown is $\overline{RS}$. Note that $\overline{RS}$ is the same as $\overline{SR}$.

Focus on rays and their symbolizations

The figure below shows a line through three points, X, Y, and Z. The line could be named $\overleftrightarrow{XY}$, $\overleftrightarrow{XZ}$, or $\overleftrightarrow{YZ}$.

The rays $\overrightarrow{XY}$ and $\overrightarrow{XZ}$ are the same. In both cases, the ray consists of endpoint X and all points to the right of it on the line.

The rays $\overrightarrow{XY}$ and $\overrightarrow{YZ}$ are different. They have different endpoints.

Unit 1 Exercise

This exercise reviews the preceding unit. The exercise is divided into three parts. Part A reviews the explanations of Unit 1. Part B offers opportunities to practice the skills and concepts of Unit 1. Part C contains problems that review your previous work in this text. All answers for Parts A and B are at the back of the book. Each problem of Part C is accompanied by a notation «CU» that refers to the Chapter (C) and Unit (U) in which that type of problem is studied.

Part A: Reviewing the foci of Unit 1.

1. Is a rectangle a symmetrical figure? If so, how many lines of symmetry does it have?

2. The capital letter H is a symmetrical figure. It has two lines of symmetry. Draw them.

3. Which of the following capital letters are symmetrical figures? B M N S T

4. What are the dimensions of a point?

5. What are the dimensions of a plane?

6. What are the dimensions of a straight line?

7. What are the dimensions of a ray?

8. What are the dimensions of a line segment?

Part B: Drill and Practice

For each description, draw a figure or state that such a figure is impossible.

1. A geometric figure with exactly two lines of symmetry.

2. A geometric figure with no line of symmetry.

3. A geometric figure with an unlimited number of lines of symmetry.

4. A geometric shape which has no points.

5. $\overleftrightarrow{AB}$ and $\overleftrightarrow{CD}$ that have no point in common.

6. $\overline{AB}$ and $\overline{AC}$ that have more than one common point.

7. $\overleftrightarrow{AB}$ and $\overleftrightarrow{AC}$ that are exactly the same ray.

8. $\overrightarrow{AB}$ and $\overline{CD}$ that have exactly one common point.

9. Two rays that have a common line segment.

10. A figure that has length and width.

11. A five-sided figure that also has five lines of symmetry.

12. A geometric figure with no dimensions.

13. A geometric figure that would not fit on a plane.

Part C: Review. Answers for the problems of Part C are not given. However, the notation «C,U» refers to the Chapter (C) and Unit (U) in which problems of the same type were presented.

1. $83,002 - 42,675$ «1,3»

2. Evaluate $-3 \cdot -7$ «6,3»

3. Simplify $-9(-x)$ «6,5»

4. Solve $13 - 6x = 1$ «6,7»

5. Evaluate $\frac{15}{7} \cdot \frac{11}{10}$ «7,1»

6. Simplify $(\frac{1}{2}x - \frac{1}{4}) + \frac{1}{3}$ «7,3»

7. Simplify $\frac{3}{4}(\frac{-5}{6}x)$ «7,4»

8. Solve $\frac{1}{x} + \frac{2}{5} = \frac{3}{x} + \frac{3}{10}$ «7,7»

9. Solve $\frac{7}{2} = \frac{5}{n}$ «8,2»

10. Round off $\frac{7}{9}$ to the nearest one percent.
    «8,4»

11. A first number is the same as 9 decreased by a second number. If s represents the second number, write an open expression for the first number.   «5,8»

12. A business spent $585 on advertising in January, $387 in February, and $715 in March. What was the average advertising costs for those three months?   «4,4»

13. Change $\frac{3}{8}$ to a decimal fraction rounded off at two decimal places.   «4,4»

14. Write an equation that shows: 17 subtracted from a number is 19.   «6,9»

15. Determine if $\frac{13}{18} = \frac{2}{3}$ is a true equality using the products of the extremes and means.
    «8,1»

16. A student gets 5 out of every 6 problems correct on a test. If there are 126 problems on the test, how many did the student answer correctly?
    «8,3»

17. _____ is 7% of 576.   «8,5»

18. A 6% commission is paid on sales of $40,000. What is the commission?   «8,6»

Unit 2:   ANGLES

Definition		Angle
A geometric figure is an **angle**	**if and only if**	the figure consists of two rays with a common endpoint.

The common endpoint of the two rays is called the **vertex** of the angle.

Frequently, a small curved arrow is placed within the arc of the two rays to indicate a direction of opening as if one of the rays were a door and the other represented the wall.

Focus on three ways to name an angle

Angles are named in three ways.

One way to name an angle is by using the name of the point that is the common vertex. Since X is the endpoint for the rays shown at the right, the angle may be named ∠ X.

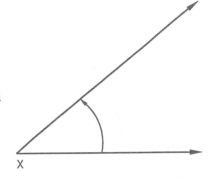

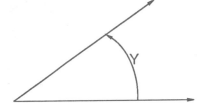

A second way for naming an angle is to use a small symbol placed next to the arrow-headed arc of rotation. The angle shown at the left may be named ∠ y.

A third way of naming an angle depends upon three points: the common endpoint of the two rays and one point from each ray. The three points are listed with the common endpoint as the second (middle) letter in the listing. The angle at the right is named ∠ DQB or ∠ BQD.

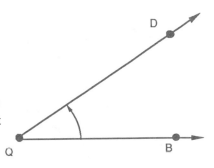

Angles Named by the Amount of Rotation

There are five types of angles depending on the amount
of rotation between the two rays.

Definition   **Straight Angle**

An angle is a   **If and**   the two rays have one-half
straight angle   **only if**   turn as their rotation.

Angle ∠ KLM is a
straight angle.

K   L   M

Definition   **Right Angle**

An angle is a   **If and**   the two rays have one-fourth
right angle   **only if**   turn as their rotation.

Angle ∠ J is a **right angle.**

The rays of a right angle are
said to be **perpendicular**.

A small box at the vertex is used
to show a right angle.

J

---

*Focus on
measuring
straight and
right angles*

The most common measuring unit
for angles is degree (symbol °).

There are 360° in a complete rotation.
Consequently, every straight angle with
one-half a rotation has one-half of 360°.
A straight angle always measures 180°.

A right angle is one-fourth of a rotation (half
of a straight angle) and every right angle measures
90°. By this fact, perpendicular lines always form
90° angles.

E   D   F

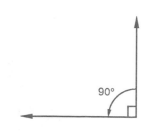

90°

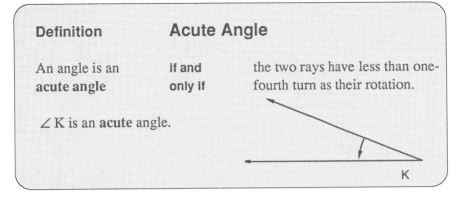

Definition   **Acute Angle**

An angle is an   **if and**   the two rays have less than one-
acute angle   **only if**   fourth turn as their rotation.

∠ K is an **acute** angle.

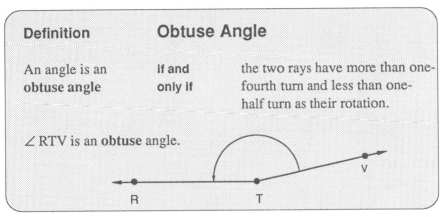

Definition   **Obtuse Angle**

An angle is an   **if and**   the two rays have more than one-
obtuse angle   **only if**   fourth turn and less than one-
                                          half turn as their rotation.

∠ RTV is an **obtuse** angle.

Focus on measuring acute and obtuse angles

An acute angle is one with less than one-fourth rotation. Since a right angle (one-fourth rotation) measures 90°, every acute angle has a measure less than 90°.

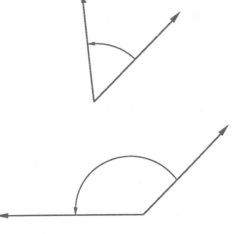

A right angle measures 90° and a straight angle measures 180° Since an obtuse angle is between a right angle and a straight angle, every obtuse angle measures between 90° and 180°.

Angles Greater than Straight Angles

The angle ∠ Z shown at the right
is formed by a rotation greater
than one-half turn. Such angles
are called **reflex angles**.

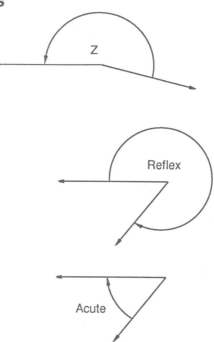

Notice that the second figure at
the right shows that the reflex
angle has the same shape as a
smaller angle. This is always
possible for a reflex angle.

In some areas of mathematics,
reflex angles are important, but
for the purposes of this text the
angles considered will not be
reflex angles.

*Focus on
reflex
angles*

The reflex angle ∠ C has the
same shape as the obtuse
angle ∠ B. ∠ C has a measure
greater than 180°. ∠ B has a
measure between 90° and
180°.

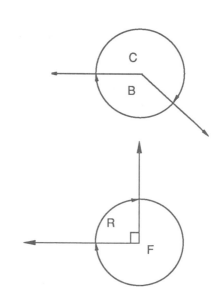

The reflex angle ∠ F has the
same shape as the right
angle ∠ R.

∠ F has a measure of 270°
which is greater than 180°.

∠ R has a measure of 90°.

Unit 2 Exercise

This exercise reviews the preceding unit. The exercise is divided into three parts. Part A reviews the explanations of Unit 2. Part B offers opportunities to practice the skills and concepts of Unit 2. Part C contains problems that review your previous work in this text. All answers for Parts A and B are at the back of the book. Each problem of Part C is accompanied by a notation «CU» that refers to the Chapter (C) and Unit (U) in which that type of problem is studied.

Part A: Reviewing the foci of Unit 2.

1. A straight angle is formed by a one-half rotation. Which of the figures below represents a straight angle?

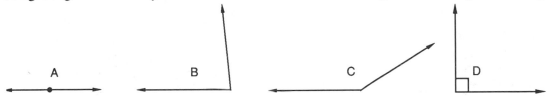

2. A right angle is formed by a one-fourth rotation. Which of the figures below represents a right angle?

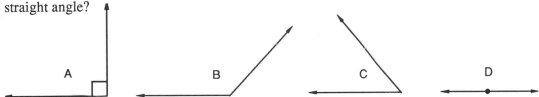

3. An acute angle is formed by less than a one-fourth rotation. Which of the figures below represents an acute angle?

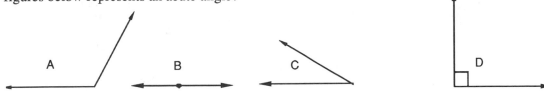

4. An obtuse angle is formed by more than a one-fourth rotation and less than a one-half rotation. Which of the figures below represents an obtuse angle?

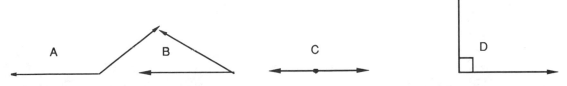

Part B: Drill and Practice

For each angle, give its letter name and its name determined by the amount of rotation.

1.
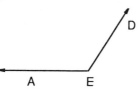

2.

3.

4.
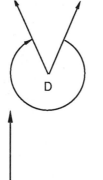

5.

6.
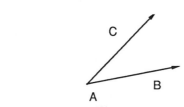

7.
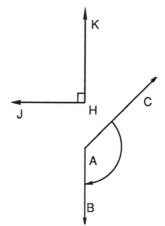

8.
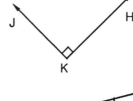

9.
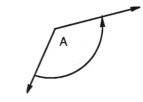

10.

11.

12.
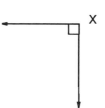

Part C: Review. Answers for the problems of Part C are not given. However, the notation «C,U» refers to the Chapter (C) and Unit (U) in which problems of the same type were presented.

1. 81,090 + 42,690 + 6,253   «1,2»

2. Find the numerator of: $\frac{4}{15} = \frac{?}{45}$ «3,2»

3. Find the difference of 68.67 and 19.4   «4,2»

4. Evaluate   17 − (-6) • -5 ÷ 3   «6,4»

5. Simplify 4(2x − 1) − (3x − 1)   «6,6»

6. Solve 9 − 4x = x − 31   «6,8»

7. Evaluate $\frac{1}{6} + (\frac{1}{2} - \frac{3}{4})$ «7,2»

8. Simplify $\frac{-3}{8}(\frac{4}{9}x)$ «7,4»

9. Solve 2x + 25 = 4x − 6   «7,6»

10. How many $\frac{1}{2}$ yards are there in 6 yards?
    «3,7»

11. Will's test scores in Computer Science were 58, 94, 97, and 78. Find his average on these four tests rounded off to the nearest whole number.
    «4,4»

12. Change $\frac{9}{13}$ to a decimal fraction rounded off at two decimal places.   «4,4»

13. One number is three times a second number. If s represents the second number, write an open expression for the first number.   «5,8»

14. Write an equation that shows: 14 less than a number is 92.   «6,9»

15. Solve $\frac{n}{5} = \frac{17}{6}$ «8,2»

16. A manufacturer found that there was a 5 cent labor cost for every 33 cents in sales. On an item selling for $250 what was the labor cost? «8,3»

17. Round off $\frac{2}{3}$ to the nearest one percent. «8,4»

18. _____% of 612 is 213.   «8,5»

19. A salesman found that he sold goods to 35% of his customers last month.   If the salesman sold to 48 customers, what is the total number of customers he has?   «8,6»

20. Draw a geometric figure with no line of symmetry.   «9,1»

Unit 3:   ANGLE MEASUREMENTS

Measuring Angles with a Protractor

The measurement of an angle cannot depend on the length of its
rays because rays continue forever. The measurement of an angle
is based on the amount of rotation between the two rays. This is
the basis for measuring angles using a protractor like the one
shown below.

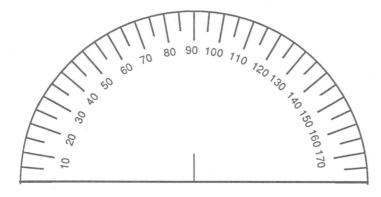

The protractor is shaped as a half-circle with its outer arc marked
off into the 180° which are in a straight angle.

To measure an angle with the protractor:

1. Place the small arrow at the half-circle's center on the vertex
    of the angle.

2. Rotate the protractor so that its straight edge lines up with one
    of the angle's rays while the protractor covers the second ray.

3. Where the second ray intersects the protractor's half-circle, there
    is a number. This number or 180° minus the number is the angle's measure
    in degrees.

4. If the angle is an acute angle, use the smaller measure. If the angle
    is an obtuse angle, use the larger measure. If the angle is a right
    angle it will have a measure of 90°.

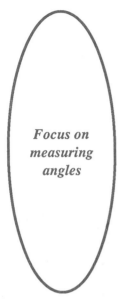

Four angles are show below with their measures indicated. Use a protractor to check your ability to measure the angles accurately.

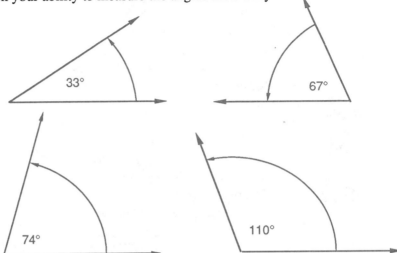

Focus on measuring angles

Adjacent and Vertical Angles

In the figure at the right, two intersecting lines have formed four angles:
> $\angle$ x, $\angle$ y, $\angle$ z, and $\angle$ w.

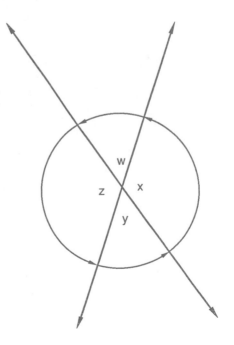

Angles $\angle$ x and $\angle$ y share a common vertex and a common ray. This pair of angles are **adjacent** angles. There are three other pairs of adjacent angles in the figure:
> $\angle$ x and $\angle$ w, $\angle$ w and $\angle$ z, and $\angle$ z and $\angle$ y.

Angles $\angle$ x and $\angle$ z share a common vertex and their rays are in opposite directions.
This pair of angles are **vertical** angles. There is one other pair of vertical angles in the figure:
> $\angle$ w and $\angle$ y.

Focus on adjacent and vertical angles

Every pair of adjacent angles has an important property in terms of their measurements. The sum of the measures of two adjacent angles is equal to the measure of the angle formed by their non-common rays.

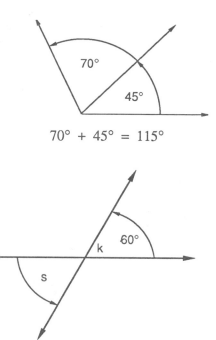

$$70° + 45° = 115°$$

Every pair of vertical angles also has an important property in terms of their measurements. The measurements of two vertical angles will always be the same. In the figure at the right, if ∠ k measures 60° then the measure of ∠ s will also be 60°.

Angles Formed by a Transversal

The figure at the right shows two lines labeled L_1 and L_2 with another line crossing them which is labeled T. In this type of situation the line T is called a **transversal** and forms the eight angles labeled in the figure.

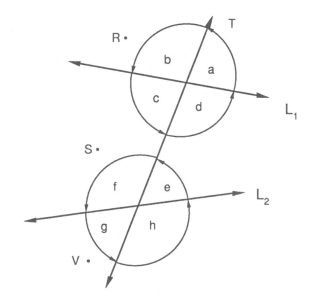

The points labeled R, S, and V all lie on the same side of the transversal T. Point S lies between lines L_1 and L_2 and is in the **interior** of those lines. Points R and V are not between lines L_1 and L_2 and are in the **exterior** of those two lines.

When two angles are on the same side of the transversal T and also in the same relative position for lines L_1 and L_2 the angles are **corresponding** angles. Angles $\angle$ a and $\angle$ e are corresponding angles. There are three other pairs of corresponding angles in the figure:

$\angle$ d and $\angle$ h, $\angle$ b and $\angle$ f, and $\angle$ c and $\angle$ g.

When two angles are on opposite sides of the transversal T and both are interior with respect to lines L_1 and L_2 the angles are **alternate interior** angles. Angles $\angle$ c and $\angle$ e are alternate interior angles. There is one other pair of alternate interior angles in the figure: $\angle$ d and $\angle$ f.

When two angles are on opposite sides of the transversal T and both are exterior with respect to lines L_1 and L_2 the angles are **alternate exterior** angles. Angles $\angle$ b and $\angle$ h are alternate exterior angles. There is one other pair of alternate exterior angles in the figure: $\angle$ a and $\angle$ g.

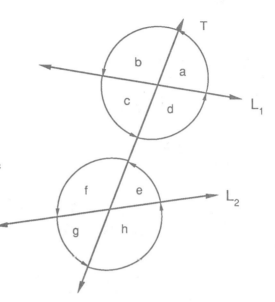

Focus on angles formed by a transversal

When a transversal intersects two parallel lines, like the ones shown in the figure at the right, a number of angles of equal measure are formed.

When L_1 and L_2 are **parallel**,

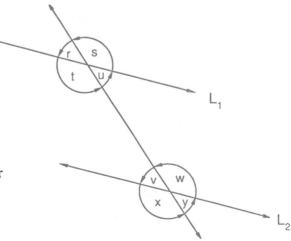

1. Every pair of corresponding angles will have the same measure. $\angle$ r = $\angle$ v

2. Every pair of alternate interior angles will have the same measure. $\angle$ u = $\angle$ v

3. Every pair of alternate exterior angles will have the same measure. $\angle$ s = $\angle$ x

Unit 3 Exercise

This exercise reviews the preceding unit. The exercise is divided into three parts. Part A reviews the explanations of Unit 3. Part B offers opportunities to practice the skills and concepts of Unit 3. Part C contains problems that review your previous work in this text. All answers for Parts A and B are at the back of the book. Each problem of Part C is accompanied by a notation «CU» that refers to the Chapter (C) and Unit (U) in which that type of problem is studied.

Part A: Reviewing the foci of Unit 3.

1.   Use a protractor to measure the four angles below.

a.   b.   c.   d.

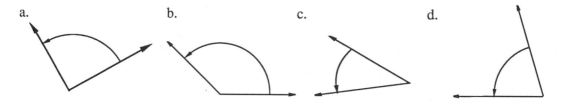

2.   Using the figure on the right, name two pairs of adjacent angles.

3.   Using the figure, name four pairs of corresponding angles.

4.   Using the figure, name two pairs of alternate interior angles.

5.   Using the figure, name two pairs of alternate exterior angles.

6.   Using the figure, if $\angle$s measures 137° find the measures of $\angle$r, $\angle$v, and $\angle$y. L_1 and L_2 are parallel.

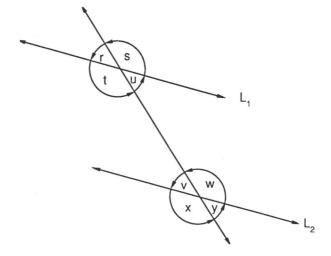

Part B: Drill and Practice

Measure the following marked angles.

1.

2.

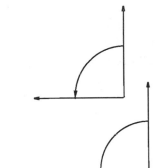

3.

4.

5.

6.

7.

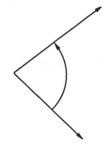

8.

9.

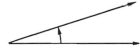

10.

11.

12.

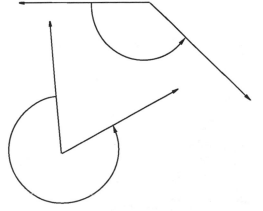

Part C: Review.   Answers for the problems of Part C are not given. However, the notation «C,U» refers to the Chapter (C) and Unit (U) in which problems of the same type were presented.

1. What is the hundreds digit of 6,248,571. «1,1»

2. Select the smaller fraction from: $\frac{5}{7}, \frac{13}{17}$ «3,2»

3. $8\frac{2}{5} - 3\frac{3}{4}$ «3,4»

4. Write 0.097 as a fraction.   «4,1»

5. Solve $-x = 5$ «5,7»

6. Simplify $-6(7x)$ «6,5»

7. Solve $7x - 13 = 29$ «6,7»

8. Evaluate $\frac{10}{21} \cdot \frac{3}{5}$ «7,1»

9. Simplify $(5x - \frac{1}{3}) + \frac{1}{6}$ «7,3»

10. Simplify $6x - 3(4x - 3)$ «7,5»

11. Solve $\frac{1}{2} - \frac{5}{4x} = \frac{1}{3} + \frac{1}{x}$ «7,7»

12. 7% of _____ is 21   «8,5»

13. A rug cleaner has completed $\frac{2}{3}$ of a job after 4 days work. How many days will the entire job require?   «3,7»

14. A first number is the same as 5 increased by a second number. If s represents the second number, write an open expression for the first number.   «5,8»

15. What is the average, to two decimal places, of 173, 154, and 159?   «4,4»

16. Write an equation that shows: the sum of a number and 38 is 95.   «6,9»

17. Determine if $\frac{0.6}{8} = \frac{0.3}{4}$ is a true equality using the products of the extremes and means.   «8,1»

18. In a ten quart cereal mixture, there are 5 parts oats for 2 parts puffed rice. How much oats are in the mixture?   «8,3»

19. Round off $\frac{4}{11}$ to the nearest one percent.   «8,4»

20. 18 girls are in a class of 42 students. What percent of the class is girls?   «8,6»

21. Draw a figure of $\overleftrightarrow{AB}$ and $\overleftrightarrow{CD}$ that have no point in common.   «9,1»

22. Name the angle shown and give its approximate measure.   «9,2»

Unit 4:   TRIANGLES

Triangles are important figures in geometry and play a significant
role in many practical building projects.

Definition	Triangle	
A geometric figure is a **triangle**	**if and only if**	it consists of three points not on the same straight line and the line segments connecting them.

The Rigidity Property of a Triangle

Physically, a triangle is a very strong figure. It will not bend
or change its shape unless one of its sides is bent or broken.

For that reason, a triangle is termed a **rigid** figure and this
property makes it an important geometric shape in constructing
bridges, buildings, and other structures where strength and
stiffness are valued.

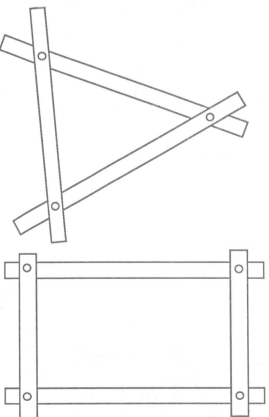

*Focus on
the rigidity
property*

The figure at the right shows
a triangle and a rectangle that
might be made from wooden
slats, which are fastened at
the corners (vertices).

Any attempt to alter the shape
of the triangle must overcome
the strength of the wooden slats
to be successful. The fasteners
on the triangle may be loose
fitting because they play little role
in the figure's strength.

Any attempt to alter the shape of
the rectangle must overcome only
the strength of the fasteners.

The Planar Property of a Triangle

A second important quality of a triangle is that it is the only two-dimensional figure that is guaranteed to remain completely on the surface of a plane.

For that reason, a triangle is termed a **planar** figure and this property makes it an important geometric shape in constructing any item that must be stable on a flat surface.

Focus on the planar property of a triangle

A milking stool is traditionally built with three legs so the farmer has a stable surface to sit on. The ends of the three legs form a triangle and will all rest on a barn floor even if that surface is not flat. A four-legged stool or chair will wobble, sometimes on a good floor, and invariably upon a rough surface. This is because three of the stool legs form a triangle that "fits" the surface, but the fourth leg will usually be off the surface and cause the stool to wobble.

Surveyors or photographers use a tripod, three legs, for instruments that need to be stable.

Classifying Triangles by Their Angles

Three special triangles are classified by the types of angles they contain.

The sum of the measures of any triangle's angles is 180°.

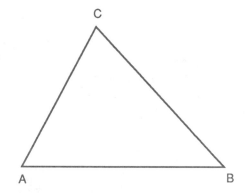

Definition   Acute Triangle

A triangle is an **acute triangle**	**if and only if**	each of its three angles is an acute angle.

Triangle $\triangle$ ABC shown above is an **acute triangle**.

Definition   **Right Triangle**

A triangle is a   **If and**   one of its angles is a
right triangle   **only If**   right angle.

Triangle Δ DEF
is a **right triangle**.
Notice that the
other two angles
are acute angles.

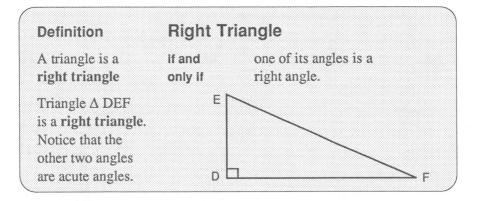

Definition   **Obtuse Triangle**

A triangle is an   **If and**   one of its angles is a
obtuse triangle   **only If**   obtuse angle.

Triangle Δ GHI
is an **obtuse triangle**.
Notice that the
other two angles
are acute angles.

*Focus on
supplementary
and
complementary
angles*

The sum of the measures of any triangle's angles is 180°. Whenever two or more angles have a sum which is 180°, those angles are said to be **supplementary angles**.

In the figure at the right, the sum of the three angles ∠ a, ∠ b, and ∠ c is 180° because they are angles of the triangle. Therefore, ∠ a, ∠ b, and ∠ c are **supplementary** angles.

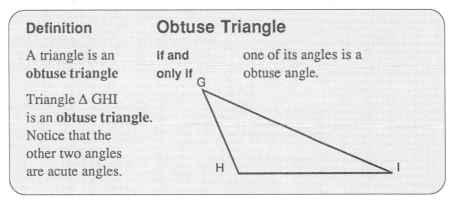

Also in the figure at the right, the adjacent angles ∠ a and ∠ d, have a sum of 180° because their uncommon rays form a straight angle. Consequently, ∠ a and ∠ d are supplementary angles.

Any two angles with a sum of 90° are called **complementary angles**. In any right triangle, the sum of the two acute angles must be 90° (the three angles have a sum of 180° and the right angle is 90°).
 Therefore, ∠ M and ∠ K in the figure at the right are complementary angles.

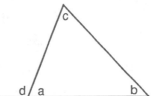

Classifying Triangles by the Lengths of Their Sides

In any triangle the sum of the lengths of any two sides
will be greater than the length of the third side.
A straight line is the shortest distance between two points.

Three special triangles are classified by the lengths of the
sides they contain.

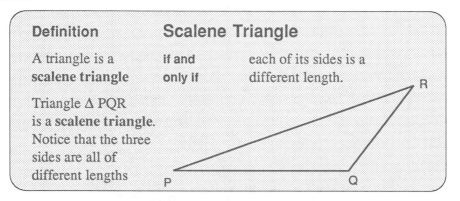

Definition   **Scalene Triangle**

A triangle is a   **If and**   each of its sides is a
scalene triangle   **only if**   different length.

Triangle Δ PQR
is a **scalene triangle**.
Notice that the three
sides are all of
different lengths

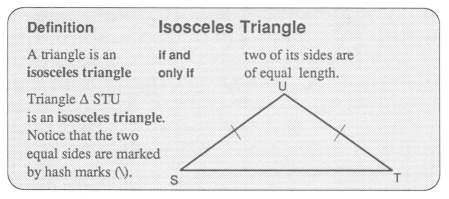

Definition   **Isosceles Triangle**

A triangle is an   **If and**   two of its sides are
isosceles triangle   **only if**   of equal length.

Triangle Δ STU
is an **isosceles triangle**.
Notice that the two
equal sides are marked
by hash marks (\).

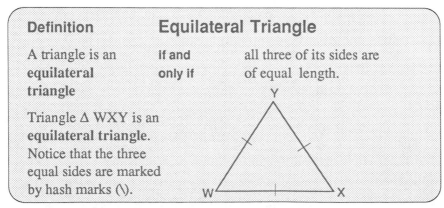

Definition   **Equilateral Triangle**

A triangle is an   **If and**   all three of its sides are
equilateral   **only if**   of equal length.
triangle

Triangle Δ WXY is an
equilateral triangle.
Notice that the three
equal sides are marked
by hash marks (\).

There is a relationship between the
lengths of sides of a triangle and the
measurements of the angles. A scalene
triangle has three different length sides
and it also has three different size angles.
**The longest side will be opposite the
greatest angle.**

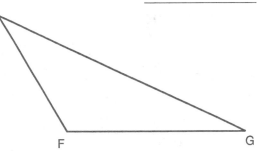

The length of side $\overline{AG}$ is shown as AG. Similarly, the length of side $\overline{AF}$ is AF.

The scalene triangle $\triangle$ AFG has side $\overline{AG}$ as its longest side. Therefore, AG > AF
and $\angle$ F > $\angle$ G.

The isosceles triangle $\triangle$ BKP has two sides, $\overline{BK}$ and $\overline{KP}$,
of equal length. Therefore, BK = KP

The angles opposite the two equal sides are also equal.
Therefore, $\angle$ B has the same measure as $\angle$ P.

An equilateral triangle has three sides of equal
length. The three angles must also have the same
measure. Since the sum of the measures is 180°
and all angles are equal, each angle of an equilateral
triangle must have a measure of 60°. For this reason
an equilateral triangle is frequently called an
equiangular triangle

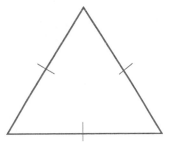

*Focus on
angle sizes
for isosceles
triangles*

Definition	Similar Triangles	
Two triangles are **similar**	if and only if	the measures of the three angles of one triangle are equal to the measures of the three corresponding angles of the second triangle.

Similar Triangles

The definition of similar triangles guarantees that two similar triangles will have identically the same shape. They may differ greatly in terms of size, but the shapes must be the same.

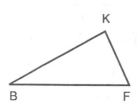 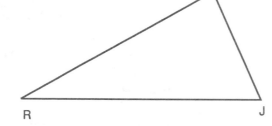

The two triangles, Δ BFK and Δ RJQ, above illustrate the idea of similarity, but any application of the concept depends upon a related idea of **corresponding parts**.

> ∠ B **corresponds** to ∠ R because the two angles have equal measures. Similarly, ∠ F corresponds to ∠ J and ∠ K corresponds to ∠ Q.

> Side $\overline{BF}$ **corresponds** to side $\overline{RJ}$ because they are between the corresponding angles, ∠ B to ∠ R and ∠ F to ∠ J. Similarly, side $\overline{BK}$ corresponds to side $\overline{RQ}$ and side $\overline{FK}$ corresponds to side $\overline{JQ}$.

Although $\overline{BF}$ **corresponds** to side $\overline{RJ}$, BF is not equal to RJ. In fact, BF < RJ. However, a valuable property of similar triangles is:

> **The lengths of corresponding sides of similar triangles are proportional.**

Therefore,
$$\frac{BF}{RJ} = \frac{FK}{JQ}, \quad \frac{BF}{RJ} = \frac{BK}{RQ}, \text{ and } \frac{FK}{JQ} = \frac{BK}{RQ}$$

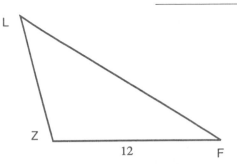

Corresponding angles of similar triangles always have the same measure.

Corresponding sides of similar triangles usually have different measures, but always are proportional. If the ratio of one pair of corresponding sides is known, then the ratio of any other pair of corresponding sides is the same.

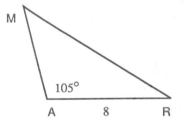

Focus on corresponding parts of similar triangles

In the figure above, △ARM and △ZFL are similar.   Since the measure of ∠ A is indicated as 105° then the measure of ∠ Z must also be 105° because they are corresponding angles and must have the same measure.

The lengths of $\overline{AR}$ and $\overline{ZF}$ are shown. AR = 8 and ZF = 12

Since the ratio of 8 to 12 is $\frac{2}{3}$, this means that the ratio of any other pair of corresponding sides must also be $\frac{2}{3}$.

$$\frac{RM}{FL} = \frac{AR}{ZF} = \frac{AM}{LZ} = \frac{2}{3} \qquad \begin{array}{l}\underline{\text{Numerators from △ARM}}\\ \text{Denominators from△ZFL}\end{array}$$

$$\frac{RM}{FL} = \frac{2}{3}$$

Therefore, the ratio of RM to FL is $\frac{2}{3}$.

Solving Problems with Similar Triangles

In the figure at the right, the △ CEK and △ JRZ are similar because they have the same angles.

Side $\overline{CE}$ corresponds to $\overline{JR}$ and $\overline{CK}$ corresponds to $\overline{JZ}$. These correspondences give the proportion

$$\frac{CE}{JR} = \frac{CK}{JZ}$$

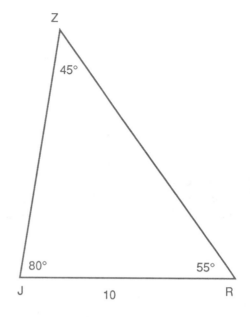

Since CE = 7, JR = 10, and CK = 11, the proportion becomes $\frac{7}{10} = \frac{11}{JZ}$

Solving the proportion gives

$$JZ = \frac{110}{7} = 15\frac{5}{7}$$

The proportionality of the sides of similar triangles often makes it possible to find a missing length.

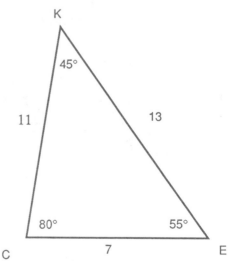

Focus on finding a missing length

The figure at the right shows similar triangles, Δ ABC and Δ ADE. Both triangles have ∠ A and a right angle. This is sufficient to force the third angles to be equal.

A proportion can be written to find the length of $\overline{AD}$.

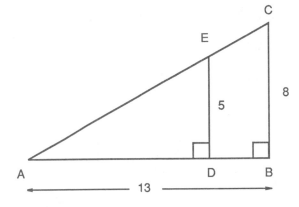

$$\frac{AB}{AD} = \frac{BC}{DE}$$

$$\frac{13}{AD} = \frac{8}{5}$$

$$8 \cdot AD = 13 \cdot 5$$
$$AD = \frac{65}{8}$$

Solving gives $AD = \frac{65}{8} = 8\frac{1}{8}$

The lengths of $\overline{AE}$ and $\overline{AC}$ cannot be found by this method because neither of these corresponding parts is known.

Unit 4 Exercise

This exercise reviews the preceding unit. The exercise is divided into three parts. Part A reviews the explanations of Unit 4. Part B offers opportunities to practice the skills and concepts of Unit 4. Part C contains problems that review your previous work in this text. All answers for Parts A and B are at the back of the book. Each problem of Part C is accompanied by a notation «CU» that refers to the Chapter (C) and Unit (U) in which that type of problem is studied.

Part A: Reviewing the foci of Unit 4.

1. Classify each triangle below as acute, right, or obtuse.

a.    b. 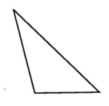   c. 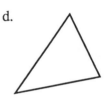   d.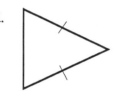

2. Classify each triangle below as scalene, isosceles, or equilateral.

a.    b. 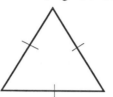   c. 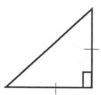   d.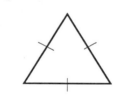

3. The two triangles at the right are similar. Make a list of the 6 corresponding parts for the triangles.

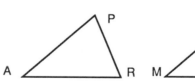

Part B: Drill and Practice

1. If a triangle is a right, isosceles triangle, find the size of all its angles.

2. If an obtuse, isosceles triangle has a 40° angle, what are the measures of the other angles?

3. Is it possible to have an acute, isosceles triangle?

4. Is it possible to have an obtuse, equilateral triangle?

5. Is it possible to have a right, scalene triangle?

6. Is it possible to have an obtuse, scalene triangle?

7. If two triangles are similar might one have a 60° angle and the other have a 70° angle?

8. If two triangles are similar might one have a 110° angle and the other have a 120° angle?

Part C: Review. **Answers for the problems of Part C are not given. However, the notation «C,U» refers to the Chapter (C) and Unit (U) in which problems of the same type were presented.**

1. Evaluate $4 + [7 + 7 \cdot 3]$ «5,2»

2. Evaluate $2x + 5y$ when $x = 3$ and $y = 6$ «5,3»

3. Simplify $(9 + 5x) + 7$ «5,4»

4. Simplify $5(x + 3) + 2(x + 8)$ «5,6»

5. Evaluate $-2 - (-4)$ «6,2»

6. Evaluate $-6 \cdot 7 - 10 \div -2$ «6,4»

7. Simplify $5(2x - 3) - (x - 3)$ «6,6»

8. Solve $4(3x + 5) - 8 = 12$ «6,8»

9. Evaluate $(\frac{5}{6} - \frac{3}{4}) - \frac{2}{3}$ «7,2»

10. Simplify $(\frac{2}{3}x + \frac{3}{4}) - \frac{3}{8}$ «7,3»

11. Simplify $\frac{4}{5}(\frac{3}{4}r)$ «7,4»

12. Simplify $5(3x + \frac{3}{5}) + 2(\frac{7}{2}x - 5)$ «7,5»

13. Solve $5x - (2x + 7) = 8 + 2x$ «7,6»

14. Solve $\frac{1}{8}x + \frac{3}{4} = \frac{7}{12}$ «7,7»

15. Solve $\frac{n}{6} = \frac{5}{16}$ «8,2»

16. The United States budgets 0.24 of its money for national defense. Bob's taxes last year were $8,450. How much of Bob's taxes go to defense?   «4,4»

17. A photograph was 2 inches high and 3 inches wide. If it is enlarged to be 5 inches high, how wide will it be?   «8,3»

18. Round off $\frac{46}{75}$ to the nearest one percent. «8,4»

19. 19% of 4,087 is _____.   «8,5»

20. 15% of a 35 member chemistry class failed the course. How many failed the course?   «8,6»

21. Draw a figure of $\overline{AB}$ and $\overline{AC}$ that have more than one common point.   «9,1»

22. Name the angle shown and give its approximate measure.   «9,2»

23. Name a pair of alternate interior angles in the figure.   «9,3»

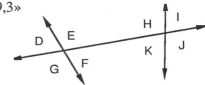

Unit 5: PERIMETERS AND CIRCUMFERENCES

Linear Measurements

The most common and simplest form of geometric measure is a **linear measure**.

Direct linear measures are made with rulers, tape measures, meter sticks, and similar tools.

Each of the line segments shown at the right could be measured by a ruler or meter stick.

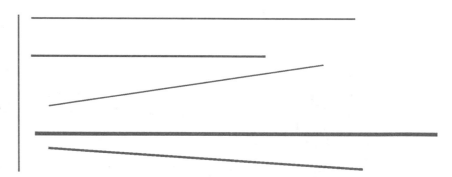

Focus on linear measures

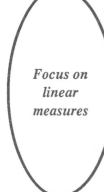

Linear measures are applied to line segments or curves.

Each of the figures shown at the right has a linear measure. Although the length of some of the figures would be difficult to measure with a ruler, if the curves were formed by pieces of string they could be easily measured by reshaping the string into a straight line.

All Linear Measures Are Approximations

Linear measures are never exact. All linear measures are limited
in their accuracy by the unit of measure used on the meter stick
or ruler.

A ruler marked off in quarter/inches has an accuracy to the nearest
one-fourth inch. If a more accurate measure is desired then a
smaller unit of measure must be used.

This process of improving the accuracy by choosing a smaller unit
of measure can, theoretically, continue forever. Therefore, each
new measure will be more accurate, but different than its predecessors.
All the measures will be approximations.

*Focus
on two
different
units of
measure*

The figure below shows two rulers and one line segment. The top ruler is
marked off in quarter/inches. Its unit of measure is $\frac{1}{4}$ inch. The bottom ruler
is marked off in eighth/inches. Its unit of measure is $\frac{1}{8}$ inch.

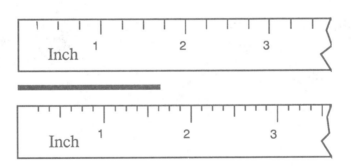

Using the $\frac{1}{4}$ inch units, the line segment is measured as $1\frac{3}{4}$ inches long
because that is the closest mark on the ruler.

Using the $\frac{1}{8}$ inch units, the line segment is measured as $1\frac{5}{8}$ inches long
because that is the closest mark on the ruler.

Both measures are correct! Both measures are approximations! The
accuracy of the measures is determined by their units of measure.

Perimeter of a Triangle

A triangle is shown at the right. Δ ABC has sides of lengths 5, 7, and 10. The linear measure of the distance from point A to point C, point C to point B, and back to A is the **perimeter** of the triangle.

Consequently, the perimeter of Δ ABC is the sum of the lengths of the three sides.

7 + 5 + 10 = 22

22 is the perimeter of Δ ABC.

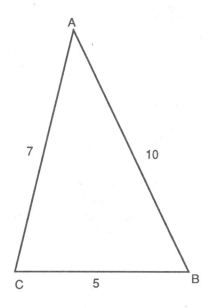

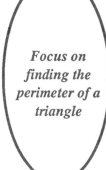

Focus on finding the perimeter of a triangle

The perimeter of a triangle is found by determining the lengths of the three sides and then finding their sum.

Δ DEF shown at the right is an isosceles triangle because it has two equal sides. The length of side $\overline{EF}$ is 16 which is the same as that of side $\overline{DE}$.

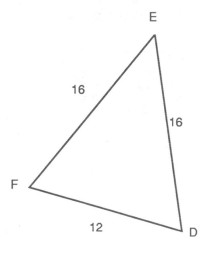

The perimeter of Δ DEF is 44 because: 16 + 16 + 12 = 44

Perimeters of Quadrilaterals

A four-sided geometric figure is
a **quadrilateral**. To find the
perimeter of a quadrilateral the
sum of the lengths of its sides
is found.

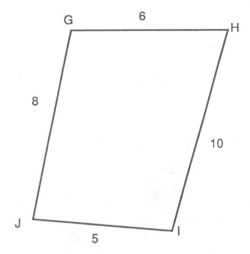

Quadrilateral GHIJ shown at the
right has sides of 5, 8, 6, and 10.
Its perimeter is:

$$5 + 8 + 6 + 10 = 29$$

The perimeter of quadrilateral GHIJ is 29.

Definition	**Trapezoid**	
A four-sided plane figure is a **trapezoid**	**if and only if**	one pair of the sides are parallel (have the same distance between them).

*Focus on
finding the
perimeter of a
quadrilateral*

The quadrilateral KLIN shown
at the right is a **trapezoid**
because it has a pair of parallel
sides, $\overline{KL}$ and $\overline{NI}$. The pair of
parallel sides are marked by
equal signs (=).

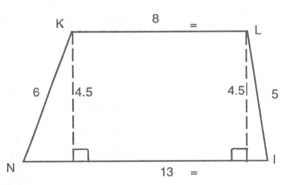

The perimeter of trapezoid KLIN
is found by adding the lengths of
its four sides.

The perimeter of trapezoid KLIN is 32 because: $6 + 8 + 5 + 13 = 32$

Definition   Parallelogram

A four-sided plane figure is **a parallelogram** **if and only if** both pairs of the opposite sides are parallel and have equal lengths.

The quadrilateral OPQR shown at the right is a **parallelogram**. Both of its pairs of opposite sides are parallel and have the same lengths.

Parallelogram OPQR has one pair of sides of length 9 and another pair of length 4.

The perimeter of the parallelogram OPQR is:

$2 \cdot 9 + 2 \cdot 4 = 18 + 8 = 26$

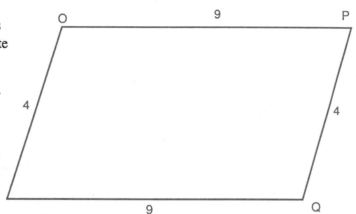

Definition   Rectangle

A four-sided plane figure is **a rectangle** **if and only if** it is a parallelogram with four right angles.

Focus on the perimeter of a parallelogram

Any rectangle, like RSTU, has its perimeter determined by the formula

$P = 2L + 2W$

where L represents the horizontal length and W represents the width or vertical length.

The perimeter of rectangle RSTU is 26 because:

$P = 2 \cdot 5 + 2 \cdot 8 = 10 + 16 = 26$

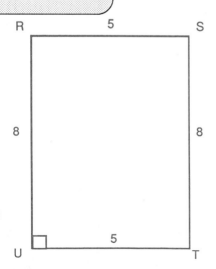

Definition	Square	
A four-sided plane figure is a square	**if and only if**	it is a rectangle with four sides of equal length.

Perimeters of Squares

The rectangle VWXY shown at the right is a **square** because all of its sides are equal in length.

The perimeter of a square is found using the formula

$$P = 4s$$

where s represents the length of one side.

The perimeter of square VWXY is:

$$P = 4 \cdot 7 = 28$$

28 is the perimeter of VWXY.

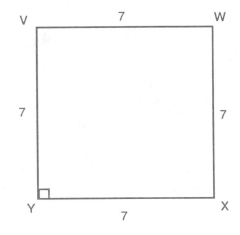

Focus on finding the perimeter of a square

The perimeter of a square, like ABCD, can be found by adding the four equal sides, but it is easier to use the formula:

$$P = 4s$$

For square ABCD the formula gives

$$P = 4 \cdot 8\frac{3}{4} = 4 \cdot \frac{35}{4} = 35$$

35 is the perimeter of ABCD.

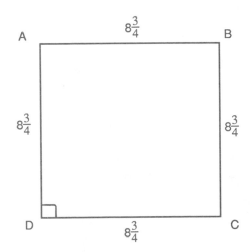

A Hierarchy of Quadrilaterals

The quadrilaterals studied in this unit can be arranged in a hierarchy where each succeeding figure possesses all of the properties of its predecessors and some new properties.

Quadrilateral	A four-sided planar figure
Trapezoid	A quadrilateral with at least one pair of parallel sides
Parallelogram	A trapezoid with two pairs of parallel sides
Rectangle	A parallelogram with four right angles
Square	A rectangle with four equal sides

Notice that the hierarchy above describes each successive figure in terms of its predecessor.

Every parallelogram is a trapezoid with two pairs of parallel sides, but this is sufficient to make:

1. Opposite sides equal in length.

2. Opposite angles equal.

3. Consecutive angles supplementary.

4. Diagonals bisect each other.

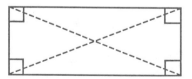

Every rectangle is a parallelogram with four right angles, but this is sufficient to make:

5. Diagonals equal in length.

Every square is a rectangle with four equal sides, but this is sufficient to make:

6. Diagonals perpendicular.

Circumferences of Circles

The distance around a circle is
called its **circumference** rather
than its perimeter.

To find the circumference of a
circle, it is necessary to know
that its diameter, d, is the distance
through the center, C, from one
point on the circle to a second
point. In the figure, $\overline{AB}$ is a
diameter. All diameters of a
circle are equal.

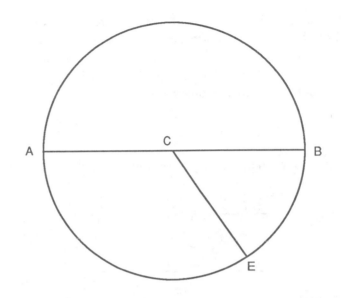

The radius, r, of a circle is the distance
from its center, C, to a point on the
circle. In the figure, $\overline{CE}$ is a
radius. All radii (plural of radius)
of a circle are equal.

*Focus on
finding the
circumference
of a circle*

The circumference of a circle is found
by one of the formulas

$$C = \pi d \quad \text{or} \quad C = 2\pi r$$

where d is the diameter and r is the radius
(the diameter is twice the radius). The
symbol π stands for the number Pi which
is approximated by 3.14 or $3\frac{1}{7}$.

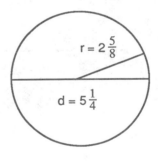

For the circle shown at the right, the circumference could be found in either
of the following ways. Both examples use the formula $C = \pi d$.

$$C = 3\tfrac{1}{7} \cdot 5\tfrac{1}{4} \qquad\qquad C = 3.14 \cdot 5\tfrac{1}{4}$$

$$= \frac{22}{7} \cdot \frac{21}{4} \qquad\qquad\quad = 3.14 \cdot 5.25$$

$$= \frac{11}{1} \cdot \frac{3}{2} \qquad\qquad\quad = 16.4850$$

$$= \frac{33}{2}$$

$$= 16\tfrac{1}{2}$$

The answers are different because different approximations for Pi were used.

Unit 5 Exercise

This exercise reviews the preceding unit. The exercise is divided into three parts. Part A reviews the explanations of Unit 5. Part B offers opportunities to practice the skills and concepts of Unit 5. Part C contains problems that review your previous work in this text. All answers for Parts A and B are at the back of the book. Each problem of Part C is accompanied by a notation «CU» that refers to the Chapter (C) and Unit (U) in which that type of problem is studied.

Part A: Reviewing the foci of Unit 5.

Determine the unit of measure for each of the following:

1.   a.

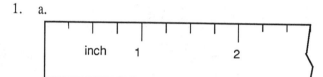

b.

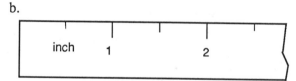

c.

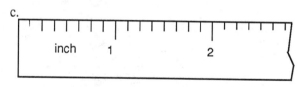

d.

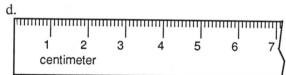

2.   Copy rulers 1a and 1b above on a folded paper and measure the segments.
   a.   b.

3.   Which of the following are approximations?
   a.   John Doe is 5 feet $10\frac{3}{4}$ inches tall.
   b.   John Doe scored 100 on a biology test.
   c.   John Doe had blood pressure of 140 over 80.
   d.   John Doe has 27 classmates in his mathematics course.

4. Give the correct name for each quadrilateral and find its perimeter.

a.
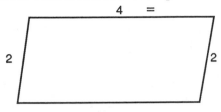

b.
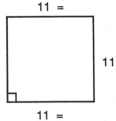

c.
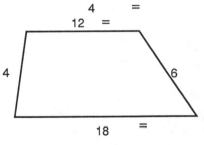

d.
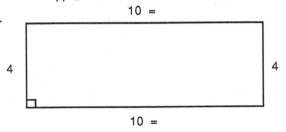

5. A circle has a radius of 17 centimeters.
   a. What length is another radius for the same circle?   b.   What length is a diameter?
   c. What is the circumference of the circle?

6. Find the perimeters or circumference for the following:

a.
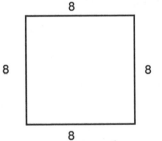

b.
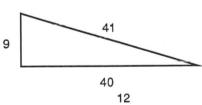

c.
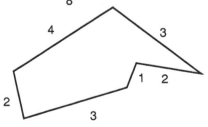

d.
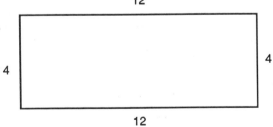

e.
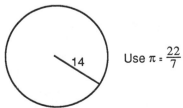

Part B: Drill and Practice

1. Use the ruler of problem 1a, page 520, to measure the segments below.

    a. ▬▬▬▬▬▬

    b.

    c. ▬▬▬▬▬▬▬▬▬

2. If a ruler has 12 divisions between each inch, what is the unit of measure for this ruler?

For problems 3 - 9, find the perimeter/circumference.

3.

    6

    10   10

    6

4.

    16

    16   16

    16

5.

    21

    19

    15

    17

6.

    20 =

    9   12

    12 =

7.

    19

    20   22

    13 15   17

8.

    9

    5   5

    9

9.

    12   Use π = 3.14

10. What is the unit of measure of the rule below?

    Inch   1   2

11. Give the name of the following:
    a. The figure in problem 3.
    b. The figure in problem 4.
    c. The figure in problem 5.
    d. A four-sided figure with one pair of parallel sides.

12. Which of the following are approximations?
    a. Sam answered 50 problems correctly on a test.
    b. Suzy jumped 21 feet.
    c. Scientists concluded that there are 2 billion molecules in a peanut.
    d. Sally ran a race in 9.823 seconds.

Part C: Review. Answers for the problems of Part C are not given. However, the notation «C,U» refers to the Chapter (C) and Unit (U) in which problems of the same type were presented.

1. $57,902 - 19,412$ «1,3»

2. Evaluate $-5 \cdot -7$ «6,3»

3. Simplify $3(-5x)$ «6,5»

4. Solve $23 + 3x = 2$ «6,7»

5. Evaluate $\frac{5}{8} \cdot \frac{11}{10}$ «7,1»

6. Simplify $(\frac{2}{3}x - \frac{1}{2}) + \frac{3}{4}$ «7,3»

7. Simplify $\frac{1}{5}(\frac{-5}{6}x)$ «7,4»

8. Solve $\frac{1}{x} + \frac{5}{8} = \frac{3}{x} + \frac{3}{4}$ «7,7»

9. Solve $\frac{8}{7} = \frac{5}{n}$ «8,2»

10. Round off $\frac{8}{11}$ to the nearest one percent. «8,4»

11. Alice earned $2,300 in January, $987 in February, and $1,811 in March. What was the average monthly earnings for those three months? «4,4»

12. Change $\frac{5}{6}$ to a decimal fraction rounded off at two decimal places. «4,4»

13. Write an equation that shows the difference when 31 is subtracted from a number is 45. «6,9»

14. A teacher finds that 19 out of every 20 students in her classes will pass. If there are 140 total students in her classes, how many will pass? «8,3»

15. _____ is 9% of 326. «8,5»

16. A 7% commission is paid on sales of $3,650. What is the commission? «8,6»

17. Draw a geometric figure with exactly two lines of symmetry. «9,1»

18. Name the angle shown and give its approximate measure. «9,2»

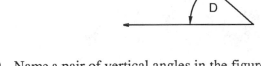

19. Name a pair of vertical angles in the figure. «9,3»

20. Is it possible to have an acute, isosceles triangle? «9,4»

Unit 6:   AREAS

Linear measures are associated with ideas of length and distance. In this unit, the topic of discussion is area measures. Area measures are associated with measuring the amount of surface enclosed by a geometric figure.

Perimeter Versus Area

The figure at the right shows a rectangle, ABCD, and a square, EFGH. Both have a perimeter of 8.

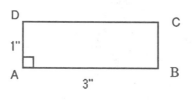

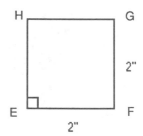

$$P = 2L + 2W \qquad P = 4s$$
$$ = 2 \cdot 3 + 2 \cdot 1 \qquad = 4 \cdot 2$$
$$ = 6 + 2 \qquad\qquad = 8$$
$$ = 8$$

Inside the rectangle there are three smaller squares, but inside the square there are four squares of the same size. This means that there is more surface area in the square than in the rectangle.

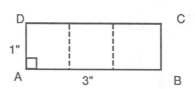

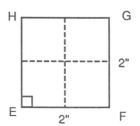

Focus on area measures

Linear measures are for distance or length. Area measures are for surface area.

Area measures are always based on the idea of fitting smaller squares inside a geometric figure. These smaller squares are the units of measurement for area and are appropriately called **square units**.

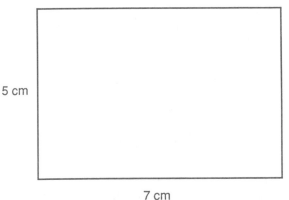

The rectangle shown at the right has length 7 centimeters (cm) and width 5 cm. To determine the rectangle's area in **square centimeters** it is necessary to find how many squares one centimeter on each side can be fitted into the rectangle.

Since 35 one centimeter squares can be fitted into the rectangle, the area of the rectangle is said to be 35 square centimeters.

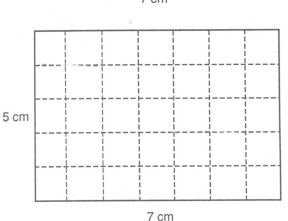

A Formula for a Rectangle's Area

The area of a rectangle can be found using the formula

$$A = LW$$

where L and W represent the linear measures of the length and width.

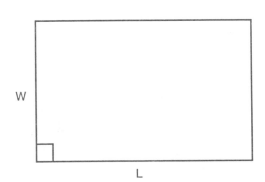

To find the area of rectangle MNOP, both length and width are first stated in terms of inches. Then the formula is applied.

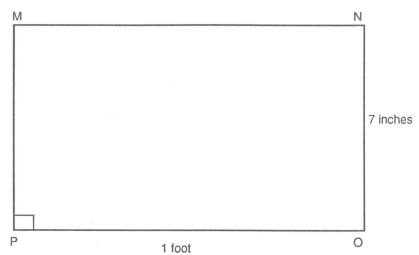

$$A = 1 \text{ foot} \cdot 7 \text{ inches}$$
$$= 12 \text{ inches} \cdot 7 \text{ inches}$$
$$= 84 \text{ square inches}$$

The area is given in square inches. This means that 84 one-inch squares could be fitted into the rectangle.

Focus on finding a rectangle's area

The rectangle RSTU at the right has length $3\frac{1}{3}$ inches and width $4\frac{1}{2}$ inches.

The area is found using the formula.

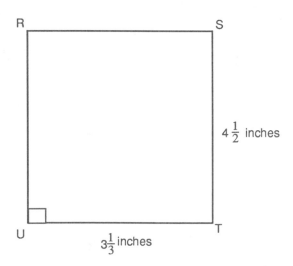

$$A = LW$$
$$A = 3\frac{1}{3} \cdot 4\frac{1}{2}$$
$$= \frac{10}{3} \cdot \frac{9}{2}$$
$$= \frac{5}{1} \cdot \frac{3}{1}$$
$$= 15$$

The area is 15 square inches. Notice that the original linear measures were given with inches as the units. This means the area measures will be square inches.

Areas of Squares

A square is a special type of
rectangle because the length
and width have the same
measure.

The rectangle formula $A = LW$
can be changed to $A = s^2$ where
s^2 means $s \bullet s$.

To find the area of square KGPX,

$A = s^2$

$\quad = 2\frac{1}{4} \bullet 2\frac{1}{4}$

$\quad = \frac{9}{4} \bullet \frac{9}{4}$

$\quad = \frac{81}{16}$

$\quad = 5\frac{1}{16}$

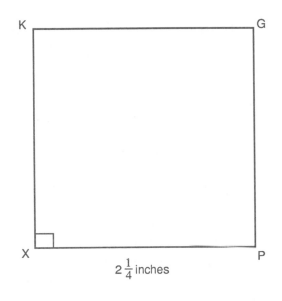

$2\frac{1}{4}$ inches

The area is $5\frac{1}{16}$ square inches.

Focus on finding the area of a figure

The figure at the right is composed
of a rectangle, BDKC, and a square,
RGHC.

To find the area of the entire figure,
find the sum of the separate areas.

Area of (BDKC) = 5 • 4

$\qquad\qquad\qquad$ = 20

Area of (RGHC) = 2^2

$\qquad\qquad\qquad$ = 4

Total area of the figure is 20 + 4
or 24 square kilometers.

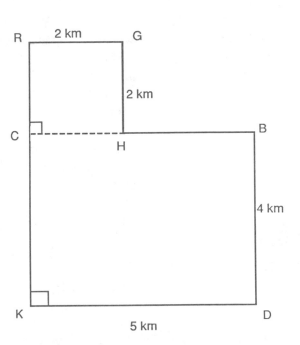

Areas of Parallelograms

Every parallelogram has the same area as an associated rectangle. In the figure below, the alteration of a parallelogram to a rectangle with the same area is shown.

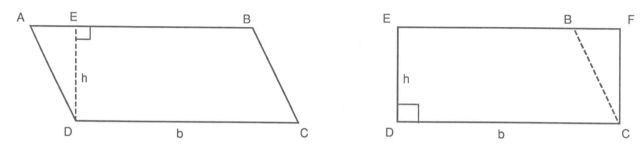

The area of parallelogram ABCD is the same as rectangle EFCD. On the parallelogram, if △ AED were detached and moved so that $\overline{AD}$ coincided with line segment $\overline{BC}$ the result would be the rectangle EFCD.

Because every parallelogram is associated with a rectangle of equal area, the rectangle formula, A = LW, can be rewritten as A = bh where b represents the base or length and h represents the vertical distance between the two lengths.

Focus on finding a parallelogram's area

Two parallelograms with lengths 10 ft and sides 6 ft are shown at the right. Parallelogram ABCD has height 5 ft and parallelogram EFGH has height 2 ft.

The perimeters of the parallelograms are the same, 32 feet, but the areas are different.

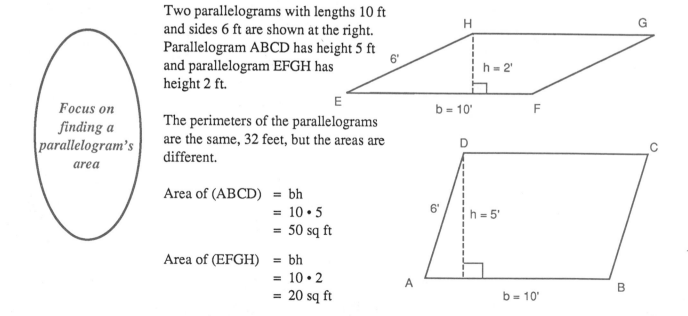

Area of (ABCD) = bh
                   = 10 • 5
                   = 50 sq ft

Area of (EFGH) = bh
                   = 10 • 2
                   = 20 sq ft

Areas of Triangles

Every triangle has an associated parallelogram and the area of the triangle is always one-half the area of its associated parallelogram. In the figure below, two triangles are shown. The dashed lines of the figure indicate a second, equal-sized triangle that could be placed to make a parallelogram.

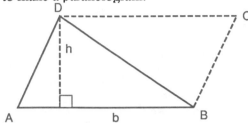 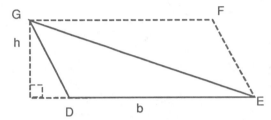

The area of △ ABD is one-half that of parallelogram ABCD. Similarly, the area of △ DEG is one-half the area of parallelogram DEFG.

Because every triangle is associated with a parallelogram of twice the area, the parallelogram formula, A = bh, can be altered to A = $\frac{1}{2}$bh.

Focus on finding a triangle's area

△ ABC is shown at the right.

If $\overline{AB}$, 8 cm, is considered the base then the height of the triangle is $\overline{CD}$, 6 cm.

If $\overline{BC}$, 12 cm, is considered the base then the height of the triangle is $\overline{AE}$, 4 cm.

The area of the triangle is found in two ways below.

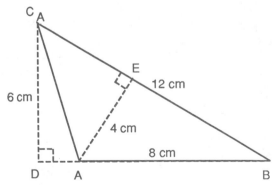

$A = \frac{1}{2}bh$ $A = \frac{1}{2}bh$

$= \frac{1}{2} \cdot 8 \cdot 6$ $= \frac{1}{2} \cdot 12 \cdot 4$

$= 4 \cdot 6$ $= 6 \cdot 4$

$= 24$ sq cm $= 24$ sq cm

Any side of a triangle may be used as its base, but the associated height must also be used. The height must always be perpendicular to the base (perpendicular lines intersect at a right angle).

Areas of Circles

The area of a circle is found by
using the formula

$$A = \pi r^2$$

where Pi is approximated by $3\frac{1}{7}$
or 3.14 and r is the length of the
radius.

The area of the circle shown at
the right is found by first using
the fact that if the diameter is 8 ft
then the radius is 4 ft.

$$
\begin{aligned}
A &= \pi r^2 \\
&= 3.14 \cdot 4^2 \\
&= 3.14 \cdot 16 \\
&= 50.24 \text{ sq ft}
\end{aligned}
$$

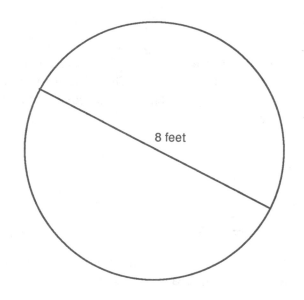

8 feet

*Focus on
finding the
area of a
geometric
figure*

The figure at the right is composed
of a rectangle, PDQR, and a half-circle
with diameter $\overline{DQ}$.

To find the area of the entire figure,
find the sum of the separate areas.

$$
\begin{aligned}
\text{Area of (PDQR)} &= LW \\
&= 7 \cdot 6 \\
&= 42
\end{aligned}
$$

$$
\begin{aligned}
\text{Area of (Half-Circle)} &= \frac{1}{2} \cdot \pi r^2 \\
&= \frac{1}{2} \cdot \frac{22}{7} \cdot 3^2 \\
&= \frac{1}{1} \cdot \frac{11}{7} \cdot 9 \\
&= \frac{99}{7} \\
&= 14\frac{1}{7}
\end{aligned}
$$

6 feet

D --- Q

7 feet

P   R

Total area of the figure is $42 + 14\frac{1}{7}$ or $56\frac{1}{7}$ square feet.

Unit 6 Exercise

This exercise reviews the preceding unit. The exercise is divided into three parts. Part A reviews the explanations of Unit 6. Part B offers opportunities to practice the skills and concepts of Unit 6. Part C contains problems that review your previous work in this text. All answers for Parts A and B are at the back of the book. Each problem of Part C is accompanied by a notation «CU» that refers to the Chapter (C) and Unit (U) in which that type of problem is studied.

Part A: Reviewing the foci of Unit 6.

Find the area and perimeter of each figure below.   Find the area of each figure below.

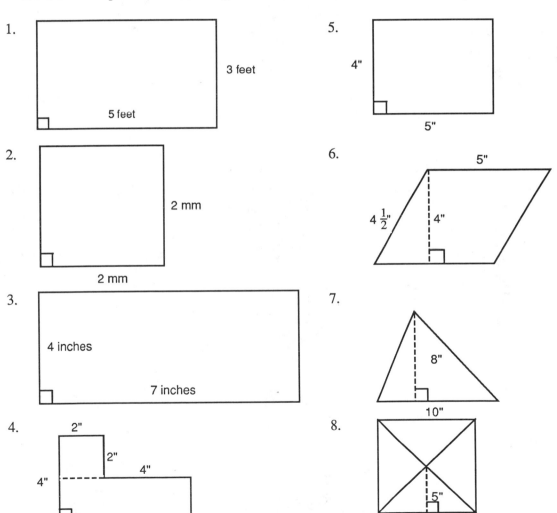

1.   3 feet
      5 feet

2.   2 mm
      2 mm

3.   4 inches
      7 inches

4.   2"
      2"
      4"
      4"

5.   4"
      5"

6.   5"
      4 ½"   4"

7.   8"
      10"

8.   5"
      10"

Find the area of each circle below.

9.

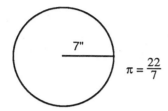

$\pi = \frac{22}{7}$

11.

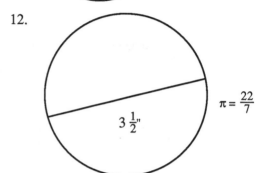

$\pi = 3.14$

10.

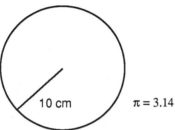

$\pi = 3.14$

12.

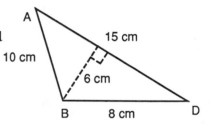

$\pi = \frac{22}{7}$

Part B: Drill and Practice

1. Find the circumference and area of the figure.

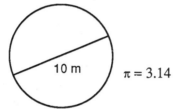

$\pi = 3.14$

4. Find the perimeter and area of the figure.

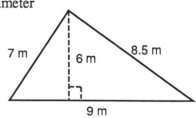

2. Find the perimeter and area of the figure.

21 ft

21 ft

5. Find the perimeter and area of the figure.

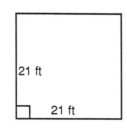

3. Find the perimeter and area of the figure.

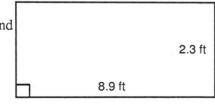

2.3 ft

8.9 ft

6. Find the circumference and area of the figure.

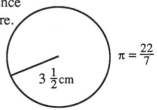

$\pi = \frac{22}{7}$

$3\frac{1}{2}$ cm

7. Find the perimeter
    and area of
    the figure.

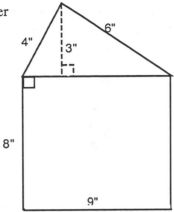

10. Find the perimeter
    and area of
    the figure.

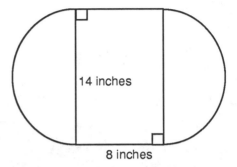

8. Find the perimeter
    and area of
    the figure.

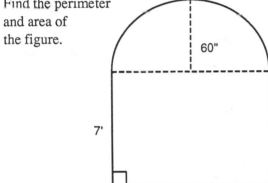

11. Find the
    area of
    the
    shaded
    portion.

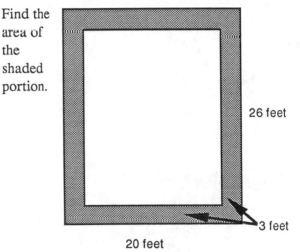

9. Find the perimeter
    and area of
    the figure.

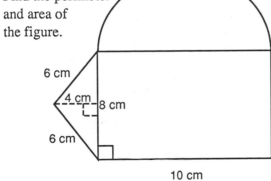

12. Find the perimeter
    and area of
    the figure.

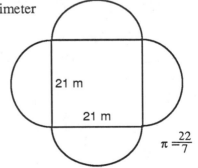

Part C: Review. Answers for the problems of Part C are not given. However, the notation «C,U» refers to the Chapter (C) and Unit (U) in which problems of the same type were presented.

1. $79,560 + 38,605 + 4,332$ «1,2»

2. Find the numerator of: $\frac{7}{15} = \frac{?}{45}$ «3,2»

3. Find the difference of 18.67 and 6.9   «4,2»

4. Evaluate $6 - (-8) \cdot -5 \div 4$ «6,4»

5. Simplify $(2x - 1) - 2(3x - 1)$ «6,6»

6. Solve $13 - 3x = x - 31$ «6,8»

7. Evaluate $\frac{7}{8} + (\frac{1}{4} - \frac{1}{2})$ «7,2»

8. Simplify $\frac{-1}{8}(\frac{4}{7}x)$ «7,4»

9. Solve $x + 17 = 4x - 4$ «7,6»

10. How many $\frac{1}{3}$ yards are there in 8 yards?.
    «3,7»

11. Frank's test scores in Earth Science were 87, 65, 91, and 78. Find his average on these four tests rounded off to the nearest whole number. «4,4»

12. Change $\frac{23}{76}$ to a decimal rounded off at two decimal places.   «4,4»

13. Solve $\frac{n}{8} = \frac{13}{9}$ «8,2»

14. In Miami 9 out of 10 days in the spring will have a temperature over 80 degrees. In a spring vacation of 14 days, how many will have a temperature over 80 degrees?   «8,3»

15. _____% of 215 is 54.   «8,5»

16. A survey found that 37% of a school's students could not run a mile in less than 12 minutes. If the school has 1,840 students, what number cannot run a mile in less than 12 minutes? «8,6»

17. Draw a figure of a geometric figure with no line of symmetry.   «9,1»

18. Name the angle shown and give its approximate measure.   «9,2»

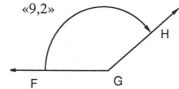

19. Name a pair of alternate exterior angles in the figure.   «9,3»

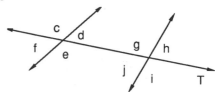

20. Is it possible to have an obtuse, scalene triangle?   «9,4»

21. Find the perimeter of a rectangle with width 3 inches and length 5 inches..   «9,5»

Unit 7:   VOLUMES

Linear measures are applied in situations where the only concern
is length. One dimensional objects like line segments are measured
by linear units. The perimeters of two dimensional objects are like
pieces of string which are reshaped as line segments so that linear
units can be used.

Square measures are applied in situations where the amount of
surface area is desired. Square units are for two dimensional
objects.

In this unit, cubic measures are presented. Cubic measures are
applied to three dimensional objects.

Cubic Measures

Objects such as cardboard boxes and tin cans are three
dimensional figures. Measuring the amount of space enclosed
by a cardboard box or tin can is equivalent to finding the volume
measures of the containers.

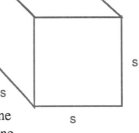

The unit of measure for three dimensional
objects is a cube. Cubes are shaped like a
child's building blocks or a pair of dice.
A cube might be called a three dimension
square because all of its sides (faces) are squares.

A cube with each side of its squares one inch long is called one
cubic inch. A cube with sides one centimeter long is called one
cubic centimeter.

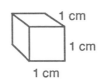

The basic idea in measuring the volume of any three dimensional
object is filling its space snugly with cubes and then counting the
number of cubes.

Focus on measuring the volume of a box

To measure the volume of a box like that shown at the right, it is necessary to find how many one inch cubes would fit in the box.

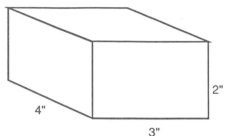

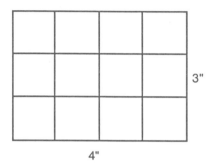

The floor of the box is a rectangle with length 4 inches and width 3 inches. It looks like the rectangle shown at the left. This means that 12 one inch cubes could be placed on the floor of the box.

One side of the box is a rectangle 4 inches long and 2 inches high. It looks like the rectangle shown at the right. It indicates that 2 layers of one inch cubes could be stacked in the box.

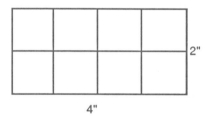

Since 12 one inch cubes could be placed in each layer in the box and there are two layers possible, the box has a capacity of 24 (12 • 2) one inch cubes. The box has a volume of 24 cubic inches.

A Formula for the Volume of a Rectangular Solid

The shape of a box is called a **rectangular solid**.

The basic idea for finding the volume of a rectangular solid is
1. determine how many cubes would cover the floor of the container (This is the area of the base).
2. determine how many layers of cubes could be placed in the box (This is the height of the box).
3. multiply the area of the base by the height.

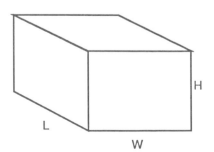

The formula for volume of a box is $V = AH$ where A is the area of the floor of the box and H is the height of the box. The formula can be written as $V = LWH$ where L is the length of the box, W its width, and H its height.

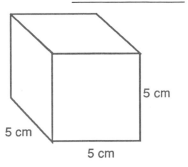

The figure at the right shows a cube which measures 5 cm on each side or edge.

The floor of the cube is a square with length 5 cm and width 5 cm. The height of the cube is 5 cm.

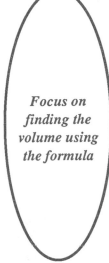

Focus on finding the volume using the formula

The volume of the cube is

$$V = LWH$$
$$= 5 \cdot 5 \cdot 5$$
$$= 25 \cdot 5$$
$$= 125$$

The volume of the cube is 125 cubic centimeters.

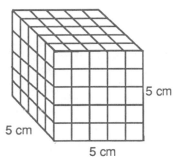

The Volume of a Cylinder

A tin can has a shape which is called a cylinder.

To find the volume of a cylinder, use the formula

$$V = Ah$$

where A is the area of the circular base and h is the height of the cylinder.

$$V = Ah$$
$$= \pi r^2 \cdot h$$
$$= 3.14 \cdot 4^2 \cdot 10$$
$$= 3.14 \cdot 16 \cdot 10$$
$$= 3.14 \cdot 160$$
$$= 502.4$$

The volume of the cylinder is 502.4 cubic inches.

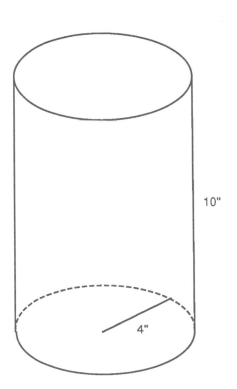

Focus on finding the volume of a cylinder

The cylinder shown at the right is lying on its side. Consider the circular ends as the bases. Then the formula V = AH is used to find its volume.

$$V = AH$$

$$= \pi r^2 H$$

$$= \frac{22}{7} \cdot (2\tfrac{1}{2})^2 \cdot 14 \qquad r = \frac{1}{2} \cdot 5$$

$$= \frac{22}{7} \cdot (\tfrac{5}{2})^2 \cdot 14$$

$$= \frac{22}{\underset{1}{\cancel{7}}} \cdot \frac{25}{4} \cdot \frac{\overset{2}{\cancel{14}}}{1}$$

$$= \frac{22}{1} \cdot \frac{25}{\underset{2}{\cancel{4}}} \cdot \frac{\overset{1}{\cancel{2}}}{1}$$

$$= \frac{\overset{11}{\cancel{22}}}{1} \cdot \frac{25}{\underset{1}{\cancel{2}}} \cdot \frac{1}{1}$$

$$= 11 \cdot 25$$

$$= 275$$

The volume of the cylinder is 275 cubic feet.

Unit 7 Exercise

This exercise reviews the preceding unit. The exercise is divided into three parts. Part A reviews the explanations of Unit 7. Part B offers opportunities to practice the skills and concepts of Unit 7. Part C contains problems that review your previous work in this text. All answers for Parts A and B are at the back of the book. Each problem of Part C is accompanied by a notation «CU» that refers to the Chapter (C) and Unit (U) in which that type of problem is studied.

Part A: Reviewing the foci of Unit 7.

1. Find the volume of a rectangular solid 6 centimeters long, 2 centimeters wide, and 5 centimeters high.

2. Find the volume of a cube with sides 4.2 feet in length.

3. Find the volume of a cylinder with radius 10 inches and height 21 inches. Use $3\frac{1}{7}$ for Pi.

4. How many cubic feet are in a cubic yard?

5. Find the volume of a cube 6.3 centimeters on a side.

6. Find the volume of a rectangle shaped swimming pool 30' by 20' by 8'.

7. Find the volume of the box shown below.

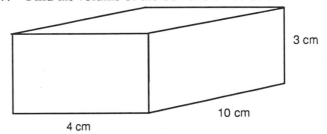

3 cm

10 cm

4 cm

8. Find the volume of the cylinder. Use $\frac{22}{7}$ for pi.

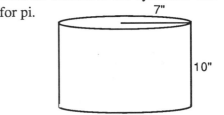

7"

10"

9. Find the volume of a cylindrical oil tank, 100 feet in diameter and 30 feet high.

Part B: Drill and Practice

Find the volume of the figures.

1.

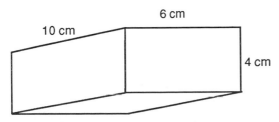

6 cm

10 cm

4 cm

2.

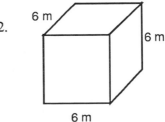

6 m

6 m

6 m

3.

12 inches

12 inches

12 inches

4.

12 inches

10 miles

28 miles

5.

5 mm

14 mm   5 mm

6.

8"

14"

7.

21 cm

30 cm

7 cm

8.

6"

4"

13"

9.

17 '

5'

11'

10.

1 m

5 m

11.

36 in

1 yd   3 ft

12.

6 in

9 in

Part C: Review.   Answers for the problems of Part C are not given. However, the notation «C,U» refers to the Chapter (C) and Unit (U) in which problems of the same type were presented.

1. What is the tens digit of 6,248,571.   «1,1»

2. $8\frac{1}{3} - 5\frac{3}{4}$ «3,4»

3. Write 0.409 as a fraction.   «4,1»

4. Solve $4x = 48$ «5,7»

5. Simplify -9(4x)   «6,5»

6. Solve $2x - 17 = 19$ «6,7»

7. Evaluate $\frac{10}{21} \cdot \frac{14}{15}$ «7,1»

8. Simplify $(3x - \frac{1}{2}) + \frac{1}{8}$ «7,3»

9. Simplify $4x - 2(5x - 7)$ «7,5»

10. Solve $\frac{3}{4} - \frac{7}{2x} = \frac{2}{3} + \frac{1}{x}$ «7,7»

11. A mechanic has completed $\frac{5}{8}$ of a job after 2 days work. How many days will the entire job require? «3,7»

12. Write an equation that shows: the sum of a number and 41 is 57.   «6,9»

13. Determine if $\frac{9}{12} = \frac{0.3}{4}$ is a true equality using the products of the extremes and means.   «8,1»

14. In a graduating class, there are 3 men for each 2 women. How many women graduate in the class of 542 students?   «8,3»

15. 8% of _____ is 412   «8,5»

16. 37 women are in a class of 54 students. What percent of the class is women?   «8,6»

17. Draw a figure of $\overleftrightarrow{AB}$ and $\overleftrightarrow{CD}$ that have more than one point in common.   «9,1»

18. Name the angle shown and give its approximate measure.   «9,2»

19. Name a pair of corresponding angles in the figure.   «9,3»

20. Is it possible to have an obtuse, equilateral triangle?   «9,4»

21. Find the perimeter of a square with one side 5 feet long.   «9,5»

22. Find the area of a rectangle with length 8 inches and width 5 inches.   «9,6»

Chapter 9 Test

«9,U» shows the unit in which this problem is found in the chapter.

1. Find the number of lines of symmetry for: «9,1»
    a. an equilateral triangle
    b. a 4 ft by 3 ft rectangle

2. Use the terms "acute," "obtuse," "right," and "straight" to identify each angle shown below.   «9,2»
    a.    b.    c. 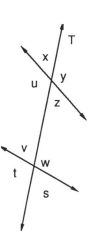   d.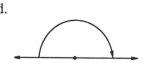

3. Use a protractor to measure the angles in problems 2 a and 2 b above.   «9,3»

4. In the figure at the right: «9,3»

    a. ∠x and _____ are alternate exterior angles.

    b. ∠v and _____ are corresponding angles.

    c. ∠u and _____ are alternate interior angles.

    d. ∠w and _____ are vertical angles.

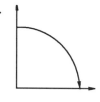

5. In an isosceles triangle, if one angle measures
    110 degrees each other angle measures _____.   «9,4»

6. Find perimeters for: «9,5»
    a. A square with one side five inches long.
    b. A parallelogram with sides 3.2 cm and
        8.1 cm.

7. Find areas for: «9,6»
    a. A triangle with base six inches and height five inches.
    b. A circle with diameter 200 yards. Use 3.14 as Pi.

8. Find volumes for: «9,7»
    a. A rectangular solid with length seven meters, width four meters, and height 2.5 meters.
    b. A cylinder with radius 7 inches and height 10 inches. Use $\frac{22}{7}$ as Pi.

9. To measure the distance across a lake, Ed used similar triangles. He measured the distances shown
    in the figure below after making $\angle$ RBS equal to $\angle$ TAS. Use proportions to find the distance from
    T to A.   «9,4»

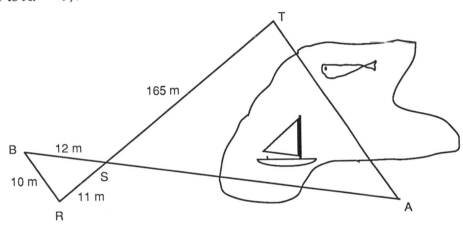

10. Draw a figure of the following situation and then find the height of the tree. A man 2 meters tall
    casts a shadow 8 meters long while standing by a tree whose shadow is 92 meters long.
    «9,4»

Answers

Chapter 1, Unit 1
Part A

1. 13
2. 23
3. 21
4. 13
5. 9 thousands
6. 3
7. hundreds
8. ten thousands
9. 10
10. 10

Chapter 1, Unit 1
Part B

1. 9
2. 15
3. 14
4. 8
5. 18
6. 12
7. 6
8. 4
9. 3
10. 8
11. 4
12. 795
13. 860
14. 978
15. 16
16. 15
17. 11
18. 14
19. 9
20. 10

Chapter 1, Unit 2
Part A

1. 17

2.
$$\begin{array}{r} 6\,3 \\ +2\,5 \\ \hline \end{array}$$

3.
$$\begin{array}{r} 7\,2\,5 \\ +\ 4\,2 \\ \hline \end{array}$$

4. 88

5.
$$\begin{array}{r} 1\ \ 1 \\ 4\,3\,8 \\ +\ 9\,2\,7 \\ \hline 1{,}3\,6\,5 \end{array}$$

6.
$$\begin{array}{r} 1\ \ 1 \\ 9\,0\,3\,6 \\ +5\,2\,8\,1 \\ \hline 1\,4{,}3\,1\,7 \end{array}$$

7.
$$\begin{array}{r} 1\ \ 1 \\ 5\,4\,3 \\ +6\,2\,8 \\ \hline 1{,}1\,7\,1 \end{array}$$

8.
$$\begin{array}{r} 1\ 1\ 1 \\ 9\,0\,5 \\ 1\,4\,8\,3 \\ +\ 3\,8\,4 \\ \hline 2{,}7\,7\,2 \end{array}$$

9.
$$\begin{array}{r} 1\ 2\ 1\ 1 \\ 6\,7\,4 \\ 8\,9\,0\,5 \\ +\ \ 6\,5\,5 \\ \hline 1\,0{,}2\,3\,4 \end{array}$$

Chapter 1, Unit 2
Part B

1. 8
2. 11
3. 9
4. 9
5. 9
6. 10
7. 15
8. 22
9. 21
10. 23
11. 16
12. 18
13. 19
14. 19
15. 17
16. 15
17. 15
18. 17
19. 14
20. 20
21. 988
22. 897
23. 98
24. 966
25. 739
26. 1,013
27. 853
28. 6,602
29. 9,596
30. 14,350
31. 15,706
32. 110,758
33. 129,287
34. 4,385,409
35. 485,771

36. 1,665,671

Chapter 1, Unit 3
Part A

1. B
2. D
3. F
4. H
5. 6
6. 5
7. 0
8. 1
9. 3
10. 3
11. 7
12. 2
13. 9
14. $1 - 4 = $ ___
15. 5
16. 6
17. 6
18. 312
19. 143
20. 230
21. 124
22. ones
23. 16
24. 419
25. no
26. yes
27. tens

28.
$$\begin{array}{r} 4 \\ 4\,\cancel{5}^{1}1\,2 \\ -2\,0\,4\,2 \\ \hline \end{array}$$

29. yes
30. tens

31.
$$4\,\cancel{6}^{1}6\,3$$
$$-4\,3\,8\,1$$

32.
$$4\,\cancel{6}^{1}6\,3$$
$$-4\,3\,8\,1$$
$$2\ 8\ 2$$

33.
$$4\,7\,\cancel{6}^{1}5$$
$$-1\,5\,3\,8$$
$$3\,2\,2\,7\ =\ 3{,}227$$

34.
$$6\,1\,\cancel{8}^{1}4$$
$$-2\,0\,4\,8$$
$$4\,1\,3\,6\ =\ 4{,}136$$

35.
$$9\,\cancel{5}\,\cancel{1}^{1}6$$
$$-4\,2\,6\,7$$
$$5\,2\,4\,9\ =\ 5{,}249$$

36.
$$\cancel{4}^{1}6\,\cancel{8}^{1}2$$
$$-3\,9\,4\,7$$
$$7\,3\,5$$

36.
$$\cancel{5}^{1}\cancel{0}^{1}\cancel{0}^{1}0$$
$$-2\,0\,4\,3$$
$$2\,9\,5\,7\ =\ 2{,}957$$

Chapter 1, Unit 3
Part B

1. yes
2. no
3. yes
4. yes
5. 3
6. 2
7. 8
8. 2
9. 5
10. 10
11. 7
12. 2
13. 9
14. 4
15. X
16. 5
17. X
18. 3
19. 2
20. X
21. 2
22. 5
23. 7
24. X
25. X
26. 8
27. 4
28. 888
29. X
30. 293
31. X
32. 2,470
33. 1,182
34. 2,903
35. 96
36. 712
37. 7,172
38. 1,745
39. 1,765
40. 3,478
41. 4,564
42. 2,696
43. 2,436
44. X
45. 34,708
46. 37,872
47. 21,554
48. 34,154
49. 30,557

Chapter 1, Unit 4
Part A

1. $599
2. no
3. no
4. yes
5. $23
6. 17 pounds
7. 4,819
8. 56
9. minuend, subtrahend
10. 527
11. yes
12. yes
13. no
14. no
15. yes

Chapter 1, Unit 4
Part B

1. 673
2. 6,715
3. 1,946
4. 17
5. 12
6. 11
7. no
8. addend
9. 305
10. 485
11. 266
12. 238
13. 7 days
14. 18 days
15. 792 pounds
16. $58
17. 133 pages
18. 291 miles
19. 563 acres
20. sum; total
21. 268 lbs.
22. $56,488
23. different
24. 30 credits
25. $447
26. 131 prescriptions
27. 75 days

Chapter 1 Test

1. 9
2. 17
3. 6
4. 1
5. 1,012
6. 1,432
7. 748
8. 1,849
9. 1,093 miles
10. $386
11. 874
12. 1,734

13. 76, 78, 139, 215, 394, 6,032
14. 4
15. 5
16. X
17. 5
18. 1,217
19. 252
20. 7,596
21. 3,308
22. 206 miles
23. 459 acres

Chapter 2, Unit 1
Part A

1. 35
2. 40
3. 5
4. 45, 50
5. 12
6. 2
7. 2, 4, 6, 8, 10, 12, 14, 16
8. 18, 21, 24, 27, 30
9. 4
10. 24
11. 24, 28, 32, 36, 40
12. 2, 4, 6, 8, 10, 12, 14, 16, 18, 20
13. 3, 6, 9, 12, 15, 18, 21, 24, 27, 30
14. 4, 8, 12, 16, 20, 24, 28, 32, 36, 40
15. 5, 10, 15, 20, 25, 30, 35, 40, 45, 50
16. 36, 42, 48, 54, 60
17. 7
18. 18

19. 4
20. 7
21. 3,906
22. 2,608
23. 16,624
24. 12,138

Chapter 2, Unit 1
Part B

1. 0
2. 0
3. 9
4. 28
5. 54
6. 63
7. 48
8. 24
9. 45
10. 36
11. 36
12. 6
13. 7
14. 9
15. 3
16. a. 2, 4, 6, 8, 10, 12, 14, 16, 18, 20
    b. 5, 10, 15, 20, 25, 30, 35, 40, 45, 50
    c. 7, 14, 21, 28, 35, 42, 49, 56, 63, 70
    d. 8, 16, 24, 32, 40, 48, 56, 64, 72, 80
17. 6
18. 9
19. 3

20. 35
21. 36
22. 40
23. 21
24. 36
25. 0
26. 56
27. 28
28. 36
29. 21
30. 32
31. 48
32. 42
33. 10,346
34. 31,752
35. 49,232
36. 18,942
37. 24,680
38. 73,431
39. 27,951
40. 19,756
41. 18,249
42. 43,947

Chapter 2, Unit 2
Part A

1. yes
2. no
3. factors
4. product

5.   874
     x 53
     2622
     4370
   46322 = 46,322

6.   537
     x 39
     4833
     1611
   20943 = 20,943

7.   2031
     x 457
     14217
     10155
     8124
   928167 = 928,167

8.   4915
     x 250
     0000
     24575
     9830
   1228750 =
   1,228,750

9. 930
10. 5,800
11. 32,000
12. 140,000

Chapter 2, Unit 2
Part B

1. 520
2. 210
3. 22,225
4. 36,684
5. 4,212
6. 14,965
7. 2,318,904
8. 1,537,650
9. 29,784
10. 34,086
11. 5,181

12. 14,715
13. 86,014
14. 37,336
15. 232,246
16. 57,133
17. 3,270
18. 78,500
19. 32,150
20. 453,000
21. 672 jars
22. $3,060

Chapter 2, Unit 3
Part A

1. 5
2. 4
3. 4
5. 8
5. 5
6. 0
7. 3
8. 8
9. 21
10. 0
11. 24
12. 30

13. $7 \overline{)6}$

14. $4 \overline{)13}$

15. $6 \overline{)9}$

16. $\begin{array}{r} 1 \text{ R4} \\ 5 \overline{)9} \\ \underline{-5} \\ 4 \end{array}$

17. $\begin{array}{r} 0 \text{ R4} \\ 6 \overline{)4} \\ \underline{-0} \\ 4 \end{array}$

18. $\begin{array}{r} 2 \\ 3 \overline{)6} \\ \underline{-6} \\ 0 \end{array}$

19. $\begin{array}{r} 0 \text{ R 6} \\ 9 \overline{)6} \\ \underline{-0} \\ 6 \end{array}$

20. $\begin{array}{r} 07 \text{ R2} \\ 5 \overline{)37} \\ \underline{-0} \\ 37 \\ \underline{-35} \\ 2 \end{array}$

Chapter 2, Unit 3
Part B

1. 2
2. 2
3. 5
4. 6
5. 1
6. 0
7. 3
8. 3
9. 9
10. 0
11. 6
12. 1
13. 4
14. 6
15. 1 R1

16. 0 R4
17. 0 R1
18. 2
19. 1 R1
20. 64 R3
21. 149 R1
22. 427 R2
23. 328 R5
24. 608 R2
25. 1,307 R2
26. 259 R2

Chapter 2, Unit 4
Part A

1. $\begin{array}{r} X \\ 92 \overline{)43871} \end{array}$

2. 3

3. $\begin{array}{r} X \\ 52 \overline{)98372} \end{array}$

4. 4

5. $\begin{array}{r} X \\ 83 \overline{)72516} \end{array}$

6. 3

7. $\begin{array}{r} 3 \\ 576 \overline{)16739} \\ 1728 \end{array}$

8. $\begin{array}{r} 1673 \\ \underline{-1728} \\ X \end{array}$

9. less

10.
$$576 \overline{)16739} \atop 2$$
$$-\underline{1152}$$

11. yes;
1,673 − 1,152 = 521
521 is less than 576
12. 16,739 ÷ 576 = 29 R35
13. 2
14. 7
15. 4
16. 5
17. 3
18. 267 R17
19. 304 R3
20. 441 R63
21. 6
22. 436
23. 62
24. quotient and divisor;
dividend
25. 306
26. 27
27. division
28. $144
29. 169 mg.
30. 13 days

Chapter 2, Unit 4
Part B

1. 125 R18
2. 497 R44
3. 214 R12
4. 522 R16
5. 337 R7
6. 111 R2
7. 198 R58
8. 264 R94
9. 609 R10
10. 232 R211

11. 186 R108
12. 184 R595
13. 287
14. 571
15. 1,232 oz.
16. 75 tires
17. 789 ornaments
18. 47 necklaces
19. 92 statues
20. 46, 912, 19, 38
21. 507 x 18 = 9,126
9,126 + 17 = 9,143
22. 51
23. $357
24. 1,563 cards

Chapter 2 Test

1. factors
2. product
3. 520
4. 22,225
5. 36,684
6. 4,212
7. 14,965
8. 2,318,904
9. 1,537,650
10. 1,200 jars
11. $5,928
12. 46 (divisor)
2,990 (dividend)
65 (quotient)
13. 903 R3
14. 887 R4
15. 243 R2
16. 186 R5
17. 987 R22
18. 1,574 R34
19. 60
20. 40 customers

Chapter 3, Unit 1
Part A

1. $\frac{1}{2}$

2.

3.

4. A

5.

6.

7. A

8.

9.

10. A

11.

12. no
13. $\frac{3}{4}$
14. no
15. no
16. no
17. $\frac{73}{140}$
18. $\frac{13}{392}$

19. $\frac{24}{35}$
20. equal
21. yes
22. yes
23. yes
24. yes
25. yes
26. no
27. 1, 2, 13, 26
28. 13
29. $\frac{3}{4}$
30. $\frac{2}{3}$

Chapter 3, Unit 1
Part B

1. $\frac{5}{7}$
2. 5
3. 10
4. $\frac{1}{5}$
5. $\frac{5}{6}$
6. $\frac{8}{45}$
7. 1, 5, 14, 653
8. $\frac{3}{8}$
9. $\frac{41}{50}$
10. $\frac{21}{22}$
11. $\frac{2}{3}$
12. $\frac{4}{5}$
13. $\frac{2}{5}$
14. $\frac{3}{10}$
15. $\frac{6}{13}$
16. $\frac{1}{5}$

17. $\frac{2}{5}$
18. $\frac{13}{15}$
19. $\frac{4}{7}$
20. $\frac{6}{11}$
21. $\frac{5}{9}$
22. $\frac{3}{4}$
23. $\frac{3}{4}$
24. yes
25. no
26. no
27. no
28. no
29. no
30. yes
31. no
32. no

Chapter 3, Unit 2
Part A

1. $\frac{27}{36}$
2. yes
3. 9
4. $\frac{5}{10}$
5. yes
6. 3
7. 2
8. 3
9. 20
10. $\frac{15}{20}$
11. $\frac{25}{30}$
12. yes
13. no
14. yes
15. no

16. is
17. no
18. yes
19. sometimes
20. 6
21. 30
22. 10
23. 20
24. 24
25. 15
26. 6
27. 35
28. no
29. yes
30. smaller
31. no
32. yes
33. no
34. 1
35. 3
36. yes
37. no
38. $1\frac{3}{7}$
39. $2\frac{3}{4}$
40. $2\frac{1}{7}$
41. $\frac{24}{8}$
42. 5
43. 7
44. 15
45. $\frac{32}{9}$
46. 15
47. 23

Chapter 3, Unit 2
Part B

1. 20
2. 8

3. 32
4. 11
5. 20
6. 10
7. 10
8. 12
9. 63
10. 18
11. 12
12. 30
13. 45
14. 12
15. $\frac{9}{11}$
16. $\frac{8}{13}$
17. $\frac{4}{15}$
18. $\frac{7}{9}$
19. $\frac{11}{16}$
20. $\frac{2}{9}$
21. $\frac{3}{5}$
22. $\frac{8}{13}$
23. proper
24. improper
25. proper
26. improper
27. $6\frac{2}{5}$
28. $5\frac{7}{8}$
29. $4\frac{3}{4}$
30. $3\frac{1}{9}$
31. $5\frac{5}{6}$
32. $5\frac{1}{8}$
33. $5\frac{5}{12}$

34. 6

35. $9\frac{3}{4}$

36. $9\frac{2}{5}$

37. 1

38. $\frac{35}{8}$

39. $\frac{41}{6}$

40. $\frac{43}{5}$

41. $\frac{59}{8}$

42. $\frac{37}{4}$

43. $\frac{62}{11}$

44. $\frac{75}{8}$

45. $\frac{21}{8}$

46. $\frac{64}{11}$

47. $\frac{19}{4}$

48. $\frac{41}{5}$

49. $\frac{29}{9}$

50. $\frac{45}{7}$

51. $\frac{29}{5}$

52. $\frac{11}{8}$

53. $\frac{18}{7}$

54. $\frac{4}{3}$

**Chapter 3, Unit 3
Part A**

1. 8 chairs
2. no
3. no

4. fourths
5. ninths
6. yes
7. yes
8. yes
9. $\frac{5}{17}$
10. yes
11. $\frac{4}{5}$
12. no
13. no
14. yes
15. $\frac{2}{5}$
16. $\frac{1}{3}$
17. 1
18. $1\frac{1}{5}$
19. $1\frac{2}{7}$
20. $1\frac{3}{5}$
21. yes
22. no
23. b
24. b
25. a
26. b
27. $\frac{19}{24}$
28. $6\frac{5}{7}$
29. $13\frac{13}{17}$
30. $13\frac{1}{2}$
31. $14\frac{9}{10}$
32. $8\frac{4}{9}$
33. $5\frac{19}{24}$

**Chapter 3, Unit 3
Part B**

1. $\frac{6}{11}$
2. $\frac{6}{7}$
3. $\frac{2}{3}$
4. $\frac{13}{17}$
5. $\frac{1}{2}$
6. $\frac{5}{7}$
7. $\frac{3}{4}$
8. $\frac{3}{5}$
9. $\frac{6}{7}$
10. $1\frac{4}{7}$
11. $1\frac{4}{13}$
12. $\frac{5}{8}$
13. $1\frac{1}{2}$
14. $1\frac{1}{5}$
15. $1\frac{2}{3}$
16. $1\frac{3}{5}$
17. $1\frac{1}{2}$
18. $1\frac{2}{13}$
19. $\frac{6}{7}$
20. $\frac{4}{5}$
21. $1\frac{1}{4}$
22. $1\frac{1}{3}$
23. $\frac{11}{15}$

24. $1\frac{1}{5}$
25. $\frac{7}{10}$
26. $\frac{7}{13}$
27. $1\frac{11}{20}$
28. $1\frac{4}{9}$
29. $1\frac{1}{2}$
30. $1\frac{1}{4}$
31. 1
32. $1\frac{2}{3}$
33. $1\frac{4}{15}$
34. $1\frac{1}{10}$
35. $\frac{2}{3}$
36. $1\frac{7}{9}$
37. $\frac{27}{40}$
38. $\frac{17}{21}$
39. $\frac{8}{9}$
40. $\frac{35}{44}$
41. $\frac{58}{65}$
42. $1\frac{3}{8}$
43. $1\frac{3}{8}$
44. $1\frac{16}{45}$
45. $1\frac{7}{24}$
46. $1\frac{1}{12}$
47. $\frac{13}{15}$
48. $1\frac{1}{30}$

49. $\frac{7}{8}$

50. $\frac{29}{35}$

51. $1\frac{5}{18}$

52. $1\frac{21}{44}$

53. $\frac{8}{9}$

54. $1\frac{1}{8}$

55. $1\frac{23}{90}$

56. $1\frac{3}{56}$

57. $\frac{31}{60}$

58. $1\frac{17}{24}$

59. $\frac{29}{35}$

60. $1\frac{71}{80}$

61. $19\frac{7}{13}$

62. $6\frac{8}{11}$

63. $9\frac{3}{4}$

64. $35\frac{3}{4}$

65. $11\frac{1}{7}$

66. $6\frac{2}{11}$

67. $14\frac{2}{5}$

68. $15\frac{1}{4}$

69. 6

70. $11\frac{11}{12}$

71. $3\frac{23}{24}$

72. $5\frac{1}{2}$

73. $10\frac{3}{8}$

74. $8\frac{5}{14}$

75. $7\frac{2}{3}$

76. $13\frac{5}{8}$

77. $3\frac{6}{7}$

78. $11\frac{4}{11}$

79. $7\frac{1}{2}$

80. $11\frac{1}{8}$

Chapter 3, Unit 4
Part A

1. no
2. no
3. yes
4. X
5. $5\frac{2}{11}$
6. $1\frac{3}{5}$
7. $2\frac{57}{88}$
8. $3\frac{2}{7}$
9. $2\frac{1}{4}$
10. $\frac{5}{16}$
11. $5\frac{1}{10}$
12. $5\frac{3}{5}$

Chapter 3, Unit 4
Part B

1. $\frac{3}{5}$
2. $\frac{3}{7}$
3. X

4. X
5. $\frac{4}{15}$
6. $\frac{1}{12}$
7. $\frac{7}{10}$
8. $\frac{1}{4}$
9. $\frac{2}{15}$
10. $\frac{19}{56}$
11. $\frac{21}{40}$
12. X
13. $\frac{5}{12}$
14. X
15. $\frac{5}{18}$
16. $\frac{7}{13}$
17. X
18. $\frac{1}{2}$
19. $\frac{9}{56}$
20. $\frac{11}{24}$
21. $\frac{3}{8}$
22. X
23. $\frac{19}{56}$
24. $\frac{1}{12}$
25. X
26. $\frac{2}{15}$
27. $\frac{29}{70}$
28. $2\frac{1}{9}$

29. $7\frac{1}{3}$
30. $4\frac{3}{11}$
31. $4\frac{15}{16}$
32. $2\frac{2}{5}$
33. $3\frac{1}{4}$
34. $\frac{3}{8}$
35. $1\frac{13}{14}$
36. $4\frac{1}{2}$
37. $3\frac{7}{12}$
38. $\frac{29}{56}$
39. $3\frac{7}{9}$
40. $6\frac{8}{11}$
41. $3\frac{3}{40}$
42. $1\frac{13}{16}$
43. $2\frac{3}{8}$
44. $7\frac{1}{16}$
45. $2\frac{3}{56}$
46. $3\frac{2}{5}$
47. $4\frac{11}{15}$
48. $3\frac{2}{63}$
49. $\frac{1}{2}$
50. $2\frac{3}{4}$
51. $4\frac{13}{18}$
52. $7\frac{11}{24}$

Chapter 3, Unit 5
Part A

1. $\frac{21}{80}$

2. $\frac{45}{88}$

3. $\frac{30}{7} = 4\frac{2}{7}$

4. 2

5. $\frac{1}{4}$

6. $\frac{16}{21}$

7. $\frac{7}{18}$

8. $\frac{5}{11}$

9. $\frac{7}{12}$

10. $\frac{1}{6}$

11. $\frac{2}{3}$

12. $\frac{3}{10}$

13. $\frac{5}{14}$

14. $\frac{1}{14}$

15. $\frac{39}{8} = 4\frac{7}{8}$

16. $\frac{9}{7} = 1\frac{2}{7}$

17. 4

18. $\frac{11}{6} = 1\frac{5}{6}$

19. 12

20. $\frac{15}{8} = 1\frac{7}{8}$

21. 2

Chapter 3, Unit 5
Part B

1. $\frac{4}{35}$

2. $\frac{30}{91}$

3. $\frac{8}{27}$

4. $\frac{9}{32}$

5. $\frac{12}{7} = 1\frac{5}{7}$

6. $\frac{9}{8} = 1\frac{1}{8}$

7. $\frac{21}{4} = 5\frac{1}{4}$

8. $\frac{10}{3} = 3\frac{1}{3}$

9. $\frac{12}{35}$

10. $\frac{15}{8} = 1\frac{7}{8}$

11. $\frac{27}{7} = 3\frac{6}{7}$

12. $\frac{16}{11} = 1\frac{5}{11}$

13. $\frac{8}{45}$

14. $\frac{15}{77}$

15. $\frac{4}{15}$

16. $\frac{18}{35}$

17. $\frac{4}{5}$

18. $\frac{18}{11} = 1\frac{7}{11}$

19. $\frac{8}{45}$

20. $\frac{27}{5} = 5\frac{2}{5}$

21. $\frac{16}{33}$

22. $\frac{16}{9} = 1\frac{7}{9}$

23. $\frac{28}{9} = 3\frac{1}{9}$

24. $\frac{27}{4} = 6\frac{3}{4}$

25. $\frac{20}{63}$

26. $\frac{10}{51}$

27. $\frac{10}{23}$

28. $\frac{7}{20}$

29. $\frac{3}{8}$

30. $\frac{6}{35}$

31. $\frac{3}{10}$

32. $\frac{9}{20}$

33. $\frac{50}{57}$

34. $\frac{10}{33}$

35. $\frac{9}{28}$

36. $\frac{3}{10}$

37. $\frac{8}{21}$

38. $\frac{15}{19}$

39. $\frac{5}{27}$

40. $\frac{5}{24}$

41. $\frac{4}{5}$

42. $\frac{5}{6}$

43. $\frac{2}{3}$

44. $\frac{3}{10}$

45. $\frac{2}{5}$

46. $\frac{15}{28}$

47. $\frac{3}{20}$

48. $\frac{27}{10} = 2\frac{7}{10}$

49. $\frac{4}{3} = 1\frac{1}{3}$

50. $\frac{7}{2} = 3\frac{1}{2}$

51. $\frac{3}{2} = 1\frac{1}{2}$

52. $\frac{3}{4}$

53. $\frac{4}{5}$

54. $\frac{6}{7}$

55. $\frac{4}{5}$

56. $\frac{24}{25}$

57. 1

58. $\frac{9}{4} = 2\frac{1}{4}$

59. $\frac{2}{3}$

60. $\frac{15}{16}$

Chapter 3, Unit 6
Part A

1. $\frac{3}{7}$

2. $\frac{9}{5}$

3. 1

4. 1

5. 1

6. $\frac{32}{45}$

7. $\frac{14}{5} = 2\frac{4}{5}$

8. $\frac{7}{40}$

9. $\frac{24}{5} = 4\frac{4}{5}$

10. $\frac{9}{28}$

11. $\frac{4}{27}$

12. $\frac{1}{2}$

13. $\frac{20}{3} = 6\frac{2}{3}$

14. $\frac{4}{21}$

15. $\frac{25}{52}$

**Chapter 3, Unit 6
Part B**

1. $\frac{21}{10} = 2\frac{1}{10}$

2. $\frac{5}{8}$

3. 10

4. $\frac{25}{6} = 4\frac{1}{6}$

5. $\frac{16}{5} = 3\frac{1}{5}$

6. $\frac{5}{8}$

7. $\frac{3}{23}$

8. $\frac{9}{8} = 1\frac{1}{8}$

9. $\frac{18}{7} = 2\frac{4}{7}$

10. $\frac{2}{5}$

11. $\frac{40}{3} = 13\frac{1}{3}$

12. $\frac{3}{7}$

13. $\frac{24}{7} = 3\frac{3}{7}$

14. 4

15. $\frac{39}{10} = 3\frac{9}{10}$

16. 2

17. 3

18. $\frac{9}{10}$

19. $\frac{9}{4} = 2\frac{1}{4}$

20. $4\frac{6}{11}$

**Chapter 3, Unit 7
Part A**

1. $2\frac{1}{4}$ lbs.
2. no
3. yes
4. multiply
5. $\frac{3}{5}$; 12
6. 20 rings
7. $\frac{7}{10}$; 21
8. 30 pills

**Chapter 3, Unit 7
Part B**

1. $5\frac{5}{6}$ days

2. $4\frac{3}{5}$ grams

3. $3\frac{3}{4}$ lbs.

4. $2\frac{3}{4}$ liters

5. $7\frac{4}{5}$ kg.

6. $4\frac{5}{12}$ hr.

7. $2\frac{3}{4}$ cups

8. $1\frac{3}{8}$

9. $3\frac{17}{20}$ hr.

10. $\frac{27}{2} = 13\frac{1}{2}$

11. 6
12. 15
13. 15 students
14. 20 hours
15. 100 pages

16. 20 ($\frac{1}{4}$ years)
17. 36 ($\frac{1}{3}$ acres)
18. 64 ($\frac{1}{2}$ ounces)
19. 12 ($\frac{1}{4}$ inches)
20. 120 ($\frac{1}{10}$ grams)
21. $\frac{35}{4} = 8\frac{3}{4}$ ($\frac{4}{5}$ liters)
22. 250 bricks
23. 70 blades
24. 20 capsules
25. 30 mortgages
26. difference
27. sum (total)
28. $4\frac{3}{8}$ inches
29. 12
30. $3\frac{1}{4}$ yards
31. 18 ($\frac{1}{3}$ years)
32. $\$20\frac{5}{8}$
33. 17 ($\frac{1}{2}$ quart) bottles
34. $9\frac{4}{5}$ kg.
35. 12 battalions
36. $1\frac{1}{2}$ gm.
37. 100 ($\frac{1}{4}$ ounce) coins
38. $3\frac{87}{100}$ meters
39. 105 blocks
40. $\frac{21}{5} = 4\frac{1}{5}$ tons
41. 16 shares

Chapter 3 Test

1. $\frac{2}{3}$

2. 5

3. 10

4. $\frac{4}{5}$

5. no

6. yes

7. $\frac{4}{5}$

8. $\frac{5}{6}$

9. $2\frac{2}{5}$

10. $9\frac{3}{8}$

11. $\frac{47}{8}$

12. $\frac{49}{13}$

13. 60

14. $\frac{21}{24}$

15. $1\frac{9}{10}$

16. $\frac{17}{21}$

17. $3\frac{9}{28}$

18. $\frac{1}{2}$

19. $\frac{27}{40}$

20. $\frac{3}{8}$

21. $4\frac{1}{4}$ kg.

22. $4\frac{3}{4}$ meters

23. $1\frac{3}{4}$ cups

24. $\frac{6}{7}$

25. $\frac{11}{10} = 1\frac{1}{10}$

26. $\frac{39}{8} = 4\frac{7}{8}$

27. $\frac{5}{6}$

28. 2

29. 40

30. $\frac{77}{12} = 6\frac{5}{12}$

31. $\frac{6}{5}, \frac{2}{3}, \frac{5}{9}$

32. 20

33. $\frac{11}{6} = 1\frac{5}{6}$ liters

34. 19 cups

35. $1,500

Chapter 4, Unit 1
Part A

1. 0.013
2. 0.00372
3. less
4. greater
5. 10.2394
6. 0.2
7. 0.12
8. 0.58
9. 1.28
10. 0.39

Chapter 4, Unit 1
Part B

1. 0.28
2. 0.000003
3. 0.01246
4. 1.52
5. 42.05
6. 5.428
7. 0.038
8. 45.125
9. 0.1
10. 3.7
11. 0.37
12. 12.5
13. 0.0007

14. 230.0681
15. $\frac{1925}{1000}$
16. $\frac{8}{10000}$
17. $\frac{204}{100000}$
18. $\frac{282}{10}$
19. $\frac{9007}{100}$
20. $\frac{900}{10000}$
21. $\frac{28}{1000}$
22. $\frac{108}{10}$
23. $\frac{514}{1000}$
24. 0.98
25. 0.9
26. 0.3
27. 51.6
28. T
29. T
30. F
31. T
32. F
33. T
34. T
35. T
36. 5.32
37. 0.33
38. 1.89
39. 0.09
40. 2.00
41. 0.47
42. 1.05
43. 1.50
44. 12.30
45. 1.89
46. 14.00
47. 0.46
48. 103.01

49. 301.99
50. 0.51
51. 14.0
52. 2.9126
53. 1.1
54. 14.0
55. 0.5
56. 243.5
57. 4.0494

Chapter 4, Unit 2
Part A

1. 0.078
2. 0.1076
3. 1.7
4. 11
5. 3.42
6. 0.04
7. 3.2
8. 0.008
   0.001
   0.007
9. 0.08
   0.02
   0.06
10. $335.31
11. $76,900
12. no

Chapter 4, Unit 2
Part B

1. 20.601
2. 1.055
3. 4.554
4. 2.2053
5. 0.35
6. 0.17
7. 1.07

8. 10.175
9. 0.6252
10. 11.25
11. 0.509
12. 0.1622
13. 7.82
14. 2.0013
15. 65.108
16. 3609.71
17. 1.0
18. 4.8
19. 0.38
20. 0.775
21. 0.021
22. 4.92
23. 13.7
24. 0.0311
25. 14.915
26. 2.625
27. 0.00195
28. 0.1989
29. 0.21
30. 0.018
31. 6.321
32. 0.44
33. 8.0562
34. 0.715
35. 4.7
36. 0.0007
37. 0.0104
38. 152.22
39. 8.0842%
40. 25.388 mg.
41. 15.736 liters
42. $.54745
43. 1.11
44. 0.0037
45. 0.37
46. 53.37
47. 1.83

48. 0.38668
49. 0.0007
50. 11.5
51. 29.52 mg.
52. 2.35 liters
53. yes
54. no
55. 48 minims
56. $83,493.41
57. 82.35 kgm.
58. 4.871 liters
59. 236.68 seconds
60. 8.511%
61. X
62. 27.9 liters
63. 8.2 mg.
64. .11%
65. $.98
66. $6.94
67. 0.0024 mm.
68. 77.9 mg.
69. 32.3 cc.
70. 0.5538
71. 22.44

Chapter 4, Unit 3
Part A

1. 4
2. 3
3. 6
4. 2
5. 3
6. 1
7. 4
8. 0.00207
9. 1.44
10. 5,000
11. $300
12. 60
13. 0.875 of 20,000

14. 0.3 of 300
15. 0.095 of $5,900

Chapter 4, Unit 3
Part B

1. 1.188
2. 0.01554
3. 0.12
4. 0.01827
5. 0.072
6. 0.306
7. 0.000078
8. 0.04875
9. 0.816296
10. 10.8
11. 0.05
12. 8.2
13. 49.7
14. 0.02781
15. 3.9
16. 3.2
17. 3.612
18. 822.0996
19. 48.26
20. 0.53613
21. 509.53
22. 114.576
23. 166.112
24. 789,804.73
25. 9.71
26. 46,310
27. 12,930
28. 0.39
29. 203.98
30. 30,000
31. 750
32. 36.34
33. 0.0776
34. 240
35. 8.137

36. 3,649.368
37. 0.05432
38. 0.51212
39. 2.8
40. 27.8
41. 3,278.4
42. 82.9
43. 516 grams
44. $7,000
45. $2.25
46. $1320
47. 869
48. 0.01702
49. multiplied
50. 700 rifles
51. 16.8 days
52. $3480
53. 1.8 hr.
54. .63 cu. m.
55. $2.46
56. 161.002 liters
57. 160 drams
58. $104
59. $237.50
60. 62.4 kgm
61. $4.50
62. $2.88
63. 0.00225 kgm
64. 2.4
65. 800
66. 1 mg.
67. $2.26
68. 174 rats

Chapter 4, Unit 4
Part A

1. 18.6
2. 1.04
3. .25
4. .1125

5. .625
6. .35
7. 3.49
8. 20.8
9. 20,000
10. .71
11. .59
12. 1.03
13. 1.85
14. 37.33
15. .43
16. .56

Chapter 4, Unit 4
Part B

1. 11.81
2. 4.3
3. 13.5
4. .002
5. .14
6. 2.5
7. .042
8. 1.2
9. .035
10. .035
11. .00016
12. 3.25
13. .3125
14. 4.25
15. .00051
16. .025
17. .2
18. .0025
19. 5.005
20. .0814
21. .59125
22. .69
23. .375
24. 2.625
25. 2,000

26. 21.28
27. 1,050
28. 4.26
29. 204
30. 1.2
31. 512.5
32. 571.2
33. 500
34. .0063
35. 3.06
36. 112.7
37. 2.86
38. 8.868
39. 5.2
40. .222
41. 1.21
42. .03
43. 6.67
44. .03
45. 1.24
46. .03
47. .04
48. 16.93
49. 937.11
50. .24
51. .57
52. .65
53. .16
54. 35 days
55. 203 coins
56. 68 tickets
57. 845 items
58. 80.6 sq. yds.
59. 192 lbs.
60. $84.08
61. 359 miles
62. 14.0 minutes
63. 26 gm/liters
64. $.83 per pkg.
65. 155 lbs.

66. 30.75 lbs.
67. $187.60 per day
68. 904 tickets
69. 1.94 meters
70. $2.57
71. $8.40 sq. yd.

Chapter 4 Test

1. 0.005
2. $\frac{47}{100}$
3. false
4. 1.27
5. 0.101
6. 2.0712
7. 0.002
8. 2.9975
9. $1,101.75
10. 6.917 kgm.
11. 3.98 liters
12. 6.72%
13. 7.52154
14. 59.17
15. .38
16. 25,800 bushels
17. 143 milligrams
18. $3,450
19. 94 cards
20. $46.37

Chapter 5, Unit 1
Part A

1. 6
2. 7 + 9
3. 4
4. 9
5. yes
6. no

7. yes
8. no
9. yes
10. no
11. yes
12. no
13. 5
14. 6
15. 3 + 8
16. 60 ÷ 2

Chapter 5, Unit 1
Part B

1. 23
2. 22
3. 30
4. 25
5. 17
6. 27
7. 20
8. 17
9. 15
10. 14
11. 5
12. 12
13. 2
14. 120
15. 42
16. 54
17. 504
18. 140
19. 25
20. 3
21. 1
22. 4
23. 2
24. 6
25. 15
26. 48

Chapter 5, Unit 2
Part A

1. 20
2. 12
3. 205
4. 12
5. no
6. 12 • 7
7. 11 + 16
8. yes
9. 5 + 4
10. 8 • 5
11. 30
12. 42
13. 3 + 1
14. 8 + [7 • 2]
15. 37
16. 31
17. 7
18. 4
19. 7
20. 26
21. 6 • 3
22. no
23. yes
24. yes
25. 17
26. 96
27. 24
28. 27
29. yes
30. no
31. yes
32. no
33. yes
34. no
35. yes
36. no
37. no
38. no

Chapter 5, Unit 2
Part B

1. 20
2. 25
3. 62
4. 28
5. 41
6. 26
7. 30
8. 19
9. 64
10. 90
11. 32
12. 70
13. 46
14. 23
15. 24
16. 23
17. 26
18. 44
19. 31
20. 66
21. 55
22. 30
23. yes
24. yes
25. no
26. no
27. yes
28. yes
29. no
30. no
31. no
32. no
33. yes
34. yes
35. yes

Chapter 5, Unit 3
Part A

1. open
2. numerical
3. number
4. 10 + 6
5. 15
6. 21
7. 35
8. 12
9. 16 + 5z
10. (w + 5) • w
11. any
12. 28
13. 70
14. 9
15. 50
16. 39
17. yes, 16
18. no
19. no
20. yes, 60

Chapter 5, Unit 3
Part B

1. open expression
2. open
3. numerical
4. open expression
5. numerical expression
6. counting
7. yes
8. yes
9. 11
10. 15
11. 32
12. 18

13. 21
14. 24
15. 81
16. 7
17. 46
18. 13
19. 34
20. 38
21. 45
22. 49
23. 37
24. 51
25. 22
26. 29
27. 24
28. 27
29. 31
30. 26
31. 27
32. 33
33. 31
34. 48
35. 19
36. 70
37. 19
38. 24
39. 18
40. 35
41. 19
42. 33
43. 27
44. 31
45. yes, 14
46. yes, 16
47. yes, 33
48. yes, 21
49. yes, 7

Chapter 5, Unit 4
Part A

1. yes
2. yes
3. 93 + x
4. 3 + a
5. y + 11
6. 3 + 17x
7. 9a + 3x
8. yes
9. Associative
10. x + (7 + p)
11. 5x + (3 + 2b)
12. yes
13. x + 0

Chapter 5, Unit 4
Part B

1. (5x + 7) + 4
2. 7 + (3 + 9b)
3. 7 + x
4. 7x + (2x + 9)
5. Commutative
6. Commutative
7. Commutative
8. Commutative
9. Commutative
10. Associative
11. Associative
12. y + x
13. x + (y + z)
14. 8x
15. (2x + 3) + 0
16. 5x + 0
17. (3x + 4)
18. b + 22
19. x + 18
20. x + 11

21. g + 17
22. r + 22
23. d + 13
24. a + 12
25. r + 12
26. a + 13
27. a + 17
28. c + 12
29. b + 38
30. 3h + 24
31. x + 12
32. y + 8
33. 3x + 11
34. 2x + 12
35. 2x + 12
36. x + 21
37. 2x + 13
38. y + 20
39. 3x + 13
40. 3x + 13

Chapter 5, Unit 5
Part A

1. yes
2. yes
3. 93x
4. 3a
5. y • 11
6. 3 • 17x
7. 9a • 3x
8. yes
9. Associative
10. x • (7 • p)
11. 5x • (3 • 2b)
12. yes
13. 1x

Chapter 5, Unit 5
Part B

1. (5x • 7) • 4
2. 7 • (3 • 9b)
3. 7x
4. 7x • (2x • 9)
5. Commutative
6. Commutative
7. Commutative
8. Commutative
9. Commutative
10. Associative
11. Associative
12. yx
13. x • (y • z)
14. 8x + 1x
15. 2x + 3
16. 1x + 5x
17. 1 • 8 + 1(3x + 4)
    or
    1[8 + (3x + 4)]
18. 105b
19. 72x
20. 28x
21. 72g
22. 85r
23. 40d
24. 32c
25. 345b
26. 384h
27. 32x
28. 15y
29. 84x
30. 45x
31. 24r
32. 48x
33. 45x

Chapter 5, Unit 6
Part A

1. yes
2. yes
3. 7 • 8 + 7 • 5
4. 6(8 + 3)
5. 15z
6. 6a
7. 9x + 2
8. 12r + 4
9. 9x + 12
10. 10x + 4
11. 3y + 15
12. 7x + 21
13. 20x + 15
14. x + 7
15. 17x + 13
16. 21x + 11

Chapter 5, Unit 6
Part B

1. 14 + 7y
2. 10m + 20
3. 7y + 63
4. 3 + 3c
5. 36 + 6x
6. 24 + 8a
7. 5 + y
8. 6m + 30
9. 18 + 9e
10. 42 + 6r
11. 3z + 18
12. 6 + 14x
13. 12x + 8
14. 4x + 7
15. 54x + 24
16. 19a
17. 31x

18. 5x
19. 13r
20. 23x
21. 2x
22. 12m
23. 11x + 13
24. 8x + 11
25. 18x + 30
26. 44x + 37
27. 8a + 7
28. 8x + 7
29. 25a
30. 17m
31. 23x
32. 5a + 7
33. 15y
34. 27t
35. 7x + 14
36. 3x + 13
37. 9x + 20
38. 5x + 19
39. 13t + 7
40. 16t + 3
41. 12x + 15
42. 8x + 17
43. 8x + 11
44. 5x + 9
45. 9x + 7
46. 6x + 16
47. 13x + 7
48. 7x + 10
49. 17x + 31
50. 23x + 11
51. 18x + 71
52. 11x + 26
53. 19x + 15
54. 5x + 17
55. x + 7
56. 10x + 5
57. 5x + 30

58. 5x + 15
59. 11x + 5

Chapter 5, Unit 7
Part A

1. equation
2. equation
3. true
4. false
5. true
6. yes
7. 4
8. 5
9. 4
10. 3
11. 4
12. 2
13. 3
14. 6
15. 6
16. 3

Chapter 5, Unit 7
Part B

1. 16
2. 11
3. 13
4. 6
5. 4
6. 5
7. 38
8. 7
9. 15
10. 9
11. 8
12. 9
13. 55
14. 175

15. 107
16. 8
17. 8
18. 16
19. 13
20. 3
21. 6
22. 3
23. 8
24. 5
25. 15
26. 8
27. 4
28. 3
29. 2
30. 3
31. 5
32. 1
33. 4
34. 9
35. 13
36. 100
37. 83
38. 4
39. 9

Chapter 5, Unit 8
Part A

1. s + 90,000
2. v + 87
3. 803 + r
4. e + 37
5. r − 17
6. e − 20
7. m − 215
8. e − 305
9. 2w
10. 6s

11. 4s
12. 3t

Chapter 5, Unit 8
Part B

1. n + 5
2. n − 9
3. e − 13
4. 5n
5. 10 − s
6. 2s
7. 11 + s
8. f + 4
9. s − 45
10. s + 5
11. 6s
12. b + 40
13. b − 40
14. d − 11,000
15. w + 42
16. 5c
17. 4a
18. n + 150
19. s − 300
20. 4c
21. s + 8
22. e − 50

Chapter 5 Test

1. 10 + 3
2. 11 • 9
3. 6 • 8
4. 4 + 24
5. 26
6. 72
7. 27
8. 24
9. x + 15

10. 35x
11. 16y + 6
12. 21 + 14x
13. 14x + 5
14. 13x + 34
15. 9
16. 7
17. 7
18. 8
19. 3
20. 6
21. n − 65
22. s + 110

Chapter 6, Unit 1
Part A

1. 27
2. 9
3. yes
4. negative
5. yes
6. yes
7. 0
8. -21
9. -8
10. 15
11. yes
12. yes
13. yes
14. yes
15. yes
16. -61
17. -39
18. 29
19. 14
20. -56
21. 0
22. -17

23.	47	17.	8	11.	9	3.	-7 • 5
24.	-84	18.	11	12.	21	4.	-9 • 3
25.	0	19.	3	13.	1	5.	-24; -24
26.	integers	20.	8	14.	31	6.	16; 16
27.	5	21.	-5	15.	-12	7.	0; 0
28.	-13	22.	-11			8.	0; 0

Chapter 6, Unit 2
Part B

29.	0	23.	8			9.	0; 0
30.	negative integers	24.	-5			10.	0; 0
31.	positive	25.	5	1.	14	11.	positive
32.	negative	26.	3	2.	-30	12.	negative
33.	-18	27.	-3	3.	0	13.	positive
34.	-29	28.	-7	4.	-2	14.	0
35.	6	29.	6	5.	18	15.	72
36.	-4	30.	33	6.	5	16.	-32
37.	-2	31.	4	7.	-28	17.	-15
38.	7	32.	-9	8.	-5	18.	0
39.	-15	33.	-15	9.	-4	19.	40
40.	3	34.	-12	10.	2	20.	120
41.	-6	35.	-7	11.	-1	21.	27
42.	-8	36.	-5	12.	1	22.	21
		37.	-21	13.	-4	23.	-24

Chapter 6, Unit 1
Part B

		38.	20	14.	-14	24.	60
		39.	-13	15.	-11	25.	-24
		40.	4	16.	-6		
1.	24	41.	0	17.	10		
2.	-15	42.	-8	18.	-17		

Chapter 6, Unit 3
Part B

3.	-32			19.	-7		
4.	17			20.	-18	1.	-45
5.	5	**Chapter 6, Unit 2**		21.	-12	2.	-60
6.	-7	**Part A**		22.	9	3.	0
7.	5			23.	11	4.	72
8.	8	1.	12 + (-5)	24.	-6	5.	-85
9.	-5	2.	3 + (-5)	25.	3	6.	-18
10.	-12	3.	-8 + (-3)			7.	-56
11.	-31	4.	12 + 4			8.	0
12.	2	5.	9 + (-12); -3			9.	-32
13.	-16	6.	10 + 5; 15	**Chapter 6, Unit 3**		10.	8
14.	49	7.	-12 + 4; -8	**Part A**		11.	-84
15.	6	8.	-4 + 10; 6			12.	-30
16.	7	9.	4	1.	-24	13.	24
		10.	8	2.	-28		

14. 32
15. -7
16. 6
17. 0
18. -71
19. 24
20. 35
21. -27
22. 24
23. 84
24. 54
25. 28
26. -28
27. -46
28. -4
29. 0
30. -20

Chapter 6, Unit 4
Part A

1. $13 \div 6$
2. $9 \cdot 3$
3. $15 \div -3$
4. any grouping and order
5. $10 - 3$
6. $-6 + 13$ or $-2 - 7$
7. $3 + 8$
8. $15 \cdot 3$
9. 8
10. -5
11. -17
12. 12
13. grouping symbol(s)
14. mult/division
15. positive
16. negative
17. 0
18. undefined

Chapter 6, Unit 4
Part B

1. -2
2. 7
3. 3
4. -4
5. 0
6. $\frac{-2}{3}$
7. $\frac{8}{5}$
8. undefined
9. $\frac{9}{2}$
10. $\frac{-4}{3}$
11. 14
12. 21
13. 62
14. 5
15. -25
16. 16
17. 18
18. -13
19. -27
20. 8
21. 6
22. 36
23. -19
24. 21
25. 7
26. -19
27. 42
28. -10
29. 1
30. -17
31. -4
32. 2
33. -72
34. 35
35. 13

36. -16

Chapter 6, Unit 5
Part A

1. -17
2. -20
3. 11
4. $-8 + x$
5. $-3x + 5$
6. $x + (-7 + 4)$
7. $(-5 + x) - 3$
8. $z + 4$
9. $x + 2$
10. $5x$
11. $3x$
12. -17
13. $-5x$
14. $-7y$
15. $8 \cdot (4z)$
16. $(5 \cdot -3)x$
17. $(7 \cdot 2)x$
18. $-20x$
19. $12x$
20. yes
21. yes
22. yes
23. yes
24. yes

Chapter 6, Unit 5
Part B

1. -7
2. 13
3. 4
4. 7
5. 7
6. 4
7. -22

8. 13
9. 5
10. 8
11. -20
12. -6
13. $8x + 4$
14. $3x + 4$
15. $3x + 2$
16. $4x$
17. $-3x + 12$
18. $x + 13$
19. $8x + 5$
20. $7x + 5$
21. $5x - 6$
22. $2x + 2$
23. $-3x - 5$
24. $3x$
25. $-18x$
26. $24x$
27. $6b$
28. $-15r$
29. $63x$
30. $x - 2$
31. $-18x$

Chapter 6, Unit 6
Part A

1. yes
2. $4x$
3. $13x$
4. $-3x + 24$
5. $-4x + 7$
6. 1
7. -1
8. $x - 24$
9. $15x + 2$
10. $-9x - 7$

Chapter 6, Unit 6
Part B

1. -2x
2. 6x
3. -9x
4. 3x
5. 0
6. 4x
7. -10x
8. -2x + 3
9. 32x − 12
10. 6x − 21
11. -45 + 18x
12. -x − 7
13. -3x
14. -3 + 5x
15. -6x − 4
16. -4x + 12
17. -5x + 7
18. -x + 14
19. x − 2
20. 5x + 8
21. -7x − 9
22. 2x + 2
23. 4x − 4
24. 19 − 3x
25. -5x + 2
26. -4x − 10
27. x − 9
28. 5x − 5
29. 2x − 3
30. 4x − 9

Chapter 6, Unit 7
Part A

1. yes
2. no
3. +7

4. -15
5. x = 4
6. x = 4
7. x = 3
8. x = 5
9. -2
10. 3
11. 2
12. -6
13. 3
14. -6
15. -35 + 13 = -20 − 2
    is true.
16. -18 + 5 = -6 − 9
    is false.

Chapter 6, Unit 7
Part B

1. 3
2. 3
3. -1
4. -2
5. -4
6. -3
7. 2
8. 5
9. 0
10. 4
11. -1
12. -6
13. 3
14. -2
15. -5
16. -3
17. -4
18. 3
19. 2
20. 3
21. 7

22. 78
23. 4
24. -47
25. -9
26. 0
27. 6
28. -2
29. -2, 7 = 3 + 4 is true
30. 1, 19 = 12 + 7 is true
31. -10, -10 + 7 = -3 is true
32. -8, 5 • -8 = -40 is true
33. 8, 24 − 6 = 18 is true

Chapter 6, Unit 8
Part A

1. 6
2. -3
3. 6
4. 10
5. -7
6. 3
7. -8
8. -3

Chapter 6, Unit 8
Part B

1. -10
2. 2
3. -2
4. -4
5. 6
6. 4
7. 3
8. 0
9. -3
10. 0
11. 1
12. 2

13. 5

14. 3

15. 2

16. 1

17. -1

18. 15

19. -10

20. 4

21. 5

22. -4

23. 8

24. 2

25. 0

26. 9

27. 2

28. -5

29. 11

30. -3

31. -13, -26 − (-13 − 4) = -9
    is true

32. 0, 5 + 2(3 − 0) = 11 is true

33. 3, 24 − 3(4 + 6) = -6 is true

34. -3, 4 + 9 = 10 + 3 is true

35. -1, 5(-2 + 4) − 17 = -7
    is true

Chapter 6, Unit 9
Part A

1. 8b − 7

2. 11h + 3

3. 2m + 18

4. 12w − 107

5. n − 9 = 12

6. 13 + n = 39

7. 2n = 20

8. 5n + 6 = 41

9. s + (s + 18) = 64

10. s + (s − 300) = 10,500

11. (3s − 13) − s = 55

12. f + 3f = 18,000

Chapter 6, Unit 9
Part B

1. 8t − 800

2. 2w + 4

3. 9c + 18

4. 13p − 100

5. 2s + 45

6. n + 8 = 21

7. n − 9 = 45

8. n + 23 = 98

9. n − 8 = 32

10. 3n − 7 = 47

11. 67 = 14 + 2n

12. 9 − n = 37

13. n + (n + 5) = 93

14. n + 4n = 130

15. 14n − 7 = 169

16. (2n + 3) + n = 105

17. z + (z + 14) = 64

18. n + (n − 300) = 850

19. (3 + 8s) − s = 52

20. t + (t − 7) = 334

21. (2h + 50) − h = 150

22. (3s + 2) + s = 93

23. (3b − 20) + b = 765

24. n + 12 = 69

25. 54 = 6n

26. n − 8 = 23

Chapter 6 Test

1. 11

2. -20

3. -27

4. 0

5. 45

6. 24

7. -70

8. -5

9. 20

10. -14

11. 3

12. -6

13. x − 3

14. -24x

15. -10x

16. -4 + 7x or 7x − 4

17. 10x − 35

18. 10 − 3x or -3x + 10

19. -x − 20

20. 9x + 6

21. -6

22. 18

23. -9

24. 2

25. 5

26. 2

27. 1

28. 2

29. 3

30. 7

31. 3d + 42

32. 62

Chapter 7, Unit 1
Part A

1. yes

2. no

3. $\frac{9}{1}$

4. $\frac{-15}{1}$

5. yes

6. 2

7. no

8. yes

9. yes

10. yes

11. yes

12. yes

13. $\frac{-4}{9}$

14. $\frac{9}{8}$

15. $\frac{7}{15}$

16. $\frac{-9}{16}$

17. $\frac{3}{8}$

18. $\frac{-8}{21}$

19. $\frac{8}{35}$

20. $\frac{27}{8}$

21. $\frac{-24}{5}$

22. 0

23. $\frac{-3}{4}$

24. $\frac{-4}{11}$

25. $\frac{-8}{3}$

26. $\frac{2}{5}$

27. $\frac{2}{7}$

28. $\frac{1}{4}$

29. $\frac{-5}{8}$

30. $\frac{1}{6}$

31. -2

32. $\frac{9}{10}$

33. $\frac{-1}{5}$

34. $\frac{15}{28}$

35. $\frac{-6}{35}$

36. $\frac{2}{11}$

37. $\frac{-3}{14}$

38. $\frac{-3}{5}$

39. $\frac{3}{7}$

40. $\frac{-11}{9}$

41. $\frac{-5}{7}$

42. $\frac{5}{16}$

43. $\frac{2}{3}$

44. $\frac{6}{7}$

45. $\frac{2}{3}$

46. 4

47. 1

**Chapter 7, Unit 1
Part B**

1. yes

2. yes

3. yes

4. yes

5. no

6. yes

7. no

8. $\frac{3}{5}$

9. 3

10. $\frac{5}{3}$

11. $\frac{2}{5}$

12. $\frac{-3}{2}$

13. $\frac{-1}{5}$

14. $\frac{1}{4}$

15. $\frac{-1}{3}$

16. $\frac{-6}{7}$

17. $\frac{3}{4}$

18. $\frac{10}{7}$

19. $\frac{-8}{5}$

20. $\frac{12}{17}$

21. $\frac{-5}{13}$

22. $\frac{9}{16}$

23. $\frac{27}{8}$

24. $\frac{-5}{9}$

25. -2

26. 9

27. $\frac{3}{5}$

28. $\frac{-4}{3}$

29. 5

30. $\frac{-3}{4}$

31. 4

32. -2

33. $\frac{3}{10}$

34. $\frac{-1}{5}$

35. $\frac{15}{28}$

36. $\frac{2}{35}$

37. $\frac{-2}{35}$

38. $\frac{-14}{45}$

39. $\frac{8}{7}$

40. $\frac{-65}{14}$

41. $\frac{-56}{15}$

42. $\frac{-12}{35}$

43. 0

44. $\frac{27}{14}$

45. $\frac{3}{10}$

46. $\frac{-9}{28}$

47. $\frac{-3}{4}$

48. 0

49. $\frac{35}{12}$

50. $\frac{-4}{9}$

51. $\frac{-27}{20}$

52. 0

53. 0

54. $\frac{10}{21}$

55. $\frac{10}{27}$

56. $\frac{-12}{35}$

57. 0

58. $\frac{28}{15}$

59. $\frac{-18}{7}$

60. $\frac{-8}{17}$

61. $\frac{-9}{7}$

62. $\frac{-11}{3}$

63. $\frac{8}{15}$

64. $\frac{8}{3}$

65. $\frac{-11}{5}$

66. $\frac{6}{7}$

67. $\frac{-10}{19}$

68. $\frac{7}{9}$

69. $\frac{14}{11}$

70. $\frac{-6}{11}$

71. $\frac{7}{4}$

72. $\frac{7}{11}$

73. $\frac{-7}{13}$

74. $\frac{5}{14}$

75. $\frac{-9}{16}$

Chapter 7, Unit 2
Part A

1. $\frac{3}{8} + \frac{-2}{5}$

2. $\frac{-5}{7} + \frac{3}{8}$

3. $\frac{-31}{24}$

4. $\frac{-41}{21}$

5. $\frac{29}{35}$

6. $\frac{23}{4}$

7. $\frac{-7}{20}$

8. $\frac{-29}{12}$

9. $\frac{1}{2}$

10. $\frac{5}{7}$

11. $\frac{-5}{4}$

12. $\frac{1}{2}$

13. $\frac{3}{7}$

14. $\frac{-17}{24}$

15. $\frac{13}{30}$

16. $\frac{34}{35}$

17. $\frac{-49}{30}$

18. $\frac{1}{60}$

19. $\frac{13}{30}$

20. $\frac{-32}{105}$

Chapter 7, Unit 2
Part B

1. $\frac{31}{21}$

2. $\frac{38}{45}$

3. $\frac{23}{7}$

4. $\frac{19}{24}$

5. $\frac{31}{30}$

6. $\frac{13}{5}$

7. $\frac{34}{35}$

8. $\frac{9}{10}$

9. $\frac{19}{30}$

10. $\frac{-8}{45}$

11. $\frac{-59}{56}$

12. $\frac{-38}{35}$

13. $\frac{-17}{6}$

14. $\frac{-43}{8}$

15. $\frac{-13}{20}$

16. $\frac{29}{36}$

17. $\frac{19}{60}$

18. $\frac{-6}{5}$

19. $\frac{-21}{20}$

20. $\frac{-29}{14}$

21. $\frac{29}{12}$

22. $\frac{-2}{3}$

23. $\frac{39}{28}$

24. $\frac{7}{15}$

25. $\frac{-11}{4}$

26. $\frac{-5}{4}$

27. $\frac{5}{6}$

28. 0

29. 0

30. $\frac{11}{21}$

31. $\frac{-23}{30}$

32. $\frac{-19}{40}$

33. $\frac{-29}{33}$

34. $\frac{31}{24}$

35. $\frac{-1}{24}$

36. $\frac{-19}{15}$

37. $\frac{2}{15}$

38. $\frac{13}{20}$

39. $\frac{-1}{14}$

40. $\frac{23}{40}$

41. $\frac{33}{20}$

42. $\frac{-34}{15}$

43. $\frac{1}{28}$

44. $\frac{-35}{24}$

45. $\frac{-23}{12}$

46. $\frac{11}{60}$

47. $\frac{-1}{12}$

48. $\frac{31}{6}$

49. $\frac{8}{15}$

50. $\frac{7}{30}$

51. $\frac{-1}{42}$

52. $\frac{41}{28}$

53. $\frac{35}{24}$

54. $\frac{-7}{20}$

55. $\frac{61}{24}$

Chapter 7, Unit 3
Part A

1. $\frac{-7}{2}$

2. $\frac{-3}{4}$

3. $\frac{1}{2}$

4. $\frac{-1}{4} + x$

5. $-3x + \frac{5}{9}$

6. $x + (\frac{-2}{3} + \frac{3}{7})$

7. $(\frac{8}{3} + x) - \frac{1}{5}$

8. $z + \frac{1}{20}$

9. $x + \frac{44}{35}$

10. $\frac{1}{4}x$

11. $\frac{-3}{10}x$

12. $\frac{-4}{9}$

13. $\frac{-3}{4}x$

Chapter 7, Unit 3
Part B

1. $\frac{17}{10}$

2.   0

3. $\frac{25}{8}$

4. $\frac{-1}{21}$

5. $\frac{-5}{3}$

6. $\frac{-19}{14}$

7. $\frac{11}{12}$

8. $\frac{31}{5}$

9. $\frac{19}{12}$

10. $\frac{31}{20}$

11. $\frac{7}{12}$

12. $\frac{17}{12}$

13. $8x + \frac{3}{8}$

14. $3x - \frac{1}{9}$

15. $\frac{1}{5}x$

16. $4x - \frac{2}{5}$

17. $\frac{61}{60} - 3x$

18. $x + \frac{47}{24}$

19. $\frac{4}{9}x + \frac{11}{36}$

20. $\frac{6}{5}x - \frac{4}{15}$

21. $\frac{1}{4}x - \frac{1}{2}$

22. $2x - \frac{11}{18}$

23. $-3x + \frac{13}{30}$

24. $3x - \frac{2}{15}$

25. $x + \frac{61}{40}$

Chapter 7, Unit 4
Part A

1. $\frac{5}{9}y$

2. $\frac{4}{3}(\frac{-1}{6}z)$

3. $(\frac{5}{13} \cdot \frac{-7}{10})x$

4. $(\frac{3}{5} \cdot \frac{5}{3})x$

5. $\frac{1}{3}x$

6. $\frac{-15}{28}x$

7.   yes

8.   yes

9.   yes

10. yes

11. yes

12. yes

13. no

14. $\frac{12}{7}$

15. $\frac{-10}{3}$

Chapter 7, Unit 4
Part B

1. $\frac{6}{35}x$

2. $5x$

3. $9x$

4. $-x$

5. $\frac{9}{14}x$

6. x

7. $\frac{-12}{7}x$

8. $3x$

9. $\frac{6}{35}x$

10. x

11. x

12. x

13. x

14. x

15. x

16. x

17. x

18. x

19. $3x$

20. $5x$

21. $\frac{-5}{12}x$

22. $\frac{-11}{18}x$

23. $\frac{4}{7}b$

24. $\frac{-8}{5}r$

25. $\frac{20}{21}x$

26. x

Chapter 7, Unit 5
Part A

1.   yes, -10

2. $8(y - \frac{3}{8})$

3. $3x - \frac{5}{6}x$

4. $-5(w + \frac{1}{3})$

5. $12y$

6. $\frac{27}{40}x$

7. $\frac{5}{6}x + 11$

8. $\frac{5}{3}x + 5$

9. $9x + \frac{17}{42}$

10. $\frac{1}{12} + \frac{59}{120}x$

11. $12x + 2$

12. $5x - 4$

13. $7x - 14$

14. $7x + 1$

Chapter 7, Unit 5
Part B

1. $8x$

2. $13x$

3. $6x$

4. $2x$

5. $-8x$

6. $-7x$

7. $\frac{7}{6}x$

8. $\frac{5}{12}x$

9. $\frac{2}{3}y$

10. $-3x + 10$

11. $2x - \frac{3}{4}$

12. $-14x - 10$

13. $8 - \frac{17}{5}x$

14. $35 - 14x$

15. $4 - 2x$

16. $-3x + 4$

17. $-12x - 3$

18. $-30 + 6x$

19. $-20 + 4x$

20. $10 - 6x$

21. $-7 + x$

22. $-32 + 16x$

23. $6 - 9x$

24. $x - 10$

25. $7x - 10$

26. $10x - 8$

27. $3x - 3$

28. $-3x + 1$

29. $2x - 12$

30. $11x$

31. $-5x - 16$

32. $16x - 19$

33. $\frac{11}{15}x + 11$

34. $x + \frac{11}{3}$

35. $4x - \frac{27}{4}$

36. $\frac{10}{x}$

37. $\frac{-3}{x}$

Chapter 7, Unit 6
Part A

1. 8

2. $\frac{1}{8}$

3. $\frac{3}{4}$

4. $\frac{4}{3}$

5. -3

6. $\frac{-1}{3}$

7. $\frac{-7}{5}$

8. $\frac{-5}{7}$

9. $\frac{-11}{9}$

10. $\frac{-3}{2}$

11. $\frac{19}{5}$

12. $\frac{-8}{3}$

Chapter 7, Unit 6
Part B

1. 16

2. 6

3. $\frac{4}{3}$

4. 9

5. $\frac{-5}{2}$

6. $\frac{15}{2}$

7. $\frac{-13}{5}$

8. $\frac{-5}{3}$

9. $\frac{35}{6}$

10. $\frac{-2}{3}$

11. $\frac{-7}{4}$

12. $\frac{-8}{3}$

13. $\frac{7}{9}$

14. $\frac{7}{5}$

15. $\frac{15}{4}$

16. $\frac{-21}{5}$

17. $\frac{17}{3}$

18. 32

19. -15

20. -20

21. 18

22. $\frac{-2}{3}$

23. $\frac{10}{3}$

24. $\frac{-7}{11}$

25. $\frac{23}{4}$

26. $\frac{8}{3}$

27. $\frac{-10}{7}$

28. $\frac{2}{3}$

29. $\frac{1}{2}$

30. 4

31. $\frac{19}{5}$

32. $\frac{1}{9}$

33. $\frac{17}{6}$

34. $\frac{20}{3}$

35. 1

36. $\frac{23}{16}$

37. 3

38. -9

Chapter 7, Unit 7
Part A

1. 6

2. $\frac{2}{5}$

3. 40

4. $\frac{25}{16}$

5. 60

6. $\frac{52}{45}$

7. 12

8. $\frac{3}{16}$

9. $7x$

10. 3

11. $4x$

12. 2

13. $6x$

14. $\frac{21}{4}$

15. $5x$

16. -10

Chapter 7, Unit 7
Part B

1. $\frac{23}{40}$

2. $\frac{9}{40}$

3. 8

4. 4

5. $\frac{29}{8}$

6. $\frac{-6}{7}$

7. $\frac{1}{10}$

8. $\frac{1}{4}$

9. $\frac{14}{9}$

10. $\frac{1}{8}$

11. $\frac{7}{2}$

12. -1

13. 1

14. $\frac{10}{7}$

15. 12

16. $\frac{5}{28}$

17. $\frac{11}{14}$

18. $\frac{-5}{11}$

19. -2

20. -6

21. 4

22. 40

23. 4

24. 8

25. $\frac{19}{4}$

26. $\frac{9}{4}$

27. $\frac{-27}{14}$

28. -6

29. 14

30. -1

31. $\frac{-9}{8}$, $\frac{-3}{4} + \frac{5}{6} = \frac{1}{12}$ is true

32. $\frac{2}{5}$, $\frac{3}{10} - \frac{1}{5} = \frac{1}{10}$ is true

33. 15, $10 - \frac{3}{8} = \frac{15}{2} + \frac{17}{8}$

    is true

34. 10, $\frac{1}{2} + \frac{1}{2} = \frac{3}{10} + \frac{7}{10}$

    is true

35. $\frac{4}{3}$, $\frac{4}{4} + \frac{1}{2} = 2 \cdot \frac{3}{4}$ is true

Chapter 7 Test

1. $\frac{-10}{7}$

2. $\frac{6}{25}$

3. 0

4. $\frac{16}{9}$

5. $\frac{16}{21}$

6. $\frac{-3}{10}$

7. $\frac{-33}{28}$

8. $\frac{19}{15}$

9. $\frac{19}{36}$

10. $\frac{13}{9}$

11. $\frac{2}{3}x + 2$

12. $\frac{1}{12}x$

13. $\frac{11}{x}$

14. $\frac{-2}{x}$

15. $\frac{-1}{5}$

16. $\frac{23}{4}$

17. $\frac{-19}{6}$

18. -12

19. $\frac{-2}{3}$

20. $\frac{7}{8}$

21. $\frac{10}{21}$

22. $\frac{-2}{25}$

23. $\frac{-5}{7}$

24. $\frac{-3}{2}$

25. $\frac{7}{5}$

26. $\frac{1}{5}$

27. $\frac{2}{7}$

Chapter 8, Unit 1
Part A

1. $\frac{7}{6}$

2. $\frac{8}{7}$

3. no

4. 6 to 9

5. 9 to 23

6. $\frac{5}{4}$

7. $\frac{30}{1}$

8. $\frac{76}{25}$

9. $\frac{16}{5}$

10. $\frac{681}{2000}$

11. no

12. yes

13. no

14. yes

15. yes

16. no

17. 48; 48

18. 12 and 1

19. 18 and 5

20. $8 \cdot n = 13 \cdot 9$

21. $47 \cdot 33 = n \cdot 28$

22. $2 \cdot 18 = 9 \cdot n$

23. $7 \cdot n = 8 \cdot 14$

Chapter 8, Unit 1
Part B

1. $\frac{10}{32} = \frac{5}{16}$

2. $\frac{28}{18} = \frac{14}{9}$

3. $\frac{2.7}{0.8} = \frac{27}{8}$

4. $\frac{18}{7.8} = \frac{30}{13}$

5. $\frac{1}{2}$

6. $\frac{7}{2}$

7. $\frac{4}{1}$

8. $\frac{1}{2}$

9. $\frac{3}{50}$

10. $\frac{15}{1}$

11. $\frac{9}{11}$

12. $\frac{261}{200}$

13. $\frac{19}{45}$

14. yes

15. $\frac{6}{13}$

16. $\frac{13}{70}$

17. $\frac{8}{3}$

18. $\frac{4}{5}$

19. $\frac{120}{1}$

20. $\frac{9}{10}$

21. $\frac{315}{1}$

22. $\frac{34}{5}$

23. $\frac{40300}{551}$

24. $\frac{29}{1150}$

25. $\frac{47560}{23}$

26. yes

27. no

28. yes

29. no

30. $n \cdot 16 = 8 \cdot 2$

31. $37 \cdot 10 = 18 \cdot n$

32. $16 \cdot n = 20 \cdot 5$

Chapter 8, Unit 2
Part A

1. 14

2. $\frac{27}{4}$ or $6\frac{3}{4}$

3. $\frac{7}{3}$ or $2\frac{1}{3}$

4. $\frac{54}{11}$ or $4\frac{10}{11}$

5. $8 \cdot 13 = n \cdot 10$

6. $7 \cdot n = 11 \cdot 5$

7. $\frac{24}{5}$ or $4\frac{4}{5}$

8. $\frac{17}{5}$ or $3\frac{2}{5}$

Chapter 8, Unit 2
Part B

1. $\frac{33}{8}$ or $4\frac{1}{8}$

2. $\frac{28}{3}$ or $9\frac{1}{3}$

3. $\frac{35}{2}$ or $17\frac{1}{2}$

4. $\frac{4}{7}$

5. $\frac{19}{4}$ or $4\frac{3}{4}$

6. $\frac{11}{2}$ or $5\frac{1}{2}$

7. $\frac{3}{4}$

8. $\frac{25}{3} = 8\frac{1}{3}$

9. 6

10. $\frac{21}{17}$ or $1\frac{4}{17}$

11. 9

12. $\frac{10}{3}$ or $3\frac{1}{3}$

13. $\frac{37}{5}$ or $7\frac{2}{5}$

14. $\frac{5}{3}$ or $1\frac{2}{3}$

15. $\frac{16}{7}$ or $2\frac{2}{7}$

16. $\frac{7}{2}$ or $3\frac{1}{2}$

17. $\frac{33}{2}$ or $16\frac{1}{2}$

18. $\frac{7}{4}$ or $1\frac{3}{4}$

19. $\frac{5}{6}$

20. $\frac{28}{3}$ or $9\frac{1}{3}$

21. $\frac{5}{6}$

22. $\frac{25}{4}$ or $6\frac{1}{4}$

23. $\frac{9}{2}$ or $4\frac{1}{2}$

24. $\frac{35}{6}$ or $5\frac{5}{6}$

25. $\frac{6}{11}$

26. $\frac{21}{5}$ or $4\frac{1}{5}$

27. $\frac{8}{15}$

Chapter 8, Unit 3
Part A

1. a. $\frac{4}{2}$ and $\frac{10}{n}$

   b. $\frac{4}{2} = \frac{10}{n}$

   c. $n = 5$; cost is $5

2. a. $\frac{6}{9}$ and $\frac{20}{n}$

   b. $\frac{6}{9} = \frac{20}{n}$

   c. $n = 30$; cost is $30

3. a. $\frac{3}{51}$ and $\frac{n}{85}$

   b. $\frac{3}{51} = \frac{n}{85}$

   c. $n = 5$; 5 runners

4. a. $\frac{7}{4}$ and $\frac{n}{11}$

   b. $\frac{7}{4} = \frac{n}{11}$

   c. $n = \frac{77}{4}$;

      $19\frac{1}{4}$ lbs. pecans

5. a. $\frac{3}{20} = \frac{n}{240}$

   b. 36 ml. medicine

6. a. $\frac{1}{9} = \frac{16,200}{n}$

   b. $145,800 cost

7. a. $\frac{5}{160} = \frac{2}{n}$

   b. $n = 64$; 64 ml. solution

8. a. $\frac{1}{3} = \frac{n}{12}$

    b. 4 cups sugar

9. a. $\frac{2}{7} = \frac{n}{35}$

    b. 10 liters of glucose

10. a. $\frac{8}{100} = \frac{18}{n}$

    b. 225 ml water

Chapter 8, Unit 3
Part B

1. $7\frac{2}{5}$

2. $3\frac{1}{4}$ l.

3. $157\frac{1}{2}$ sq. m.

4. $36 discount

5. 3 will die

6. 5 accidents

7. 15¢ profit

8. $1\frac{2}{5}$ tablespoons of drug

9. 150 problems

10. $4 advertising cost

11. $11\frac{2}{3}$ inches wide

12. 100 grams of gold

13. $19.50

14. $11\frac{7}{11}$ cm. wide

Chapter 8, Unit 4
Part A

1. one-hundredths
2. one-hundredths
3. one-hundredths
4. one-hundredths
5. percent
6. percent

7. same
8. 9%
9. %
10. 56%; $\frac{56}{100}$
11. .3%; $\frac{3}{1000}$
12. 133%; $\frac{133}{100}$
13. .34
14. .91
15. 4.13
16. .049
17. .49
18. .05
19. .02
20. two
21. left
22. $\frac{73}{100}$; 73%
23. $\frac{14}{100}$; 14%
24. $\frac{3.9}{100}$; 3.9%
25. $\frac{370}{100}$; 370%
26. 18%
27. 41%
28. 6%
29. two; right
30. $\frac{3}{11} = \frac{n}{100}$
31. $r = 27\frac{3}{11}$
32. 27%
33. $\frac{9}{46} = \frac{r}{100}$
34. $r = 19\frac{13}{23}$
35. 20%
36. 36%
37. 53%
38. 72%
39. more
40. 180%
41. 188%

Chapter 8, Unit 4
Part B

1. 150%
2. 4%
3. 1%
4. $\frac{17}{100}$
5. $\frac{49}{100}$
6. $\frac{57}{100}$
7. $\frac{143}{100}$
8. $\frac{7}{100}$
9. $\frac{47}{1000}$
10. $\frac{3195}{100}$
11. 19%
12. 81%
13. 3%
14. 181%
15. 67%
16. 5.9%
17. 615%
18. .95
19. .57
20. .64
21. .08
22. .03
23. 1.4
24. 1.07
25. .026
26. .579
27. .78
28. .02
29. 1.1
30. .001
31. .016
32. .0003
33. .01
34. .27

35. .85
36. .56
37. .04
38. .37
39. 9%
40. 71%
41. 156%
42. 50%
43. 9%
44. 620%
45. 1%
46. .7%
47. 101%
48. 8%
49. .1%
50. .9%
51. 162%
52. 90%
53. 38%
54. 75%
55. 64%
56. 32%
57. 7%
58. 188%
59. 256%
60. 38%
61. 56%
62. 80%
63. 33%
64. 30%

Chapter 8, Unit 5
Part A

1. base
2. rate
3. part
4. base, rate, part
5. of
6. 44
7. 40
8. 40
9. 50
10. 50
11. 18
12. 150
13. rate
14. 80, 55, 44
15. 500, 17, 85
16. 90, 150, 135

Chapter 8, Unit 5
Part B

1. base, rate, part
2. of
3. %
4. 2, 200, 4
5. 8, 3, 0.24
6. 70, 20, 14
7. 500, 53, 265
8. 20, 85, 17
9. 60, 125, 75
10. 180, 80, 144
11. 5.52
12. 187.04
13. 336
14. 175
15. 32.52
16. 78.47
17. 750.96
18. 50.4
19. 92.95
20. 176.76
21. 22%
22. 9%
23. 72%
24. 7%
25. 165%
26. 55%
27. 15%
28. 23%
29. 39%
30. 111%
31. 300
32. 600
33. 80
34. 150
35. 140
36. 872
37. 540
38. 438
39. 212
40. 35

Chapter 8, Unit 6
Part A

1. $71,052.63
2. 30
3. rate
4. ?, 20, 8
5. 40%
6. base
7. 3, b, 36
8. 1200 kg.
9. part
10. 234
11. 18%
12. 700
13. b = $2,000,000,
     r = 6; $120,000
14. r = 87, p = 2.61; 3 liters

Chapter 8, Unit 6
Part B

1. 544.04
2. 5%
3. 149

4. 26%
5. 2.82
6. 60
7. .312 liters
8. 147.2 lbs.
9. $18.75
10. $45,000
11. 88,000,000
12. 40%
13. 6.696 parts per million
14. 280 w.
15. $2400
16. 75 games
17. 43%
18. 32
19. 2.25 kg.
20. $30,000

Chapter 8 Test

1. 4
2. 15
3. 30
4. 48
5. 40 minutes
6. 8 days
7. 68 days
8. $2,000
9. 325 particles
10. 56 kg.
11. $\frac{1}{50}$; .02
12. $\frac{11}{10}$; 1.1
13. $\frac{11}{20}$; .55
14. $\frac{9}{100}$; .09
15. 50%
16. 20%
17. 101%
18. 58%

19. s, r, t
20. 50%
21. 300
22. 20
23. $6,000
24. 5%
25. 2,300 kg.
26. 69.92
27. 8

Chapter 9, Unit 1
Part A

1. yes, two
2.
3. B, M, T
4. no dimensions
5. length and width
6. length
7. length
8. length

Chapter 9, Unit 1
Part B

1. For example, a
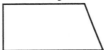

2. For example, a

3. For example, a

4. none

5. parallell lines
    A   B

    C   D

6. A   B   C

7. A   B   C

8.   D
    A   B

        C

9. $\overrightarrow{AB}$ and $\overrightarrow{BA}$.
    A   B

10. For example, a

11. A regular five-sided figure

12. point

13. For example, a ball.

Chapter 9, Unit 2
Part A

1. D
2. D
3. C
4. A

Chapter 9, Unit 2
Part B

1. ∠AED or ∠E, obtuse
2. ∠ABC or ∠B, straight
3. ∠R, acute
4. ∠D, reflex
5. ∠F, right
6. ∠CAB, or ∠A, acute
7. ∠JHK or ∠H, right
8. ∠BAC, or ∠A, obtuse
9. ∠JKH or ∠K, right
10. ∠A, obtuse
11. ∠M, acute
12. ∠X, right

Chapter 9, Unit 3
Part A

1.   a. 90°
    b. 137°
    c. 38°
    d. 75°
2. ∠r and ∠s, ∠r and ∠t, ∠s and ∠u, ∠u and ∠t, ∠v and ∠w, ∠w and ∠y, ∠y and ∠x, ∠x and ∠v
3. ∠r and ∠v, ∠s and ∠w, ∠t and ∠x, ∠u and ∠y
4. ∠t and ∠w, ∠u and ∠v
5. ∠r and ∠y, ∠s and ∠x
6. ∠r = 43°, ∠v = 43°, ∠y = 43°

Chapter 9, Unit 3
Part B

1. 45°
2. 90°
3. 130°

4. 224°
5. 44°
6. 175°
7. 81°
8. 61°
9. 20°
10. 90°
11. 136°
12. 293°

Chapter 9, Unit 4
Part A

1.   a. right
    b. acute
    c. obtuse
    d. acute
2.   a. scalene
    b. isosceles
    c. equilateral
    d. isosceles
3. ∠A and ∠M
   ∠R and ∠Q
   ∠P and ∠N
   $\overline{AR}$ and $\overline{MQ}$
   $\overline{AP}$ and $\overline{MN}$
   $\overline{PR}$ and $\overline{NQ}$

Chapter 9, Unit 4
Part B

1. 90°, 45°, 45°
2. 40°, 100°
3. Yes
4. No
5. Yes
6. Yes
7. Yes
8. No

Chapter 9, Unit 5
Part A

1.   a. $\frac{1}{4}$ inch
    b. $\frac{1}{2}$ inch
    c. $\frac{1}{8}$ inch
    d. $\frac{1}{10}$ cm
2.   a. $1\frac{3}{4}$ inches, 2 inches
    b. $1\frac{1}{4}$ inches, $1\frac{1}{2}$ inches
3. a and c
4.   a. parallelogram, 12
    b. square, 44
    c. trapezoid, 40
    d. rectangle, 28
5.   a. 17 centimeters
    b. 34 centimeters
    c. 34π, 106.76 cm or $106\frac{6}{7}$ cm
6.   a. 32
    b. 90
    c. 15
    d. 32
    e. 88

Chapter 9, Unit 5
Part B

1.   a. $1\frac{1}{2}$ inches
    b. $1\frac{3}{4}$ inches
    c. $2\frac{1}{4}$ inches
2. $\frac{1}{12}$ inch
3. 32
4. 64
5. 72
6. 53
7. 106
8. 28

9.   75.36
10. $\frac{1}{16}$ inch
11. a. rectangle
     b. square
     c. quadrilateral
     d. trapezoid
12. b, c, d

Chapter 9, Unit 6
Part A

1.   15 sq ft; 16 ft
2.   4 sq mm; 8mm
3.   28 sq in; 22 inches
4.   16 sq in; 20 inches
5.   20 sq in
6.   20 sq in
7.   40 sq in
8.   100 sq in
9.   154 sq in
10. 314 sq cm
11. 50.24 sq m
12. $9\frac{5}{8}$ sq in

Chapter 9, Unit 6
Part B

1.   C = 31.4 m
     A = 78.50 sq m
2.   P = 84 ft
     A = 441 sq ft
3.   P = 22.4 ft
     A = 20.47 sq ft
4.   P = 33 cm
     A = 45 sq cm
5.   P = 24.5 m
     A = 27 sq m
6.   C = 22 cm
     A = $38\frac{1}{2}$ sq cm

7.   P = 35 in
     A = $85\frac{1}{2}$ sq in
8.   P = 39.7 ft
     A = 109.25 sq ft
9.   P = 45.7 cm
     A = 135.25 sq cm
10. P = 60 in
     A = 266 sq in
11. A = 240 sq ft
12. P = 132 m
     A = 1134 sq m

Chapter 9, Unit 7
Part A

1.   60 cubic cm
2.   74.088 cubic ft
3.   6,600 cubic in
4.   27 cubic ft
5.   250.047 cubic cm
6.   4,800 cubic ft
7.   120 cubic cm
8.   1,540 cubic in
9.   235,500 cubic ft

Chapter 9, Unit 7
Part B

1.   240 cubic cm
2.   216 cubic m
3.   8 cubic ft
4.   6,160 cubic miles
5.   350 cubic mm
6.   2,816 cubic in
7.   4,410 cubic cm
8.   312 cubic in
9.   935 cubic ft
10. 78.5 cubic m
11. 27 cubic ft or 1 cubic yd
12. 1526.04 cubic in

Chapter 9 Test

1.   a. 3
     b. 2
2.   a. acute
     b. obtuse
     c. right
     d. straight
3.   a. 60°
     b. 160°
4.   a. $\angle$s
     b. $\angle$x
     c. $\angle$w
     d. $\angle$t
5.   35°
6.   a. 20 in
     b. 22.6 cm
7.   a. 15 sq in
     b. 31,400 sq yds
8.   a. 70 cubic m
     b. 1,540 cubic in
9.   150 m
10. 23 m

Index

Z